Mathematical Approaches to Molecular Structural Biology

Mathematical Approaches to Molecular Structural Biology

Subrata Pal

ELSEVIER

ACADEMIC PRESS
An imprint of Elsevier

Academic Press is an imprint of Elsevier
125 London Wall, London EC2Y 5AS, United Kingdom
525 B Street, Suite 1650, San Diego, CA 92101, United States
50 Hampshire Street, 5th Floor, Cambridge, MA 02139, United States
The Boulevard, Langford Lane, Kidlington, Oxford OX5 1GB, United Kingdom

Copyright © 2023 Elsevier Inc. All rights reserved.

No part of this publication may be reproduced or transmitted in any form or by any means, electronic or mechanical, including photocopying, recording, or any information storage and retrieval system, without permission in writing from the publisher. Details on how to seek permission, further information about the Publisher's permissions policies and our arrangements with organizations such as the Copyright Clearance Center and the Copyright Licensing Agency, can be found at our website: www.elsevier.com/permissions.

This book and the individual contributions contained in it are protected under copyright by the Publisher (other than as may be noted herein).

Notices
Knowledge and best practice in this field are constantly changing. As new research and experience broaden our understanding, changes in research methods, professional practices, or medical treatment may become necessary.

Practitioners and researchers must always rely on their own experience and knowledge in evaluating and using any information, methods, compounds, or experiments described herein. In using such information or methods they should be mindful of their own safety and the safety of others, including parties for whom they have a professional responsibility.

To the fullest extent of the law, neither the Publisher nor the authors, contributors, or editors, assume any liability for any injury and/or damage to persons or property as a matter of products liability, negligence or otherwise, or from any use or operation of any methods, products, instructions, or ideas contained in the material herein.

ISBN: 978-0-323-90397-4

For Information on all Academic Press publications
visit our website at https://www.elsevier.com/books-and-journals

Publisher: Andre G. Wolff
Acquisitions Editor: Michelle Fisher
Editorial Project Manager: Tracy I. Tufaga
Production Project Manager: Swapna Srinivasan
Elsevier Cover Designer: Vicky Pearson
Cover Image Designer: Sharmishtha Pal

Typeset by MPS Limited, Chennai, India

Working together
to grow libraries in
developing countries
www.elsevier.com • www.bookaid.org

Dedication

To my parents

Contents

About the author

Subrata Pal

Bachelor of Science in Physics, Calcutta University, Kolkata, India, 1972.
Master of Science in Physics, Calcutta University, Kolkata, India, 1974.
Ph.D. in Molecular Biology, Calcutta University, Kolkata, India, 1982.
Assistant Professor, Physics, in a Calcutta University-affiliated college, 1979–82.
Visiting Fellow, National Institutes of Health, National Cancer Institute, Bethesda, MD, United States, 1984–87.
Fellow/Claudia Adams Barr Investigator, Harvard Medical School, Dana-Farber Cancer Institute, Boston, MA, United States, 1988–92.
Assoc. Professor, Molecular Biology, Jadavpur University, Kolkata, India, 1993–2001.
Professor, Molecular Biology, Genomics and Proteomics, Jadavpur University, Kolkata, India, 2001–15.
Claudia Adams Barr special investigator award for basic contribution to cancer research at the Dana-Farber Cancer Institute (Harvard Medical School), Boston, MA, United States, 1991.

Preface

With the discovery of the double-helical structure of DNA in 1953 and, a few years later, the structures of the proteins, myoglobin and hemoglobin, molecular biology rose to the status of molecular structural biology (MSB). Since then, scientific literature and databases have been flooded with biomolecular structural information for which three different physics-based techniques, namely, X-ray crystallography (XRC), nuclear magnetic resonance spectroscopy, and cryoelectron microscopy (cryoEM), have been essentially responsible. It is now possible to mechanistically explain and predict functions of biomolecules based on their structures.

As technologies improved and molecules became discernible at the atomic scale, it was further revealed that biological macromolecules are intrinsically flexible and naturally exist in multiple conformations. The conformational flexibility is important for their function. Biomolecular functional dynamics has become amenable to investigations by the application of physical principles aided by mathematical and statistical tools. Furthermore, rapid developments in computer hardware and software have remarkably facilitated mathematical analysis of biomolecular structure and dynamics. Needless to say, mathematical approaches are not a "total substitute" for experimental investigations in MSB but, beyond any doubt, an unchallenged complement.

Concomitantly, molecular biology courses are also going through a desirable "transformation"—they are no longer restricted to mere description of biological phenomena but adopting a more mechanistic format. Further, some of the interdisciplinary courses at the graduate level, such as structural biology, macromolecular machines, protein folding, molecular recognition, interaction, etc., being built on structural dynamics of biomolecules, are becoming intensely dependent on mathematics and statistics.

It is true that most of these courses have mathematics as their prerequisites. However, it is also not unexpected that in such "exclusive" mathematics courses, students are exposed to a variety of topics which are important for different areas of science and technology but may not all be required for MSB. Consequently, it is likely that the students may not be able to do equal justice to all the topics and, in the process, those who would be going for MSB-related courses later discount the importance of the topics which, subsequently, turn out to be indispensable.

The book has attempted to sort out and succinctly offer some topics in mathematics and statistics which build the foundation of a discourse in MSB and orient the reader in the appropriate direction. Further, in order to reinforce the "orientation," the last three chapters of the book have illustrated how this mathematical background can be applied to a few of the most important MSB problems.

It is well known that both XRC and cryoEM are capable of solving biomacromolecular structures at atomic resolution. Invariably, they are both dependent on rigorous mathematical and computational analysis. Chapter 7 has discussed the basic physics of X-ray diffraction and gone on to describe how diffraction data from macromolecular crystals are channelized into structural models based on the mathematical theory of Fourier transform and convolution.

CryoEM is based on the scattering of electron waves by the object under investigation. In Chapter 8, the quantum physics of electron scattering has been briefly reviewed. Further, the chapter has discussed how the issues of heterogeneity and noise in cryoEM have been addressed by statistical approaches—principal component analysis and multivariate statistical analysis.

Chapter 9 highlights the essential mathematics concerning some selected and widely used computational techniques to investigate biomolecular structure and dynamics. These include quaternion approach to biomolecular structure comparison, molecular dynamics simulations to predict the movement of each atom in a molecular system, and normal mode analysis which considers the constituent atoms of a biomolecule as a set of simple harmonic oscillators.

Undoubtedly, each of the above stated topics in MSB deserves an entire book for the coverage of all its aspects. Nonetheless, the intention of the presentation has been to restrict to the fundamentals and care has been taken to avoid overloading.

The book should be useful for the students at different levels, advanced undergraduate, graduate, or research, in "regular" molecular (structural) biology or related courses, who would not like to restrict themselves to the narrative aspects of biomolecular phenomena but take a keen interest in the mechanistic and mathematical aspects of MSB.

Subrata Pal

Acknowledgments

The book is dedicated to the memory of my father who introduced me to mathematics in my early childhood and my mother in whose tenderness I grew up.

I respectfully express my indebtedness to late Professor Binayak Dutta-Roy from whom I learnt quantum mechanics.

Plenty of thanks to Michelle Fisher, Megan Ashdown, Tracy Tufaga, and the entire Elsevier team for their unhesitating help and cooperation in addressing my queries and difficulties right from the inception of the project to the publication of the book.

My wife has been very supportive of the effort in all possible ways.

And, last but not least, I am extremely proud of my daughter Sharmishtha who has designed the cover page of the book.

Table of symbols

$x, y, z, a, b, c, \alpha, \beta, \gamma, \lambda$	scalars
$\mathbf{x}, \mathbf{y}, \mathbf{z}, \mathbf{u}, \mathbf{v}, \mathbf{r}$	vectors
$\mathbf{A}, \mathbf{B}, \mathbf{C}$	matrices
$\mathbf{x}^T$	transpose of a vector
$\mathbf{A}^T$	transpose of a matrix
$\mathbf{A}^{-1}$	inverse of a matrix
$\mathbb{Z}$	integers
$\mathbb{N}$	natural numbers
$\mathbb{R}$	real numbers
$\mathbb{C}$	complex numbers
$\mathbb{R}^n$	n-dimensional vector space of real numbers
$\mathbb{R}^{m \times n}$	$m \times n$ ordered array of real numbers
$a := b$	a defined as b
$a =: b$	b defined as a
$\gg$	much greater than
$>$	greater than
$\geq$	greater than or equal to
$\approx$	approximately equal to
$\leq$	less than or equal to
$<$	less than
$\ll$	much less than
$\Rightarrow$	implies
$\Leftrightarrow$	if and only if

CHAPTER

Mathematical preliminaries 1

Structural dynamics-function correlation of biomacromolecules is perhaps the most important theme of molecular structural biology. The last three chapters of the book have attempted to illustrate the mathematical approaches to investigate this structure-function paradigm currently overwhelming the field of theoretical and computational molecular biology. This chapter reviews the fundamental mathematical concepts that are needed for the said purpose. To begin with, functions—algebraic, trigonometric, exponential and logarithmic, and complex—which are used to describe physical systems (including biological systems), have been recollected. This has been followed by the introduction of vectors and matrices that are of paramount importance in the study on cryo-electron microscopy of biomolecules, in particular, and biomolecular structure and dynamics, in general. The nature of changes in a function, with respect to a variable(s) on which it depends, has been discussed in the section on calculus.

1.1 Functions

In biological literature, we have seen that amino acids are often symbolized by single alphabets. For example, the letter G denotes the amino acid glycine, K denotes lysine, and so on. Clearly, we have two sets, one consisting of the alphabets and the other the amino acids, and a defined "correspondence" so that "corresponding" to an alphabet in the first set, a unique amino acid can be identified in the second set. This special kind of correspondence between two sets is called a ***function***—the first set is called the ***domain*** while the second set the ***range*** of the function.

Definition 1.1.1 A function is a correspondence between one set, called the domain, with a second set, called the range, such that each member of the domain corresponds to exactly one member of the range.

1.1.1 Algebraic functions

Physical systems (including biological systems) and their dynamics are quantitatively described in terms of "observables" or entities which can be measured. In

Mathematical Approaches to Molecular Structural Biology. DOI: https://doi.org/10.1016/B978-0-323-90397-4.00009-3
© 2023 Elsevier Inc. All rights reserved.

a specific physical problem, we may have an entity, for example, the distance of an object (represented geometrically by a point P) with respect to a reference point O and the value of the entity is denoted by a variable x. Now, it may happen that there is another entity (say, a force on the object), denoted by a variable y, which is related to x. The relation between the two variables may be expressed as

$$y = f(x) \tag{1.1}$$

Formally, it is said that if there is a unique value of y for each value of x, then y is a ***function*** of x. The set of all permitted values of x is called a ***domain*** and that of all permitted values of y, a ***range***. Being a function of a single variable, $f(x)$ is also called a ***univariate*** function which can be used to describe and analyze a one-dimensional system. Graphically, it is represented by a line. Examples of two univariate functions are shown in Fig. 1.1.

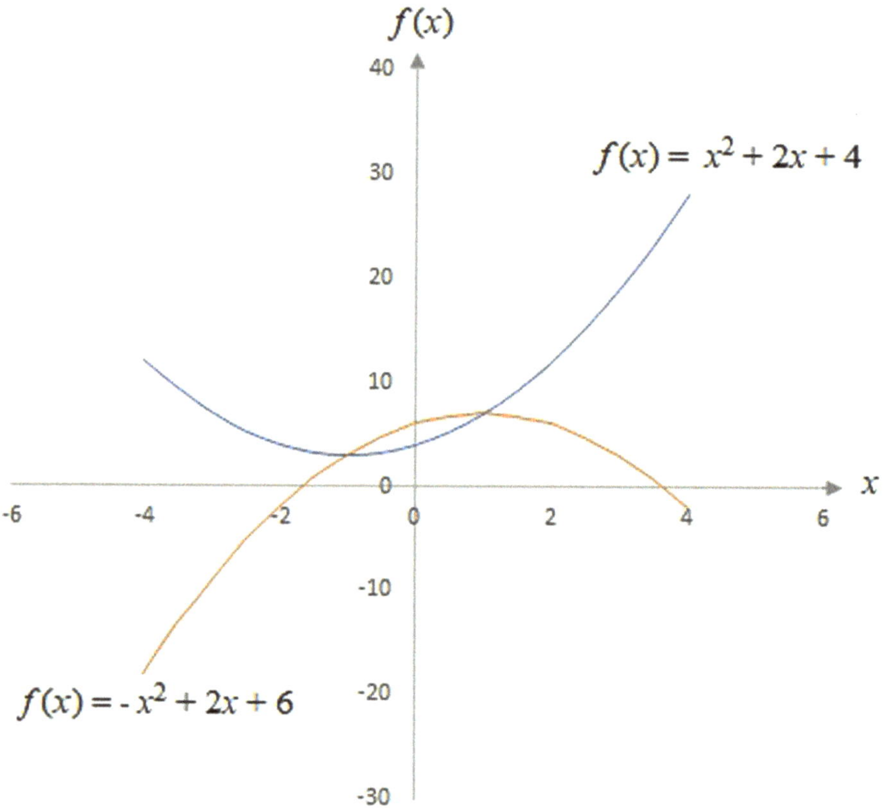

FIGURE 1.1

Graphical representation of univariate algebraic functions.

On the other hand, if there be a unique value of x for each value of y, we can write

$$x = g(y) \tag{1.2}$$

which is defined as the reverse function. $g(y)$ is also a univariate function.

However, we may have a function $f(x, y)$ which depends on two variables x and y. In this case, it is required that for any pair of values (x, y), $f(x, y)$ has a well-defined value. The notion can be extended to a function $f(x_1, x_2, \ldots, x_n)$ that depends on n number of variables $x_1, x_2, \ldots, x_n$. Functions of two variables can be represented by a surface in a three-dimensional space; functions with higher number of variables are usually difficult to visualize. Functions involving more than one variable are called ***multivariate***.

In certain physical problems, a function can be expressed as a polynomial in x.

$$f(x) = a_0 + a_1 x + a_2 x^2 + a_3 x^3 + \cdots + a_{n-1} x^{n-1} + a_n x^n \tag{1.3}$$

When $f(x)$ is set equal to zero, the polynomial equation

$$a_0 + a_1 x + a_2 x^2 + a_3 x^3 + \cdots + a_{n-1} x^{n-1} + a_n x^n = 0 \tag{1.4}$$

is satisfied by specific values of x known as the roots of the equation. $n > 0$ is an integer known as the degree of the polynomial and the equation. The coefficients $a_0, a_1, \ldots, a_n$, $(n \neq 0)$ are real quantities determined by the physical properties of the system under study.

In (1.4), if $n = 1$, the equation takes the form of a linear equation

$$a_0 + a_1 x = 0 \tag{1.5}$$

whose solution (root) is given by $\alpha_1 = -a_0/a_1$

For $n = 2$, (1.4) becomes a quadratic equation

$$a_0 + a_1 \mathrm{x} + a_2 x^2 = 0 \tag{1.6}$$

the roots of which are

$$\alpha_{1,2} = \frac{-a_1 \pm \sqrt{a_1^2 - 4 a_0 a_2}}{2a_2} \tag{1.7}$$

$n = 3$ gives a cubic equation.

1.1.2 Trigonometric functions

Physical systems that involve the periodic motion of a point can be represented by trigonometric functions that are also periodic in nature. Here, the point in question is considered as the projection of another point moving along a circle. To illustrate, let us consider a circle in the xy-plane (Fig. 1.2). The radius of the circle is r and its center is at the origin of xy-coordinate system. P' is a point on the circle, the coordinates of which are given by (x, y) and makes an angle ϕ with the x-axis. P is the projection of P' on the x-axis.

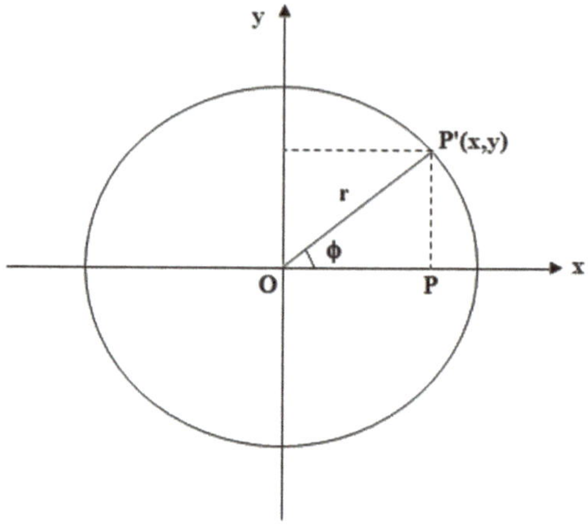

FIGURE 1.2

Position of point P' indicated by Cartesian coordinates (x, y) and polar coordinates (r,).

As P' moves counterclockwise along the circle starting from the x-axis, ϕ increases from 0 to 2π radians or 360 degrees. With P' (x, y) associated with angle ϕ, we have the following definitions:

$$\begin{aligned} \cos\phi &= \frac{\mathrm{OP}}{\mathrm{OP'}} = \frac{x}{r} \\ \sin\phi &= \frac{\mathrm{PP'}}{\mathrm{OP'}} = \frac{y}{r} \\ \tan\phi &= \frac{\mathrm{PP'}}{\mathrm{OP}} = \frac{y}{x} \end{aligned} \tag{1.8}$$

and the reciprocal relations

$$\begin{aligned} &\sec\phi = 1/\cos\phi = r/x \\ &\mathrm{cosec}\ \phi = 1/\sin\phi = r/y \\ &\cot\phi = 1/\tan\phi = x/y \end{aligned} \tag{1.9}$$

If the circle be of unit radius, that is, $r = 1$

$$x = \cos\phi, \quad y = \sin\phi, \quad y/x = \tan\phi$$

$$1/x = \sec\phi, \quad 1/y = \csc\phi, \quad x/y = \cot\phi$$

All these relations are collectively called trigonometric functions.

Referring to Fig. 1.2, one can see that the point P' can be represented also by polar coordinates r (the radial coordinate) and ϕ (the angular coordinate). The relations between the polar coordinates and Cartesian coordinates are given by

$$\begin{aligned} x &= r\cos\phi \\ y &= r\sin\phi \end{aligned} \tag{1.10}$$

Further, from (1.8), we have $\cos 0 = \cos 2\pi = 1$ and $\cos \pi = -1$. Therefore it can easily be visualized that as P' moves along the circle from 0 to 2π, its projection P moves along the x-axis from $+1$ to -1 and back to $+1$. The cycle is repeatable between 2π and 4π and so on. It can be seen that

$$f(\phi + 2n\pi) = f(\phi) \tag{1.11}$$

where $n = 1, 2, 3, \ldots$ $f(\phi) = \cos \phi$ is, therefore, a periodic function, 2π being the period. Similarly, $\sin \phi$ is also a periodic function. Fig. 1.2 also shows that

$$\sin^2 \phi + \cos^2 \phi = 1 \tag{1.12}$$

The basic trigonometric functions, as given previously, are periodic. Hence, their inverses are not single-valued. However, by restricting the domain to an appropriate interval, each of the inverses can be defined as a (single-valued) function.

$$\begin{aligned}
&\sin^{-1} x = y \text{ if and only if } x = \sin y \text{ and } -\pi/2 \le y \le \pi/2 \\
&\cos^{-1} x = y \text{ if and only if } x = \cos y \text{ and } 0 \le y \le \pi \\
&\tan^{-1} x = y \text{ if and only if } x = \tan y \text{ and } -\pi/2 \le y \le \pi/2 \\
&\cot^{-1} x = y \text{ if and only if } x = \cot y \text{ and } 0 \le y \le \pi \\
&\sec^{-1} x = y \text{ if and only if } x = \sec y \text{ and } 0 \le y \le \pi, y \ne \pi/2 \\
&\csc^{-1} x = y \text{ if and only if } x = \csc y \text{ and } -\pi/2 \le y \le \pi/2, y \ne 0
\end{aligned} \tag{1.13}$$

1.1.3 Exponential and logarithmic functions

We have seen that a polynomial contains terms like $a_n x^n$ where x is the variable and n is a fixed number (an integer). However, there can be occasions when a function will contain terms like n^x where n is a fixed number ($n > 0$ and $n \ne 1$) and the variable x is in the exponent.

As for example, in a steadily growing and dividing bacterial culture, the number of cells after x generations is given by $N = N_0 2^x$, where N_0 is the initial number of cells. Such functions are called exponential functions.

Definition 1.1.2

An exponential function f is given by

$$f(x) = a^x \tag{1.14}$$

where x is a real number and the base $a > 0$, $a \ne 0$.

Example 1.1.1: Commonly used exponentials are 10^x and 2^x, where the bases are, respectively, 10 and 2.

However, the most commonly used exponential is e^x, where the number e (≈ 2.718281828459), which can be defined as a limiting sum of an infinite series (see Section 1.5), is called the ***natural base***.

All exponential functions follow the same rules of manipulations

$$n^{x+y} = n^x \cdot n^y \tag{1.15a}$$

and

$$n^{xy} = (n^x)^y = (n^y)^x \tag{1.15b}$$

The exponential function $f(x) = a^x$ is "one-to-one" with domain $(-\infty, \infty)$ and range $(0, \infty)$. Hence, it does have an inverse function, defined as the logarithmic function with base b.

Definition 1.1.3

If $x = a^y$, then $y = \log_a x$, $a > 0$, $a \neq 1$

The number a is called the logarithmic base.

The logarithmic function that uses e as its base is called the natural logarithm and denoted by

$$\log_e x \quad \text{or} \quad \ln x$$

So, we have

$$\ln(e^x) = x, \quad x \in \mathbb{R}; \quad e^{\ln x} = x, \quad x > 0 \tag{1.16a}$$

and

$$\log_{10}(10^x) = x, \quad x \in \mathbb{R} \quad 10^{\log_{10} x} = x, \quad x > 0 \tag{1.16b}$$

The rule for conversion of $\log_{10}$ and $\ln$ is

$$\ln x = (\ln 10)\left(\log_{10} x\right) = (2.3026\ldots)\log_{10} x \tag{1.17}$$

1.1.4 Complex number and functions

Mathematical analysis of a periodic system has been well facilitated by the introduction of what are known as complex numbers and complex functions. To understand what a complex number is, let us consider a function

$$f(z) = z^2 - 3z + 2 \tag{1.18}$$

For $f(z) = 0$, (1.18) becomes

$$z^2 - 3z + 2 = 0 \tag{1.19}$$

which is a quadratic equation whose roots (solutions) are given, in accordance to (1.7), as

$$z_1 = 1 \text{ and } z_2 = 2 \tag{1.20}$$

However, for another function

$$g(z) = z^2 - 2z + 2 = 0 \tag{1.21}$$

the solutions will be given as

$$z_1, z_2 = 1 \pm \sqrt{-1} \tag{1.22}$$

The problem can be understood by plotting the functions $f(z)$ and $g(z)$ against z (Fig. 1.3). It is seen that $f(z)$ intersects the z-axis (a real line) at 1 and 2 and the roots are, therefore, real. However, $g(z)$ does not intersect the z-axis.

We may note here that that the real line or continuum can be considered to be composed of an infinite number of points, each of the points is represented by a real number. The set $\mathbb{R}$ of real numbers includes both rational and irrational numbers. It is obvious that while the first term in (1.22) is real, the second term is not. The problem has been solved by introducing the concept of an ***imaginary number*** (or the basic complex number). An imaginary number has been defined as

$$i = \sqrt{-1} \tag{1.23}$$

so that a generalized complex number can be written as

$$z = x + iy \tag{1.24}$$

where x and y are real numbers. x is the real part denoted by Re $(z) = x$, and y is the imaginary part denoted by Im $(z) = y$. Just as a real number is visualized as a

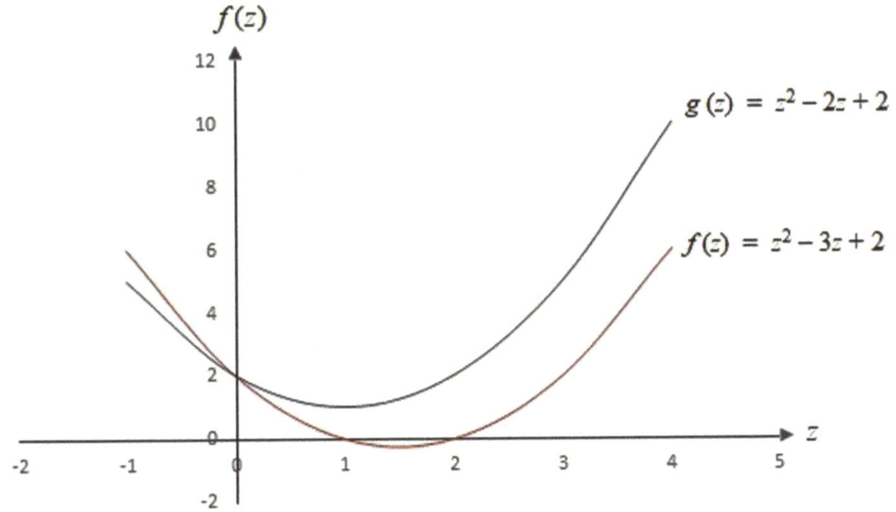

FIGURE 1.3

Graphical solution of equations.

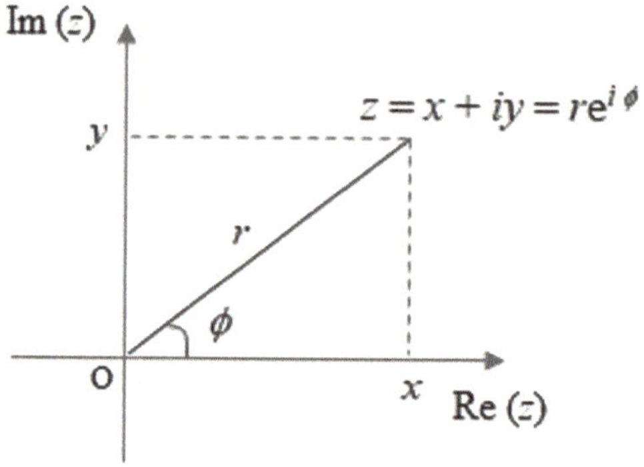

FIGURE 1.4

Representation of a complex number z.

point on an infinite straight line, a complex number can be considered a point on an infinite (complex) plane (Fig. 1.4).

$z^* = x - i\,y$ is the called the complex conjugate of $z = x + i\,y$. The magnitude or modulus $|z|$ of a complex number is given by

$$|z|^2 = z\,z^* = x^2 + y^2 \tag{1.25}$$

and the argument of the complex number is denoted by

$$\arg = \tan^{-1}\left(\frac{y}{x}\right) \tag{1.26}$$

With reference to Fig. 1.4, (1.24) can be written as

$$z = r(\cos\phi + i\sin\phi) \tag{1.27}$$

Later in this chapter we shall see that

$$e^{i\phi} = \cos\phi + i\sin\phi$$

so that a complex number can be represented in the polar form

$$z = re^{i\phi} \tag{1.28}$$

where r and ϕ can be identified with $|z|$ and $\arg z$, respectively. Polar representation of a complex number is easier to manipulate.

1.2 Vectors

Mathematical analysis of biomolecular structure requires that each atom is described by a position vector written as $\mathbf{r}_i = (x_i, y_i, z_i)^T$ while the conformation of the molecule, containing N atoms, is represented by a $3N$-dimensional vector $\mathbf{r} = (x_1, y_2, z_3, \ldots, x_N, y_N, z_N)^T$. An electron microscopic image (or measurement) of a biomacromolecule is considered a point or (vector) in a multidimensional space—the coordinates of the point are defined by the intensities of the pixel (or voxel).

1.2.1 Concept of vector in physics

Vectors can be looked at from different perspectives. In physics, we know that certain entity can be completely specified by a single number, called its magnitude, together with the unit in which it is measured—such entities are called scalars. Examples are mass, temperature, energy, and so on. However, several entities, called vectors, require both magnitude and direction to be specified completely.

A vector is represented graphically using an arrow—the length of the arrow, drawn to scale, specifies the magnitude while the direction of the arrowhead specifies the direction of the vector quantity. Mathematicians call this a "geometric vector" —the tail and the tip of the arrow being the "initial point" and "terminal point," respectively.

If a vector **v** in a plane or two-dimensional space (2-space) is positioned with its initial point at the origin of a rectangular coordinate system, it is completely (both in magnitude and direction) determined by the coordinates (v_1, v_2) of the terminal point. These coordinates are called the components of **v** with respect to the coordinate system.

The arrow can be moved around, but as long as its length and the direction it is pointing in remain the same, it is the same vector. The arrow from an initial point (a, b) to a terminal point ($a + v_1$, $b + v_2$) is equivalent to that from the origin to (v_1, v_2).

A vector can be denoted by **a**; its magnitude will be denoted by |**a**|. A unit vector, that is, a vector whose magnitude is unity, in the direction of **a** will be

$$\hat{\mathbf{a}} = \frac{\mathbf{a}}{|\mathbf{a}|} \tag{1.29}$$

Multiplication of a vector by a scalar, say λ, gives a vector $\lambda\mathbf{a}$ in the same direction as the original but with the magnitude changed λ times.

The scalar product (or dot product) of two vectors **a** and **b**, denoted by $\mathbf{a} \cdot \mathbf{b}$, is given by

$$\mathbf{a} \cdot \mathbf{b} = |\mathbf{a}||\mathbf{b}|\cos\phi, \quad 0 \le \phi \le \pi \tag{1.30}$$

where ϕ is the angle between the two vectors with their tails at the same point.

Evidently, **a** and **b** are perpendicular (or ***orthogonal***) if $\mathbf{a} \cdot \mathbf{b} = 0$. A set of vectors S is said to be ***orthonormal*** if every vector in S has magnitude unity and the vectors are mutually orthogonal.

In the three-dimensional Cartesian coordinate system, **i, j, and k** are considered to be orthonormal unit vectors in the directions of x, y, and z, respectively.

$$\mathbf{i} \cdot \mathbf{i} = \mathbf{j} \cdot \mathbf{j} = \mathbf{k} \cdot \mathrm{k} = 1$$

and

$$\mathbf{i} \cdot \mathbf{j} = \mathbf{j} \cdot \mathbf{k} = \mathbf{k} \cdot \mathbf{i} = 0$$

Any vector in the three-dimensional space can be considered a linear combination of the unit vectors.

$$\mathbf{a} = \mathbf{i}a_x + \mathbf{j}a_y + \mathbf{k}a_z \tag{1.31}$$

(a_x, a_y, a_z) are the components of **a**.

In contrast to the scalar product, the vector product (also called cross product) of two vectors **a** and **b** is a third vector denoted as

$$\mathbf{c} = \mathbf{a} \times \mathbf{b} \tag{1.32}$$

whose magnitude is given by

$$|\mathbf{c}| = |\mathbf{a}||\mathbf{b}|\sin\phi \quad 0 \le \phi \le \pi \tag{1.33}$$

The vector **c** is orthogonal to both **a** and **b** and its direction can be conceptualized in analogy to the motion of a corkscrew. As the corkscrew turns **a** up to **b**, it advances in the direction of **c** (Fig. 1.5).

Example 1.2.1: If we have two vectors written in component form

$$\mathbf{a} = (a_x, a_y, a_z) \quad \text{and} \quad \mathbf{b} = (\mathrm{b}_x, \mathrm{b}_y, \mathrm{b}_z)$$

then, in the component form, the dot product of the two vectors is given by

$$\mathbf{a} \cdot \mathbf{b} = a_x b_x + a_y b_y + a_z b_z \tag{1.34}$$

and the cross product by

$$\begin{aligned} \mathbf{a} \times \mathbf{b} &= (a_y b_z - a_z b_y,\ a_z b_x - a_x b_z,\ a_x b_y - a_y b_z) \\ &= \begin{bmatrix} a_y b_z - a_z b_y \\ a_z b_x - a_x b_z \\ a_x b_y - a_y b_x \end{bmatrix} \end{aligned} \tag{1.35}$$

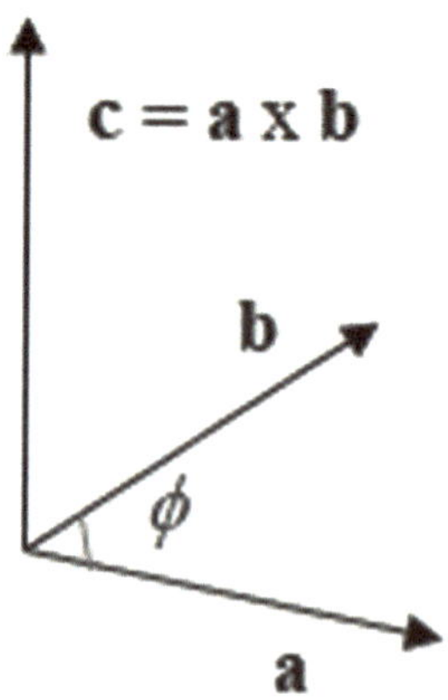

FIGURE 1.5

The vector product.

1.2.2 Vector as an ordered set of numbers

From another perspective, a vector **v** in two dimensions (2-space, $\mathbb{R}^2$) is an ordered pair of real numbers denoted by

$$\begin{pmatrix} v_1 \\ v_2 \end{pmatrix} \quad \text{or} \quad \langle v_1 \quad v_2 \rangle.$$

Synonymously it can be considered a point in $\mathbb{R}^2$. Vectors in $\mathbb{R}^2$ follow certain basic algebraic rules:

1. *Vector equality*: Two vectors

$$\mathbf{u} = \begin{pmatrix} u_1 \\ u_2 \end{pmatrix} \quad \text{and} \quad \mathbf{v} = \begin{pmatrix} v_1 \\ v_2 \end{pmatrix}$$

   are equal if and only if $u_1 = v_1$ and $u_2 = v_2$.
2. *Vector addition*: The sum **w** of two vectors $\mathbf{u} = \begin{pmatrix} u_1 \\ u_2 \end{pmatrix}$ and $\mathbf{v} = \begin{pmatrix} v_1 \\ v_2 \end{pmatrix}$ is defined by

$$\mathbf{w} = \mathbf{u} + \mathbf{v} = \begin{pmatrix} u_1 + v_1 \\ u_2 + v_2 \end{pmatrix} = \begin{pmatrix} w_1 \\ w_2 \end{pmatrix} \tag{1.36}$$

   (1.36) can be demonstrated by the "parallelogram rule for vector addition" (Fig. 1.6A).
3. *Scalar multiplication*: The product of multiplication of a vector $\mathbf{v} = \begin{pmatrix} v_1 \\ v_2 \end{pmatrix}$ with a scalar $\lambda \in \mathbb{R}$ is defined by

$$\lambda \mathbf{v} = \lambda \begin{pmatrix} v_1 \\ v_2 \end{pmatrix} = \begin{pmatrix} \lambda v_1 \\ \lambda v_2 \end{pmatrix} \tag{1.37}$$

and shown in Fig. 1.6B. It is to be noted that for $\mathbf{u}, \mathbf{v} \in \mathbb{R}^2$, the sum $\mathbf{w} \in \mathbb{R}^2$ and also the scaled product $\lambda\, \mathbf{v} \in \mathbb{R}^2$.

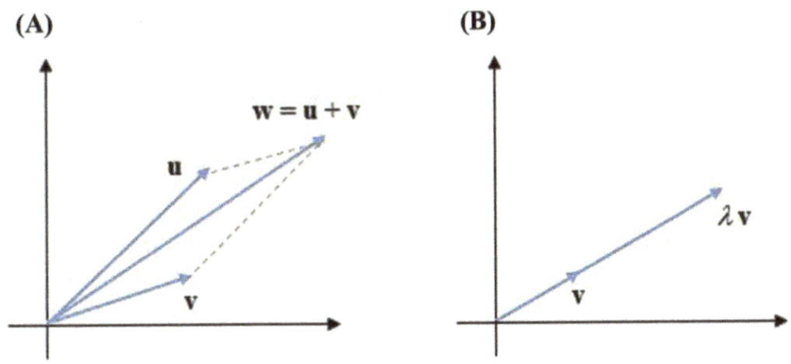

FIGURE 1.6

Vector addition (A) and scalar multiplication (B).

Similarly, a vector **v** in three-dimensional space (3-space, $\mathbb{R}^3$) can be represented by an arrow in a three-dimensional Cartesian coordinate system. It can also be denoted as a 3-tuple

$$\begin{pmatrix} v_1 \\ v_2 \\ v_3 \end{pmatrix} \quad \text{or} \quad \{ v_1 \ v_2 \ v_3 \}$$

For **u**, **v** $\in \mathbb{R}^3$ the same rules of vector equality, vector addition, and scalar multiplication are valid, and both $\mathbf{u} + \mathbf{v} = \mathrm{w}$, $\lambda\,\mathbf{v} \in \mathbb{R}^3$.

Extending the idea of a vector in 2-space or 3-space, a "generalized vector" can be represented by a "generalized arrow" or a "generalized point" in n-dimensional space (n-space, $\mathbb{R}^n$), where n is a positive integer. Alternatively, an n-dimensional vector is an ordered list of n real numbers (n-tuple). It can be written in the **comma-delimited** form

$$\mathbf{v} = (v_1, v_2, \ldots, v_n)$$

or, in the **row-vector** form as

$$\mathbf{v} = \begin{bmatrix} v_1 & v_2 & \ldots & v_n \end{bmatrix}$$

or, in the **column-vector** form as

$$\mathbf{v} = \begin{bmatrix} v_1 \\ v_2 \\ \vdots \\ v_n \end{bmatrix}$$

The numbers in the n-tuple (v_1, v_2,. . ., v_n) can be considered as either the coordinates of the generalized point or the components of the generalized vector.

Introducing here the term "transpose," it can be said that for a vector $\mathbf{v} = [v_1 \ v_2 \ldots v_n]$, its transpose is denoted by

$$\mathbf{v}^T = \begin{bmatrix} v_1 \\ v_2 \\ \vdots \\ v_n \end{bmatrix}$$

Alternatively, for

$$\mathbf{v} = \begin{bmatrix} v_1 \\ v_2 \\ \vdots \\ v_n \end{bmatrix}$$

$$\mathbf{v}^T = \begin{bmatrix} v_1 & v_2 & \ldots & v_n \end{bmatrix},$$

In other words, transposing a row-vector results in a column-vector while transposing a column-vector results in a row-vector.

The basic rules of vector algebra as followed in 2-space and 3-space can also be extended to n-space as well. So, for vectors **u**, **v** $\in \mathbb{R}^n$ and a scalar

$\lambda \in \mathbb{R}$, $\mathbf{u}$ $(u_1, u_2, \ldots, u_n) = \mathbf{v}$ $(v_1, v_2, \ldots, v_n)$ (that is, $\mathbf{u}$ and $\mathbf{v}$ are equal or equivalent) if and only if

$$u_1 = v_1, \quad u_2 = v_2, \ldots, \quad u_n = v_n$$

Further,

$$\mathbf{u} + \mathbf{v} = \begin{bmatrix} u_1 + v_1 \\ u_2 + v_2 \\ \vdots \\ u_n + v_n \end{bmatrix} = \begin{bmatrix} w_1 \\ w_2 \\ \vdots \\ w_n \end{bmatrix} = \mathbf{w} \quad \mathbf{w} \in \mathbb{R}^n \tag{1.38}$$

and

$$\lambda\mathbf{u} = \begin{bmatrix} \lambda u_1 \\ \lambda u_2 \\ \vdots \\ \lambda u_n \end{bmatrix} \lambda\mathbf{v} = \begin{bmatrix} \lambda v_1 \\ \lambda v_2 \\ \vdots \\ \lambda v_n \end{bmatrix} \quad \lambda\mathbf{u}, \lambda\mathbf{v} \in \mathbb{R}^n \tag{1.39}$$

1.2.3 Mathematical viewpoint of vector

Thus it can be seen that vectors $\mathbf{u}$ and $\mathbf{v}$ in $\mathbb{R}^n$ can be added together to produce $\mathbf{w}$, and multiplied by a scalar to yield $\lambda\mathbf{u}$ and $\lambda\mathbf{v}$, respectively, all $\mathbf{w}$, $\lambda\mathbf{u}$ and $\lambda\mathbf{v}$ being in $\mathbb{R}^n$. Based on these two properties of addition and scalar multiplication, an abstract mathematical viewpoint generalizes the concept of vectors—these are "special objects" that can be added together or multiplied by scalars to produce "objects of the same kind."

Besides the geometric vectors as described previously, the following are also examples of vector objects:

1. *Polynomials*. Two polynomials, when added together, will result in another polynomial; any of the two, if multiplied by a scalar, will also produce a polynomial. Hence, the polynomials can be considered vectors.
2. *Color*. In RGB color model, a 3-vector represents a color. For example, (255, 0, 0) is red, (0, 255, 0) is green and (0, 0, 255) is blue. These colors are added and scaled up or down to produce different colors.
3. *Signals*. A time series or signal (such as an audio or a video signal) can be represented by an n-vector. When two audio signals are added, the result is a new audio signal. If an audio signal is scaled, a new audio signal is obtained as well.

Evidently, addition and scalar multiplication can be used in combination to form new vectors. A vector $\mathbf{w} \in \mathbb{R}^n$ is said to be a linear combination of vectors $\mathbf{v}_1, \mathbf{v}_2, \ldots, \mathbf{v}_k$ in $\mathbb{R}^n$ if it can be expressed in the form

$$\mathbf{w} = \mathbf{v}_1 + \lambda_2\,\mathbf{v}_2 + \cdots + \lambda_k\,\mathbf{v}_k \tag{1.40}$$

where the scalars $\lambda_1, \lambda_2, \ldots, \lambda_k$ are called the coefficients of the linear combination.

Scalar multiplication of a vector by the number 0 results in a special vector called the zero vector

$$\mathbf{0} = \begin{bmatrix} 0 \\ 0 \\ \vdots \\ 0 \end{bmatrix} = 0\mathbf{v} = 0 \begin{bmatrix} v_1 \\ v_2 \\ \vdots \\ v_n \end{bmatrix}$$

It has been earlier said that the length of a vector, represented by an arrow drawn to scale, specifies its magnitude. A common mathematical synonym for length is the term "norm" (also called Euclidean norm). For a vector $\mathbf{v} = (v_1, v_2, \ldots, v_n) \in \mathbb{R}^n$, the norm (or length or magnitude) is denoted by $\|\mathbf{v}\|$, and defined by the expression

$$\|\mathbf{v}\| = \sqrt{v_1^2 + v_2^2 + \cdots + v_n^2} \tag{1.41}$$

It follows from the above equation that

a. $\|\mathbf{v}\| \geq 0$
b. $\|\mathbf{v}\| = 0$ if and only if $\mathbf{v} = 0$

Further, if k is a scalar, then

$$\|k\mathbf{v}\| = |k| \, \|\mathbf{v}\|$$

A vector having norm 1 is called a ***unit vector***. If $\mathbf{u} \in \mathbb{R}^n$ be a nonzero vector, a unit vector that is in the same direction as $\mathbf{u}$ is defined as

$$\hat{\mathbf{u}} = \frac{1}{\|\mathbf{u}\|} \mathrm{u} \tag{1.42}$$

It can be shown that $\|\hat{\mathbf{u}}\| = 1$. Hence it appears that multiplication of a nonzero vector by the reciprocal of its magnitude produces a unit vector. The process is called normalizing a vector.

1.3 Matrices and determinants

Mathematical analysis of a problem, whether in physics, chemistry, economics, or molecular structural biology, can often be reduced to solving a system of linear equations. If the system is organized in the form of a rectangular array of numbers, called a "matrix," its solution can be obtained by performing appropriate operations on this matrix.

1.3.1 Systems of linear equations

A linear equation, in two dimensions, is written as

$$ax + by = c, \quad \text{both } a, b \neq 0, \tag{1.43}$$

and, in three dimensions, as

$$ax + by + cz = d, \quad \text{all} \quad a, b, c \neq 0, \tag{1.44}$$

In (1.43), a and b are, respectively, the coefficients of the (unknown) variables x and y while c is the constant term; in (1.44), a, b, and c are, respectively, the coefficients of the (unknown) variables x, y, and z, d being the constant term.

More generally, a linear equation in n dimensions can be expressed in the form

$$a_1 x_1 + a_2 x_2 + \cdots + a_n x_n = b \tag{1.45}$$

where $a_1, a_2, \ldots, a_n$ denote real numbers called the **coefficients** of the (unknown) variables $x_1, x_2, \ldots, x_n$, respectively, and b is also a number called the **constant** term of the equation. Such an equation does not contain any products or roots of variables which appear only to the first power and not as arguments of trigonometric, logarithmic, or exponential functions.

In the special case where $b = 0$, the equation takes the form

$$a_1 x_1 + a_2 x_2 + \cdots + a_n x_n = 0 \tag{1.46}$$

called a **homogeneous linear equation** in variables $x_1, x_2, \ldots, x_n$.

A sequence $s_1, s_2, \ldots, s_n$ of n numbers is called a **solution** of (1.45) if

$$a_1 s_1 + a_2 s_2 + \cdots + a_n s_n = b$$

that is, the equation is satisfied when substitutions $x_1 = s_1$, $x_2 = s_2, \ldots, x_n = s_n$ are made.

A **system of linear equations**, or briefly a linear system, is a finite set of linear equations involving the same variables. The generalized form of a linear system can be written as

$$\begin{aligned} a_{11} x_1 + a_{12} x_2 + \cdots + a_{1n} x_n &= b_1 \\ a_{21} x_1 + a_{22} x_2 + \cdots + a_{2n} x_n &= b_2 \\ &\cdots \\ a_{m1} x_1 + a_{m2} x_2 + \cdots + a_{mn} x_n &= b_m \end{aligned} \tag{1.47}$$

Eq. (1.47) can be compactly represented by

$$\begin{bmatrix} a_{11} & a_{12} & \cdots & a_{1n} \\ a_{21} & a_{22} & \cdots & a_{2n} \\ \vdots & \vdots & \cdots & \vdots \\ a_{m1} & a_{m2} & \cdots & a_{mn} \end{bmatrix} \begin{bmatrix} x_1 \\ x_2 \\ \vdots \\ x_n \end{bmatrix} = \begin{bmatrix} b_1 \\ b_2 \\ \vdots \\ b_n \end{bmatrix} \tag{1.48}$$

along with the rule of multiplication

$$\begin{bmatrix} a_{11} & a_{12} & \cdots & a_{1n} \\ a_{21} & a_{22} & \cdots & a_{2n} \\ \vdots & \vdots & \cdots & \vdots \\ a_{m1} & a_{m2} & \cdots & a_{mn} \end{bmatrix} \begin{bmatrix} x_1 \\ x_2 \\ \vdots \\ x_n \end{bmatrix} = \begin{bmatrix} a_{11} x_1 + a_{12} x_2 + \cdots + a_{1n} x_n \\ a_{21} x_1 + a_{22} x_2 + \cdots + a_{2n} x_n \\ \cdots \\ a_{m1} x_1 + a_{m2} x_2 + \cdots + a_{mn} x_n \end{bmatrix} \tag{1.49}$$

We already know that

$$\mathbf{x} = \begin{bmatrix} x_1 \\ x_2 \\ \vdots \\ x_n \end{bmatrix} \quad \text{and} \quad \mathbf{b} = \begin{bmatrix} b_1 \\ b_2 \\ \vdots \\ b_m \end{bmatrix}$$

are the column vectors.

$$\mathbf{A} = \begin{bmatrix} a_{11} & a_{12} & \cdots & a_{1n} \\ a_{21} & a_{22} & \cdots & a_{2n} \\ \vdots & \vdots & \cdots & \vdots \\ a_{m1} & a_{m2} & \cdots & a_{mn} \end{bmatrix}$$

is defined as an $m \times n$ matrix. It can also be written in a compact notation as

$$[a_{ij}]_{m \times n} \quad \text{or} \quad [a_{ij}]$$

The numbers a_{ij} are called the elements of the matrix **A**. Eq. (1.48) can also be written as

$$\mathbf{A\,x} = \mathbf{b} \tag{1.50}$$

A is called the coefficient matrix.

A sequence of numbers is called **a solution to a system** of equations such as (1.47) if it is a solution to every equation in the system. A system may have no solution at all, a unique solution, or an infinite family of solutions.

For example, the system $x + y = 5$, $x + y = -3$ has no solution since the sum of two numbers cannot be 5 and -3 simultaneously.

A system without a solution is called **inconsistent** while a system with at least one solution is called **consistent**.

Usually, a system of equations with a greater number of variables than equations has infinitely many solutions. For example, the linear system

$$2x - 3y + 4z = 5$$

$$-x + 2y + 3z = 3$$

can be solved by setting

$$z = s\,(\text{arbitrary})$$

as

$$x = 19 - 17s$$

$$y = 11 - 10s$$

$$z = s$$

The equations are satisfied for all choices of s. Similarly, the equation

$$3x - y + 2z = 6$$

can be solved as

$$x = s$$
$$y = 3s + 2t - 6$$
$$z = t$$

(s and t arbitrary)

The quantities s and t in the above examples are called **parameters**. The set of solutions in each case, said to be in **parametric form**, is called the **general solution** to the system.

When only two or three variables are involved, a system can be solved graphically or by simple algebraic operations. However, if more than three variables are involved a graphical solution is not possible and further, with an increasing number of unknowns, the complexity of the algebra also increases. The computation in such cases necessitates suitable notations and standardized procedures. For this purpose, the system in (1.48) can be further abbreviated as

$$\begin{bmatrix} a_{11} & a_{12} & \cdots & a_{1n} & b_1 \\ a_{21} & a_{22} & \cdots & a_{2n} & b_2 \\ \vdots & \vdots & \cdots & \vdots & \vdots \\ a_{m1} & a_{m2} & \cdots & a_{mn} & b_m \end{bmatrix} \tag{1.51}$$

which is called the ***augmented matrix*** for the system. In this matrix the constant b_i's form the last column.

Having this notation for a system of linear equations (1.48), the standardized procedure for solving the system is based on the following principle: Two such systems are considered ***equivalent*** if they have same set of solutions. Therefore a system is solved by deriving a series of systems, each from and equivalent to the preceding, to end up with a system that is easy to solve. The operations carried out are simple and do not change the set of solutions.

To illustrate, let us see how a specific system of linear equations is solved by (elementary) operations carried out on the equations in each step (left column). In each step the corresponding augmented matrix is displayed in the right column.

$$\begin{aligned} 2x - 3y + 5z &= 16 \\ -x + 2y - 2z &= -6 \\ 3x + y - z &= 4 \end{aligned} \qquad \begin{bmatrix} 2 & -3 & 5 & 16 \\ -1 & 2 & -2 & -6 \\ 3 & 1 & -1 & 4 \end{bmatrix}$$

The first equation is added to 2 times the second equation to obtain

The first row is added to 2 times the second row to obtain

$$\begin{aligned} 2x - 3y + 5z &= 16 \\ y + z &= 4 \\ 3x + y - z &= 4 \end{aligned} \qquad \begin{bmatrix} 2 & -3 & 5 & 16 \\ 0 & 1 & 1 & 4 \\ 3 & 1 & -1 & 4 \end{bmatrix}$$

3 times the first equation is added to -2 times the third equation to obtain

3 times the first row is added to -2 times the third row to obtain

$$\begin{aligned} 2x - 3y + 5z &= 16 \\ y + z &= 4 \\ -11y + 17z &= 40 \end{aligned}$$

$$\begin{bmatrix} 2 & -3 & 5 & 16 \\ 0 & 1 & 1 & 4 \\ 0 & -11 & 17 & 40 \end{bmatrix}$$

11 times the second equation is added to the third equation to obtain

11 times the second row is added to the third row to obtain

$$\begin{aligned} 2x - 3y + 5z &= 16 \\ y + z &= 4 \\ 28z &= 84 \end{aligned}$$

$$\begin{bmatrix} 2 & -3 & 5 & 16 \\ 0 & 1 & 1 & 4 \\ 0 & 0 & 28 & 84 \end{bmatrix}$$

Multiply the third equation by 1/28 to obtain

Multiply the third row by 1/28 to obtain

$$\begin{aligned} 2x - 3y + 5z &= 16 \\ y + z &= 4 \\ z &= 3 \end{aligned}$$

$$\begin{bmatrix} 2 & -3 & 5 & 16 \\ 0 & 1 & 1 & 4 \\ 0 & 0 & 1 & 3 \end{bmatrix}$$

3 times the second equation is added to the first equation to obtain

3 times the second row is added to the first row to obtain

$$\begin{aligned} 2x + 8z &= 28 \\ y + z &= 4 \\ z &= 3 \end{aligned}$$

$$\begin{bmatrix} 2 & 0 & 8 & 28 \\ 0 & 1 & 1 & 4 \\ 0 & 0 & 1 & 3 \end{bmatrix}$$

−8 times the third equation is added to the first equation and −1 time the third equation is added to the second equation to obtain

−8 times the third row is added to the first row and −1 time the third row is added to the second row to obtain

$$\begin{aligned} 2x &= 4 \\ y &= 1 \\ z &= 3 \end{aligned}$$

$$\begin{bmatrix} 2 & 0 & 0 & 4 \\ 0 & 1 & 0 & 1 \\ 0 & 0 & 1 & 3 \end{bmatrix}$$

The first equation is multiplied by ½ to obtain

The first row is multiplied by ½ to obtain

$$\begin{aligned} x &= 2 \\ y &= 1 \\ z &= 3 \end{aligned}$$

$$\begin{bmatrix} 1 & 0 & 0 & 2 \\ 0 & 1 & 0 & 1 \\ 0 & 0 & 1 & 3 \end{bmatrix}$$

So, the solution is $x = 2$, $y = 1$, $z = 3$.

Let us now generalize the concept of elementary operations carried out to generate a series of equivalent systems in the following definition.

Definition 1.3.1 The following algebraic operations are called ***elementary operations***:

1. Interchanging two equations.
2. Multiplying an equation by a nonzero number.
3. Adding a multiple of one equation to another equation.

The above definition is associated with the following theorem:

Theorem 1.3.1 If a sequence of elementary operations is performed on a system of linear equations, the resulting system has the same set of solutions as the original; hence the two systems, initial and final, are equivalent.

The corresponding operations to manipulate the augmented matrix are called ***elementary row operations***.

Definition 1.3.2 Elementary row operations on a matrix are the following:

1. Interchanging two rows.
2. Multiplying a row by a nonzero number.
3. Adding a multiple of one row to another row.

Gaussian elimination

In the above illustration, a linear system in the unknowns x, y, and z has been solved by reducing the augmented matrix to the form

$$\begin{bmatrix} 1 & 0 & 0 & * \\ 0 & 1 & 0 & * \\ 0 & 0 & 1 & * \end{bmatrix}$$

where the asterisks represent arbitrary numbers. It appears to be in a "suitable" form from where the corresponding equations are easy to solve. Such matrices have the following properties:

1. Zero rows (consisting entirely of zeros), if any, are grouped together at the bottom.
2. In each nonzero row the first nonzero entry from the left is a 1. This is called the **leading 1** for that row.
3. Each leading 1 is to the right of all leading 1 s in the rows above it.

   A matrix satisfying these three conditions is said to be in **row-echelon form**. A row-echelon matrix is said to be in **reduced row-echelon form** if it satisfies the following additional condition:
4. Each column that has a leading 1, contains zeros everywhere else.

Examples of matrices in row-echelon form

$$\begin{bmatrix} 1 & * & * \\ 0 & 0 & 1 \end{bmatrix} \quad \begin{bmatrix} 1 & * & * & * \\ 0 & 1 & * & * \\ 0 & 0 & 0 & 1 \end{bmatrix} \quad \begin{bmatrix} 1 & * & * \\ 0 & 1 & 0 \\ 0 & 0 & 1 \end{bmatrix}$$

Examples of matrices in reduced row-echelon form

$$\begin{bmatrix} 1 & * & 0 \\ 0 & 0 & 1 \end{bmatrix} \quad \begin{bmatrix} 1 & 0 & * & 0 \\ 0 & 1 & * & 0 \\ 0 & 0 & 0 & 1 \end{bmatrix} \quad \begin{bmatrix} 1 & 0 & 0 \\ 0 & 1 & 0 \\ 0 & 0 & 1 \end{bmatrix}$$

It is to be noted that every matrix can be brought to (reduced) row-echelon form by a sequence of elementary row operations.

Example 1.3.1: Let us consider the matrix

$$\mathbf{A} = \begin{bmatrix} 1 & -2 & -1 & 3 \\ 2 & -4 & 1 & 0 \\ 1 & -2 & 2 & -3 \end{bmatrix}$$

By a couple of row operations, the matrix A can be brought to the row-echelon form

$$\mathbf{A}_{\text{re}} = \begin{bmatrix} 1 & -2 & -1 & 3 \\ 0 & 0 & 1 & -2 \\ 0 & 0 & 0 & 0 \end{bmatrix}$$

and, in another step, to the reduced form

$$\mathbf{A}_{\text{rre}} = \begin{bmatrix} 1 & -2 & 0 & 1 \\ 0 & 0 & 1 & -2 \\ 0 & 0 & 0 & 0 \end{bmatrix}$$

It can be seen that the number r of leading 1s is the same in each of these row-echelon matrices. In general, it is true that r is dependent only on the original matrix **A** and not on the way **A** is brought to row-echelon form. Hence the following definition.

Rank. The ***rank*** of a matrix **A** is the number of leading 1s in any row-echelon matrix to which **A** can be brought by row operations.

1.3.2 Matrices

As seen previously, rectangular arrays of numbers are often used to abbreviate systems of linear equations. However, as we shall see later in the book, they are used in other important contexts as well. For example, it may be necessary to align two biomacromolecule to compare their structures. It has been mentioned earlier that the conformation of a molecule can be represented by a vector. Hence, the problem boils down to the alignment of two vectors. Here, we have a simple example to illustrate the alignment of two vectors.

Example 1.3.2: Let us consider two vectors $\mathbf{v}_1 = (2, 3)$ and $\mathbf{v}_2 = (-3, 2)$ in 2-space with their initial point at the origin. Clearly, $\mathbf{v}_1$ needs to be rotated by 90 degrees counterclockwise to align with $\mathbf{v}_2$. This can be achieved by taking the following 2×2 array

$$\begin{bmatrix} a & b \\ c & d \end{bmatrix}$$

where $a = 0$, $b = -1$, $c = 1$ and $d = 0$, along with a rule of multiplication

$$\begin{bmatrix} a & b \\ c & d \end{bmatrix} \begin{bmatrix} x \\ y \end{bmatrix} = \begin{bmatrix} ax + by \\ cx + dy \end{bmatrix}$$

and carrying out the operation

$$\begin{bmatrix} 0 & -1 \\ 1 & 0 \end{bmatrix} \begin{bmatrix} 2 \\ 3 \end{bmatrix} = \begin{bmatrix} -3 \\ 2 \end{bmatrix}$$

Here, we have a formal mathematical definition of a matrix:

Definition 1.3.3

A real-valued $m \times n$ ***matrix*** **A** is an $m \cdot n$-tuple of elements a_{ij}, $i = 1, 2, \ldots, m$, $j = 1, 2, \ldots, n$ ($m, n \in \mathbb{N}$) arranged in a rectangular array consisting of m rows and n columns. The numbers in the array, a_{ij} are called the ***entries*** in the matrix. The set of all real-valued $m \times n$ matrices is denoted by $\mathbb{R}^{m \times n}$ or $\mathbb{R}^{m\,n}$.

Example 1.3.3: Matrices of different sizes

$$\underset{3\times 2}{\begin{bmatrix} 3 & 2 \\ 5 & 6 \\ -4 & 1 \end{bmatrix}} \quad \underset{1\times 4}{\begin{bmatrix} 5 & 2 & 4 & 1 \end{bmatrix}} \quad \underset{4\times 4}{\begin{bmatrix} 3 & 0 & \pi & 1 \\ e & \sqrt{7} & 0 & 4 \\ -0.5 & 4 & 2 & 6 \\ 0 & -4 & 2 & 5 \end{bmatrix}} \quad \underset{3\times 1}{\begin{bmatrix} -2 \\ 0 \\ 4 \end{bmatrix}}$$

Evidently, just as a vector is a one-dimensional array of numbers, a matrix is defined as a two-dimensional array of numbers. In other words, *a vector is an array of scalars, while a matrix is an array of vectors.* A $1 \times n$ matrix, with only one row, is equivalent to a row-vector as shown earlier—hence it is also called a row matrix. Similarly, an $m \times 1$ matrix, with just one column, is called either a column-vector or a column matrix.

Square matrices. **A**, as in Definition 1.3.3, will be a square matrix if $m = n$ and called a square matrix of order m (or n). This type of matrices is of particular importance in matrix applications and analysis. A fourth-order square matrix is shown in the above example.

A square matrix is called a **diagonal matrix** if all the entries off the main diagonal are zero. An $n \times n$ diagonal matrix can be written as

$$\mathbf{D} = \begin{bmatrix} d_1 & 0 & \cdots & 0 \\ 0 & d_2 & \cdots & 0 \\ \vdots & \vdots & \ddots & \vdots \\ 0 & 0 & \cdots & d_n \end{bmatrix} \tag{1.52}$$

A square matrix in which all the entries above the main diagonal are zero is called a ***lower triangular*** matrix and a square matrix wherein all the entries below the main diagonal are zero is called an ***upper triangular*** matrix. Either is a ***triangular*** matrix. The following are representations of a lower triangular and an upper triangular matrix.

$$\underset{\text{lower triangular}}{\begin{bmatrix} a_{11} & 0 & \cdots & 0 \\ a_{21} & a_{22} & \cdots & 0 \\ \vdots & \vdots & \ddots & \vdots \\ a_{n1} & a_{n2} & \cdots & a_{nn} \end{bmatrix}} \quad \underset{\text{upper triangular}}{\begin{bmatrix} a_{11} & a_{12} & \cdots & a_{1n} \\ 0 & a_{22} & \cdots & a_{2n} \\ \vdots & \vdots & \ddots & \vdots \\ 0 & 0 & \cdots & a_{nn} \end{bmatrix}}$$

It is worthwhile to look into the rules of operations on matrices for discussion on their applications in later chapters.

Equality. Two matrices $\mathbf{A} = [a_{ij}]$ and $\mathbf{B} = [b_{ij}]$ are defined to be equal if they have the same size and their corresponding entries are equal, that is,

$$a_{ij} = b_{ij} \tag{1.53}$$

the equalities holding for all values of i and j.

Addition and subtraction. The sum (or difference) of two matrices $\mathbf{A} \in \mathbb{R}^{m\times n}$ and $\mathrm{B} \in \mathbb{R}^{m\times n}$ is defined as the sum (or difference) of their corresponding entries, that is,

$$\mathbf{A} \pm \mathbf{B} = \begin{bmatrix} a_{11} & \cdots & a_{1n} \\ \vdots & & \vdots \\ a_{m1} & \cdots & a_{mn} \end{bmatrix} \pm \begin{bmatrix} b_{11} & \cdots & b_{1n} \\ \vdots & & \vdots \\ b_{m1} & \cdots & b_{mn} \end{bmatrix} = \begin{bmatrix} a_{11} \pm b_{11} & \cdots & a_{1n} \pm b_{1n} \\ \vdots & & \vdots \\ a_{m1} \pm b_{m1} & \cdots & a_{mn} \pm b_{mn} \end{bmatrix} \tag{1.54}$$

Scalar product. If $\mathbf{A} = [a_{ij}]$ is a matrix and λ a scalar, the scalar product $\lambda\mathbf{A}$ is obtained by multiplying each entry of the matrix $\mathbf{A}$ by λ. $\lambda\mathbf{A}$ is also called a scalar multiple of $\mathbf{A}$. That is, for $\mathbf{A} \in \mathbb{R}^{m\times n}$

$$\lambda\mathbf{A} = \lambda[a_{ij}] = [\lambda a_{ij}] = \begin{bmatrix} \lambda a_{11} & \cdots & \lambda a_{1n} \\ \vdots & & \vdots \\ \lambda a_{m1} & \cdots & \lambda a_{mn} \end{bmatrix} \tag{1.55}$$

Product of matrices. For matrices $\mathbf{A} \in \mathbb{R}^{m\times l}$, $\mathbf{B} \in \mathbb{R}^{l\times n}$, each element c_{ij} of the product $\mathbf{C} = \mathbf{A}\,\mathbf{B} \in \mathbb{R}^{m\times n}$ is equal to the vector dot product of row i of $\mathbf{A}$ with column j of $\mathbf{B}$. That is,

$$c_{ij} = \mathbf{a}_i \cdot \mathbf{b}_j = \begin{bmatrix} a_{i1} & a_{i2} & \cdots & a_{il} \end{bmatrix} \begin{bmatrix} b_{1j} \\ b_{2j} \\ \vdots \\ b_{lj} \end{bmatrix}$$

$$c_{ij} = \sum_{k=1}^{l} a_{ik}\, b_{kj}, \quad i = 1, 2, \ldots, m, \quad j = 1, 2, \ldots, n \tag{1.56}$$

Matrices can be multiplied only if their neighboring dimensions match. For matrices $\mathbf{P} \in \mathbb{R}^{k\times l}$, $\mathbf{Q} \in \mathbb{R}^{l\times m}$, $\mathbf{V} \in \mathbb{R}^{m\times k}$, the products $\mathbf{P}\,\mathbf{Q} \in \mathbb{R}^{k\times m}$, $\mathbf{Q}\,\mathbf{V} \in \mathbb{R}^{l\times k}$ and $\mathbf{V}\,\mathbf{P} \in \mathbb{R}^{m\times l}$ are defined while $\mathbf{V}\,\mathbf{Q}$, $\mathbf{Q}\,\mathbf{P}$, and $\mathbf{P}\,\mathbf{V}$ are not defined.

Following the rules of matrix multiplication, we can write from (1.49)

$$\mathbf{A}\,\mathbf{x} = \begin{bmatrix} a_{11}x_1 + a_{12}x_2 + \cdots + a_{1n}x_n \\ a_{21}x_1 + a_{22}x_2 + \cdots + a_{2n}x_n \\ \cdots \\ a_{m1}x_1 + a_{m2}x_2 + \cdots + a_{mn}x_n \end{bmatrix} = x_1 \begin{bmatrix} a_{11} \\ a_{21} \\ \vdots \\ a_{m1} \end{bmatrix} + x_2 \begin{bmatrix} a_{12} \\ a_{22} \\ \vdots \\ a_{m2} \end{bmatrix} + \cdots x_n \begin{bmatrix} a_{1n} \\ a_{2n} \\ \vdots \\ a_{mn} \end{bmatrix} \tag{1.57}$$

From the above equation, it can be said that if a matrix $\mathbf{A} \in \mathbb{R}^{m\times n}$, and a vector $\mathbf{x} \in \mathbb{R}^{n\times 1}$, the product $\mathbf{A}\,\mathbf{x}$ can be expressed as a linear combination of the column vectors of $\mathbf{A}$ where the coefficients are the entries of $\mathbf{x}$.

Zero matrices. A matrix whose entries are all zero is known as a zero matrix. It is usually denoted by **0**. Some examples of zero matrices are

Example 1.3.4: Zero matrices

$$\begin{bmatrix} 0 & 0 \\ 0 & 0 \\ 0 & 0 \\ 0 & 0 \end{bmatrix} \quad \begin{bmatrix} 0 & 0 & 0 & 0 \end{bmatrix} \quad \begin{bmatrix} 0 & 0 & 0 & 0 \\ 0 & 0 & 0 & 0 \end{bmatrix}$$

Apparently, the zero matrix plays same role in a matrix equation as the number 0 in a numerical equation. However, it should be noted that a zero matrix is not the same as the number 0. The following matrix addition

$$\mathbf{A} + \mathbf{0} = \mathbf{0} + \mathbf{A} = \mathbf{A} \tag{1.58}$$

will be valid only if **A** and **0** are matrices with the same size. For the sake of clarity, sometimes an $m \times n$ zero matrix is denoted by $\mathbf{0}_{m \times n}$.

Multiplication of two nonzero matrices can result in a zero matrix. For example, if

$$\mathbf{A} = \begin{bmatrix} 0 & 2 \\ 0 & 3 \end{bmatrix} \quad \text{and} \quad \mathbf{B} = \begin{bmatrix} 5 & 3 \\ 0 & 0 \end{bmatrix}$$

$$\mathbf{AB} = \begin{bmatrix} 0 & 0 \\ 0 & 0 \end{bmatrix}$$

Identity matrix. A square matrix with ones on the diagonal and zeroes elsewhere is defined as an identity matrix.

Example 1.3.5: A 4×4 identity matrix

$$\mathbf{I}_4 = \begin{bmatrix} 1 & 0 & 0 & 0 \\ 0 & 1 & 0 & 0 \\ 0 & 0 & 1 & 0 \\ 0 & 0 & 0 & 1 \end{bmatrix} \in \mathbb{R}^{4 \times 4}$$

For a matrix $\mathbf{A} \in \mathbb{R}^{m \times n}$

$$\mathbf{A}\mathbf{I}_n = \mathbf{A} \quad \text{and} \quad \mathbf{I}_m\mathbf{A} = \mathbf{A} \tag{1.59}$$

Transpose. For any $m \times n$ matrix, **A**, the transpose, denoted by $\mathbf{A}^T$, is defined to be the $n \times m$ matrix obtained by interchanging the rows and columns of **A**. In other words, for $\mathbf{A} \in \mathbb{R}^{m \times n}$, the matrix $\mathbf{B} \in \mathbb{R}^{n \times m}$ with $b_{ij} = a_{ji}$ is known as the transpose of **A**. That is, $\mathbf{B} = \mathbf{A}^T$.

For

$$\mathbf{A} = \begin{bmatrix} a_{11} & a_{12} & \cdots & a_{1n} \\ a_{21} & a_{22} & \cdots & a_{2n} \\ \vdots & \vdots & \cdots & \vdots \\ a_{m1} & a_{m2} & \cdots & a_{mn} \end{bmatrix}, \quad \mathbf{A}^T = \begin{bmatrix} a_{11} & a_{21} & \cdots & a_{m1} \\ a_{12} & a_{22} & \cdots & a_{m2} \\ \vdots & \vdots & \cdots & \vdots \\ a_{1n} & a_{2n} & \cdots & a_{mn} \end{bmatrix}$$

Example 1.3.6:

$$\text{For } \mathbf{B} = \begin{bmatrix} 3 & 2 \\ 5 & 6 \\ -4 & 1 \end{bmatrix}, \quad \mathbf{B}^T = \begin{bmatrix} 3 & 5 & -4 \\ 2 & 6 & 1 \end{bmatrix}$$

$$\text{For } \mathbf{C} = \begin{bmatrix} 6 & -2 & 1 & 0 \end{bmatrix}, \quad \mathbf{C}^T = \begin{bmatrix} 6 \\ -2 \\ 1 \\ 0 \end{bmatrix}$$

Transposition of a matrix can also be shown as

$$\begin{bmatrix} - & \mathbf{u} & - \\ - & \mathbf{v} & - \\ - & \mathbf{w} & - \end{bmatrix}^T = \begin{bmatrix} | & | & | \\ \mathbf{u} & \mathbf{v} & \mathbf{w} \\ | & | & | \end{bmatrix}$$

Here are some properties of the transpose

Theorem 1.3.2 Provided the sizes of the matrices permit the stated operations,

1. $(\mathbf{A}^T)^T = \mathbf{A}$
2. $(\mathbf{A} + \mathbf{B})^T = \mathbf{A}^T + \mathbf{B}^T$
3. $(\mathbf{A} - \mathbf{B})^T = \mathbf{A}^T - \mathbf{B}^T$
4. $(k\,\mathbf{A})^T = k\,\mathbf{A}^T$

In addition, we have the following important law for transposition.

Theorem 1.3.3 Reverse order law for transposition
Provided the matrix product **A B** is defined,

$$(\mathbf{AB})^T = \mathbf{B}^T \mathbf{A}^T \tag{1.60}$$

Symmetric matrix.

A matrix **A** is called **symmetric,** if $\mathbf{A} = \mathbf{A}^T$, that is, $a_{ij} = a_{ji.}$

A matrix **A** is called **skew-symmetric,** if $\mathbf{A} = -\mathbf{A}^T$, that is, $a_{ij} = -a_{ji}.$

If **A** is an $m \times n$ matrix, $\mathbf{A}^T$ is an $n \times m$ matrix. Evidently, $\mathbf{A} = \mathbf{A}^T \Rightarrow n = m$. Hence, A symmetric (or skew-symmetric) matrix **A** is necessarily a square matrix.

It is to be noted that for every $m \times n$ matrix **A**,

$$\left(\mathbf{A}^T\mathbf{A}\right)^T = \mathbf{A}^T\mathbf{A}^{T^T} = \mathbf{A}^T\mathbf{A} \tag{1.61a}$$

and

$$\left(\mathbf{A}\mathbf{A}^T\right)^T = \mathbf{A}^{T^T}\mathbf{A}^T = \mathbf{A}\mathbf{A}^T \tag{1.61b}$$

which gives the following theorem.

Theorem 1.3.4
For every matrix $\mathbf{A} \in \mathbb{R}^{m \times n}$, the products $\mathbf{A}^T\mathbf{A}$ and $\mathbf{A}\mathbf{A}^T$ are symmetric matrices.

Example 1.3.7: Let

$$\begin{bmatrix} 3 & 5 & -4 \\ 2 & 6 & 1 \end{bmatrix}$$

Then,

$$\mathbf{A}^T\mathbf{A} = \begin{bmatrix} 3 & 2 \\ 5 & 6 \\ -4 & 1 \end{bmatrix} \begin{bmatrix} 3 & 5 & -4 \\ 2 & 6 & 1 \end{bmatrix} = \begin{bmatrix} 13 & 27 & -10 \\ 27 & 61 & -14 \\ -10 & -14 & 17 \end{bmatrix}$$

and

$$\mathbf{A}\mathbf{A}^T = \begin{bmatrix} 3 & 5 & -4 \\ 2 & 6 & 1 \end{bmatrix} \begin{bmatrix} 3 & 2 \\ 5 & 6 \\ -4 & 1 \end{bmatrix} = \begin{bmatrix} 50 & 32 \\ 32 & 41 \end{bmatrix}$$

Evidently, both $\mathbf{A}^T\mathbf{A}$ and $\mathbf{A}\mathbf{A}^T$ are symmetric matrices.

Inverse. Let us consider a square matrix $\mathbf{A} \in \mathbb{R}^{n \times n}$. If there exists a matrix $\mathbf{B} \in \mathbb{R}^{n \times n}$ such that

$$\mathbf{AB} = \mathbf{BA} = \mathbf{I}_n$$

then $\mathbf{A}$ is said to be ***regular***, ***invertible***, or ***nonsingular***, and $\mathbf{B}$ is known as the ***inverse*** of $\mathbf{A}$. In the absence of any such matrix $\mathbf{B}$, the matrix $\mathbf{A}$ is ***noninvertible*** or ***singular***.

If $\mathbf{A}$ is invertible and $\mathbf{B}$ is the inverse of $\mathbf{A}$, it is also true that $\mathbf{B}$ is invertible and $\mathbf{A}$ is the inverse of $\mathbf{B}$.

Let there be two matrices

$$\mathbf{A} = \begin{bmatrix} a_{11} & a_{12} \\ a_{21} & a_{22} \end{bmatrix} \in \mathbb{R}^{2\times 2}, \quad \text{and} \quad \mathbf{B} = \frac{1}{\lambda}\begin{bmatrix} a_{22} & -a_{12} \\ -a_{21} & a_{11} \end{bmatrix} \in \mathbb{R}^{2\times 2}$$

where λ is a scalar.

Multiplying $\mathbf{A}$ with $\mathbf{B}$, we obtain

$$\mathbf{AB} = \frac{1}{\lambda}\begin{bmatrix} a_{11}a_{22} - a_{12}a_{21} & 0 \\ 0 & a_{11}a_{22} - a_{12}a_{21} \end{bmatrix} = \frac{1}{\lambda}(a_{11}a_{22} - a_{12}a_{21})\mathbf{I}_2$$

If $\lambda = (a_{11}a_{22} - a_{12}a_{21}) \neq 0$, $\mathbf{A}\,\mathbf{B} = \mathbf{I}_2 = \mathbf{A}\,\mathbf{A}^{-1}$so that

$$\mathbf{A}^{-1} = \frac{1}{\lambda}\begin{bmatrix} a_{22} & -a_{12} \\ -a_{21} & a_{11} \end{bmatrix} = \frac{1}{a_{11}a_{22} - a_{12}a_{21}}\begin{bmatrix} a_{22} & -a_{12} \\ -a_{21} & a_{11} \end{bmatrix} \tag{1.62}$$

if and only if $(a_{11}a_{22} - a_{12}a_{21}) \neq 0$.

Example 1.3.8:

$$\begin{bmatrix} -1 & 1 \\ 1 & 0 \end{bmatrix} \text{ is inverse of } \begin{bmatrix} 0 & 1 \\ 1 & 1 \end{bmatrix}$$

$$\begin{bmatrix} 1 & 1 \\ 2 & 2 \end{bmatrix} \text{ is singular}$$

In general, for any square matrix $\mathbf{A}$, the inverse $\mathbf{A}^{-1}$ (if it exists) is a square matrix of the same size as $\mathbf{A}$ with the following property:

$$\mathbf{A}\mathbf{A}^{-1} = \mathbf{I} = \mathbf{A}^{-1}\mathbf{A} \tag{1.63}$$

The reverse order law also holds for inverses of matrix products.

Theorem 1.3.5

If matrices $\mathbf{A}$ and $\mathbf{B}$ are of the same size and both are invertible, then $\mathbf{A}\,\mathbf{B}$ is invertible and

$$(\mathbf{AB})^{-1} = \mathbf{B}^{-1}\mathbf{A}^{-1} \tag{1.64}$$

Eq. (1.64) is extendible to cases where three or more matrices are involved.

Theorem 1.3.6

A product of any number of invertible matrices of the same size is invertible, and the inverse of the product is equal to the product of the inverses in reverse order.

Applying (1.60) to (1.64), we have

$$\left(\mathbf{A}\mathbf{A}^{-1}\right)^T = \mathbf{I}^T = \mathbf{I} = \left(\mathbf{A}^{-1}\right)^T\mathbf{A}^T \tag{1.65}$$

and the following theorem.

Theorem 1.3.7

For an invertible matrix $\mathbf{A}$, the transpose of the inverse equals the inverse of the transpose, that is $(\mathbf{A}^{-1})^T$ is the inverse of $\mathbf{A}^T$.

(1.62) leads to the definition of the determinant as we shall see next.

1.3.3 Determinants

The quantity $(a_{11}\ a_{22} - a_{12}\ a_{21})$ is defined as the determinant of the matrix $\mathbf{A} \in \mathbb{R}^{2\times 2}$. That is

$$\det(\mathbf{A}) = \begin{vmatrix} a_{11} & a_{12} \\ a_{21} & a_{22} \end{vmatrix} = a_{11}\,a_{22} - a_{12}\,a_{21} \tag{1.66}$$

The relationship between the determinant and invertibility of a 2×2 matrix as evident in (1.62) holds for $n \times n$ matrices as well.

Theorem 1.3.8

A matrix $\mathbf{A} \in \mathbb{R}^{n \times n}$ is invertible if and only if det $(\mathbf{A}) \neq 0$.

Determinants are defined for only square matrices. For a matrix $\mathbf{A} \in \mathbb{R}^{3 \times 3}$, the determinant, of order 3, is denoted by

$$\det(\mathbf{A}) = \begin{vmatrix} a_{11} & a_{12} & a_{13} \\ a_{21} & a_{22} & a_{23} \\ a_{31} & a_{32} & a_{33} \end{vmatrix}$$

and, in general, for $\mathbf{A} \in \mathbb{R}^{n \times n}$,

$$\det(\mathbf{A}) = \begin{vmatrix} a_{11} & a_{12} & \cdots & a_{1n} \\ a_{21} & a_{22} & \cdots & a_{2n} \\ \vdots & \vdots & \ddots & \vdots \\ a_{n1} & a_{n2} & \cdots & a_{nn} \end{vmatrix}$$

To calculate the determinant, it is necessary to introduce the notions of the minor and the cofactor of a matrix. Using cofactors, we can write a 3×3 determinant in terms of 2×2 determinants, a 4×4 determinant in terms of 3×3 determinants,..., and an $n \times n$ determinant in terms of $(n-1) \times (n-1)$ determinants.

Minor and cofactor: For a square matrix $\mathbf{A} \in \mathbb{R}^{n \times n}$, the minor M_{ij} of the element a_{ij} is defined to be the determinant of the $(n-1) \times (n-1)$ submatrix that remains after removing the ith row and jth column from $\mathbf{A}$. The associated cofactor, $C_{ij} = (-1)^{i+j} M_{ij}$.

For a 2×2 matrix

$$\mathbf{A} = \begin{bmatrix} a_{11} & a_{12} \\ a_{21} & a_{22} \end{bmatrix}$$

it is very easy to see that

$$M_{11} = a_{22}; \quad C_{11} = a_{22}$$

$$M_{12} = a_{21}; \quad C_{12} = -a_{21}$$

$$M_{21} = a_{12}; \quad C_{21} = -a_{12}$$

$$M_{22} = a_{11}; \quad C_{22} = a_{11}$$

Then, (1.66) can also be written as

$$\begin{aligned} \det(\mathbf{A}) &= a_{11} C_{11} + a_{12} C_{12} \\ &= a_{21} C_{21} + a_{22} C_{22} \\ &= a_{11} C_{11} + a_{21} C_{21} \\ &= a_{12} C_{12} + a_{22} C_{22} \end{aligned}$$

It can be noted that in each of the four equations the determinant is expressed as the sum of the products of the elements of any row or column and their corresponding cofactors—the elements and cofactors all correspond to the same row or same column of **A**.

Such a rule of cofactor expansion, which is also called ***Laplace expansion***, can be extended to higher order determinants as well. For $\mathbf{A} \in \mathbb{R}^{3\times 3}$, there are six expansions such as $a_{11}C_{11} + a_{12}C_{12} + a_{13}C_{13}$ (expanded along the first row), or $a_{13}C_{13} + a_{23}C_{23} + a_{33}C_{33}$ (expanded along the third column). Each of these expansions can be further elaborated as

$$\begin{aligned}\det(\mathbf{A}) &= a_{11}\,C_{11} + a_{12}\,C_{12} + a_{13}\,C_{13}\\ &= a_{11}\,(a_{22}\,a_{33} - a_{23}\,a_{32}) - a_{12}\,(a_{21}\,a_{33} - a_{23}\,a_{31}) + a_{13}\,(a_{21}\,a_{32} - a_{22}\,a_{31})\\ &= a_{11}\,a_{22}\,a_{33} + a_{12}\,a_{23}\,a_{31} + a_{13}\,a_{21}\,a_{32} - a_{13}\,a_{22}\,a_{31} - a_{12}\,a_{21}\,a_{33} - a_{11}\,a_{23}\,a_{32}\end{aligned}$$

and so on. All six expansions give the same value for det (**A**).

In general, for a matrix $\mathbf{A} \in \mathbb{R}^{n\times n}$, the number obtained by multiplying the entries in any row or column of **A** by the corresponding cofactors and adding the resulting products is called the determinant of **A**, and the sums themselves are called cofactor expansions of **A**. That is,

$$\det(\mathbf{A}) = a_{1j}\,C_{1j} + a_{2j}\,C_{2j} + \cdots + a_{nj}\,C_{nj}$$

[cofactor expansion along the *j*th column]

$$\det(\mathbf{A}) = a_{i1}\,C_{i1} + a_{i2}\,C_{i2} + \cdots + a_{in}\,C_{in}$$

[cofactor expansion along the *i*th row]

It is to be noted that whichever row or column is chosen, the same result is obtained.

Definiteness of a symmetric matrix

As we shall see later, in Chapter 4, the concept of definiteness of a symmetric matrix is important in laying down the rules for optimization. Here, we consider how a specific property of the determinant is related to the concept.

Definition 1.3.4 Principal submatrix.
In a matrix $\mathbf{A} \in \mathbb{R}^{n\times n}$, the *k*th principal submatrix is the submatrix consisting of the first *k* rows and columns of **A**.

Example 1.3.8: In the 4×4 matrix

$$\mathbf{A} = \begin{bmatrix} 3 & 0 & \pi & 1\\ e & \sqrt{7} & 0 & 4\\ -0.5 & 4 & 2 & 6\\ 0 & -4 & 2 & 5 \end{bmatrix}$$

$$[3] \quad \text{is the first principal submatrix}$$

$$\begin{bmatrix} 3 & 0 \\ e & \sqrt{7} \end{bmatrix} \quad \text{is the second principal submatrix}$$

$$\begin{bmatrix} 3 & 0 & \pi \\ e & \sqrt{7} & 0 \\ -0.5 & 4 & 2 \end{bmatrix} \quad \text{is the third principal submatrix}$$

The definiteness of a symmetric matrix is related to the determinants of its submatrices in accordance with the following theorem.

Theorem 1.3.9 A symmetric matrix $\mathbf{A} \in \mathbb{R}^{n \times n}$ is

1. positive definite if and only if the determinants of all its principal submatrices are positive;
2. negative definite if and only if the determinants of its principal submatrices alternate between negative and positive values, beginning with a negative value for the determinant of the first principal submatrix; and
3. indefinite if and only if it is neither positive definite nor negative definite and at least one principal submatrix has a positive determinant and at least one a negative determinant.

1.4 Calculus

Most often, it is not enough to just describe a physical system by a function of one or more variables. To understand the dynamics of the system, it is important to determine how the function changes with respect to the variable/variables on which it depends. The branch of mathematics that is concerned with changes in a function in a precise manner is calculus. It consists of two subbranches—differential calculus, which computes the change in the function (derivative) brought about by an infinitesimal (approaching zero) change in a variable and integral calculus, which is concerned with the summation (integration) of the products of the derivative and the infinitesimal changes in the variable within defined "limits."

1.4.1 Differentiation

Differentiation, as it has been indicated previously, is a limiting process and hence it is restricted by the rule of limit. Therefore, before moving on to the definition of a derivative that is obtained in the process of differentiation, it will be reasonable to have a basic understanding of the concept of limit. We have here the formal definition of the limit of a function.

Definition 1.4.1

As x approaches a, the limit of $f(x)$ is l, written as

$$\operatorname*{Lim}_{x \to a} f(x) = l \tag{1.67}$$

is defined as follows: given any real number $\varepsilon > 0$, there exists another real number $\delta > 0$ such that for $0 < |x-a| < \delta$, $|f(x)-l| < \varepsilon$. The limit must be a unique real number.

Now, if x changes to $x + \Delta x$, where Δx is a very small amount, leading to a change Δf in the value of the function, that is, $\Delta f = f(x + \Delta x) - f(x)$, then the (first) derivative of $f(x)$ is defined as follows.

Definition 1.4.2

The derivative of a function $f(x)$ at x is defined by

$$f'(x) = \frac{df(x)}{dx} = \lim_{\Delta x \to 0} \frac{\Delta f}{\Delta x} = \lim_{\Delta x \to 0} \frac{f(x + \Delta x) - f(x)}{\Delta x} \tag{1.68}$$

provided that the limit exists and is independent of the direction from which ε approaches zero. If the limit exists at $x = a$, the function is said to be differentiable at a.

The derivative of $f(x)$ can be visualized graphically as the gradient or slope of the function at x. It can be seen from Fig. 1.7 that as $\Delta x \to 0$, $\Delta f/\Delta x$ approaches the gradient given by tan ϕ.

Since $f'(x)$ is also a function of x, its derivative with respect to x, that is, the second derivative of $f(x)$ can also be defined.

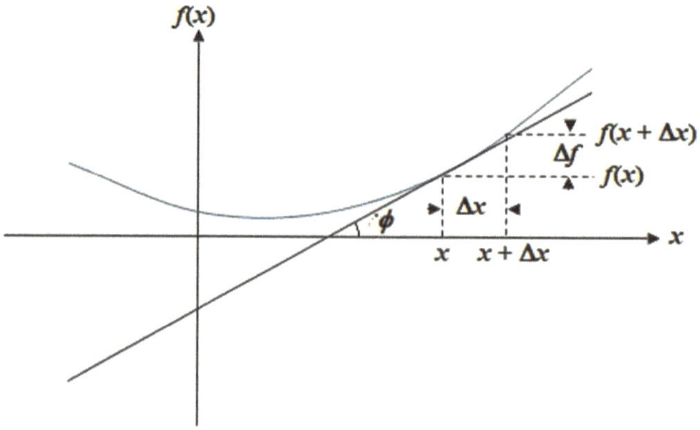

FIGURE 1.7

The gradient or slope of a function $f(x)$ at x.

Definition 1.4.3

The second derivative of a function $f(x)$ at x is defined by

$$f''(x) = \frac{d}{dx}\frac{df(x)}{dx} = \frac{d^2f(x)}{dx^2} = \lim_{\Delta x \to 0} \frac{f'(x+\Delta x) - f'(\Delta x)}{\Delta x} \tag{1.69}$$

provided the limit exists.

For most applications of the differentiation, one does not need to start from the first principles (1.68). Here, we have presented the rules and techniques, in the form of theorems which can nevertheless be deduced from the first principles, to evaluate the derivatives of simple to complicated functions.

Simple algebraic functions

Theorem 1.4.1 For any real number k, if $f(x) = x^k$, then

$$\frac{d}{dx}f(x) = \frac{d}{dx}x^k = k \cdot x^{k-1}$$

Example 1.4.1:

$$\frac{d}{dx}x^2 = 2x, \quad \frac{d}{dx}x^5 = 5x^4, \quad \frac{d}{dx}\left(\frac{1}{x}\right) = -\frac{1}{x^2}$$

Theorem 1.4.2 The derivative of a constant is zero, that is

$$\frac{d}{dx}c = 0$$

Example 1.4.2:

$$\frac{d}{dx}(x^3 - 5) = 3x^2 - 0 = 3x^2$$

Simple algebraic functions are not sufficient to describe and analyze atomic and molecular structure and dynamics. In most cases, the application of trigonometric, exponential, and logarithmic functions is useful and necessary. Here, the derivatives of these functions have been given (without proof). All these derivatives can be deduced from the first principles.

Theorem 1.4.3 ***The derivatives of trigonometric functions***

$$\frac{d}{dx}(\sin x) = \cos x \quad \frac{d}{dx}(\text{cosec}\, x) = -\text{cosec}\, x \cot x$$

$$\frac{d}{dx}(\cos x) = -\sin x \quad \frac{d}{dx}(\sec x) = \sec x \tan x$$

$$\frac{d}{dx}(\tan x) = \sec^2 x \quad \frac{d}{dx}(\cot x) = -\text{cosec}^2 x$$

Theorem 1.4.4 ***The derivative of exponential function***

For the exponential function $f(x) = e^x$,

$$\frac{d}{dx}e^x = e^x, \quad \text{or} \quad f'(x) = f(x)$$

Theorem 1.4.5 ***The derivative of the natural logarithmic function***

If $x > 0$ and $f(x) = \ln x$, then

$$f'(x) = \frac{d}{dx}\ln x = \frac{1}{x}$$

In many cases, one has to deal with a function which is the product of two functions, quotient of two functions, or a function "within" another function. Here below are the rules of differentiation of such functions.

Theorem 1.4.6 ***The product rule***

If $F(x)$ be a product of two functions $f(x)$ and $g(x)$, that is $F(x) = f(x) \cdot g(x)$, then

$$F'(x) = \frac{d}{dx}[f(x) \cdot g(x)] = f(x) \cdot \left[\frac{d}{dx}g(x)\right] + g(x) \cdot \left[\frac{d}{dx}f(x)\right]$$

also expressed as

$$F'(x) = f(x) \cdot g'(x) + g(x) \cdot f'(x) \tag{1.70}$$

Example 1.4.3

a. $f = \text{constant} = 15$, $g = g(x) = x^4$, $F(x) = f \cdot g = 15x^4$

$$\frac{d}{dx}F(x) = 15\frac{d}{dx}(x^4) + x^4\frac{d}{dx}(15) = 60x^3 + 0 = 60x^3$$

b. $f(x) = x^5 + 5,\ g(x) = x^4 + 3x^2 + 7,\ F(x) = f(x) \cdot g(x) = (x^5 + 5)(x^4 + 3x^2 + 7)$

$$\frac{d}{dx}F(x) = (x^5 + 5)\frac{d}{dx}(x^4 + 3x^2 + 7) + (x^4 + 3x^2 + 7)\frac{d}{dx}(x^5 + 5)$$

$$= (x^5 + 5)(4x^3 + 6x) + (x^4 + 3x^2 + 7)(5x^4)$$

c. $f(x) = 5x^3,\ g(x) = \sin x,\ F(x) = f(x) \cdot g(x) = 5x^3 \sin x$

$$\frac{d}{dx}F(x) = 5x^3 \cdot \frac{d}{dx}(\sin x) + \sin x \cdot \frac{d}{dx}(5x^3)$$

$$= 5x^3 \cos x + 15x^2 \sin x$$

Theorem 1.4.7 ***The quotient rule***

$$\text{If } F(x) = \frac{f(x)}{g(x)}, \quad \text{then } F'(x) = \frac{g(x) \cdot f'(x) - f(x) \cdot g'(x)}{[g(x)]^2} \tag{1.71}$$

Example 1.4.4:

$$f(x) = e^x, \quad g(x) = x^3, \quad F(x) = \frac{f(x)}{g(x)} = \frac{e^x}{x}$$

$$\frac{d}{dx}F(x) = \frac{d}{dx}\left(\frac{e^x}{x^3}\right) = \frac{x^3 \cdot e^x - e^x \cdot 3x^2}{x^6} = \frac{e^x(x-3)}{x^4}$$

Theorem 1.4.8 ***The chain rule***

If $F(x) = g\{f(x)\}$, then

$$\frac{dF}{dx} = \frac{dg}{df}\frac{df}{dx} \tag{1.72}$$

also written as

$$F'(x) = g'(f(x)) \cdot f'(x)$$

Example 1.4.5:

$$f(x) = x^2, \quad F(x) = g\{f(x)\} = e^{x^2}$$

$$\frac{d}{dx}F(x) = \frac{d}{dx}e^{x^2} = e^{x^2} \cdot 2x = 2x\,e^{x^2}$$

Example 1.4.6: Differentiation using the quotient rule and the chain rule combined.

$$F(x) = \frac{e^{x^2}}{x^2}$$

$$\frac{d}{dx}F(x) = \frac{d}{dx}\left(\frac{e^{x^2}}{x^2}\right) = \frac{x^2 \cdot \frac{d}{dx}e^{x^2} - e^{x^2} \cdot \frac{d}{dx}(x^2)}{x^4}$$

$$= \frac{x^2 \cdot 2x\, e^{x^2} - e^{x^2} \cdot 2x}{x^4} = \frac{2\, e^{x^2}(x^2 - 1)}{x^3}$$

1.4.2 Integration

An integral has been conceived as the area under a curve in an $f(x)$ versus x plot. There are two kinds of integral—definite and indefinite. The definite integral of a function $f(x)$ sets the limits for integration and denoted as

$$\int_{a}^{b} f(x)dx \tag{1.73}$$

where $x = a$ is the lower limit and $x = b$ is the upper limit of integration.

To arrive at a formal definition of the definite integral, let us consider the graph of $f(x)$ as in Fig. 1.8. Let the finite interval $[a, b] = [x_0, x_n]$ be subdivided into a large number (say, n) of subintervals $[x_{i-1}, x_i]$, $i = 1, 2, \ldots, n$, each of width

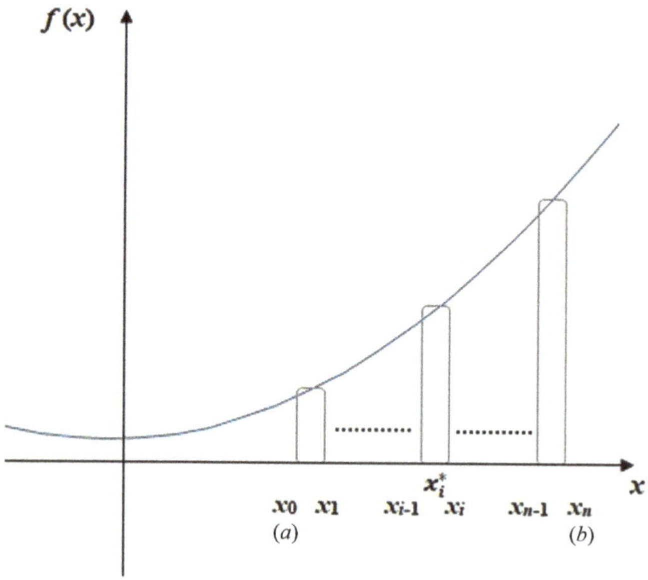

FIGURE 1.8

Defining a definite integral by subdividing the interval $[a, b]$.

Δx. Let x_i^* be any point in $[x_{i-1}, x_i]$, so that the height of the corresponding rectangle is $f(x_i^*)$. The total area of n subintervals is given by the ***Riemann sum***

$$\sum_{i=1}^{n} f(x_i^*)\Delta x \tag{1.74}$$

The area under the curve $f(x)$ on $[a, b]$ is then given by

$$A = \lim_{n \to \infty} \sum_{i=1}^{n} f(x_i^*)\Delta x \tag{1.75}$$

So, we have the following definition of the definite integral.

Definition 1.4.4

If a function $f(x)$ be defined on an interval $[a, b]$, the definite integral of f from a to b is given by

$$\int_a^b f(x)\,dx = \lim_{n \to \infty} \sum_{i=1}^{n} f(x_i^*)\Delta x \tag{1.76}$$

provided the limit exists. In such case, $f(x)$ is said to be integrable on $[a, b]$ and called the integrand.

The indefinite integral can be considered an inverse of the derivative—hence, ***antidifferentiation*** is the process of differentiation in reverse. Thus if $f(x)$ is the derivative of $F(x)$ with respect to x, that is,

$$\frac{dF(x)}{dx} = f(x) \tag{1.77}$$

then $F(x)$ is the indefinite integral of $f(x)$ expressed as

$$F(x) = \int^{x} f(x)\,dx \tag{1.78}$$

where the lower limit is arbitrary. It should be noted that the differentiation does not have a unique inverse—any function $F(x)$ obeying (1.78) can be an indefinite integral of $f(x)$. However, any two such integrals will differ only by an arbitrary constant.

Definition 1.6.4

An indefinite integral of $f(x)$ can be formally written as

$$\int f(x)\,dx = F(x) + c \tag{1.79}$$

where c is known as the constant of integration. c can be evaluated from what are known as boundary conditions.

Some useful rules of antidifferentiation are listed in the following theorem.

Theorem 1.4.9 ***Rules of antidifferentiation***

1. Constant rule

$$\int m\,dx = mx + c, \quad \text{since} \quad \frac{d}{dx}(mx + c) = m$$

2. Power rule

$$\int x^n dx = \frac{1}{n+1}x^{n+1} + c, \quad n \neq -1, \quad \text{since} \quad \frac{d}{dx}\left(\frac{x^{n+1}}{n+1} + c\right) = x^n$$

3. Natural logarithm rule

$$\int \frac{1}{x}dx = \ln x + c, \quad x > 0, \quad \text{since} \quad \frac{d}{dx}(\ln x + c) = \frac{1}{x}$$

4. Exponential (base e) rule

$$\int e^{\alpha x}dx = \frac{1}{a}e^{\alpha x} + c, \quad \alpha \neq 0, \quad \text{since} \quad \frac{d}{dx}\left(\frac{1}{a}e^{\alpha x} + c\right) = e^{\alpha x}$$

Example 1.4.7: Some integrations result in inverse trigonometric functions.

1. $\int \frac{dx}{\sqrt{a^2 - x^2}} = \sin^{-1}\frac{x}{a} + c$
2. $\int \frac{dx}{a^2 + x^2} = \frac{1}{a}\tan^{-1}\frac{x}{a} + c$
3. $\int \frac{dx}{x\sqrt{x^2 - a^2}} = \frac{1}{a}\sec^{-1}\frac{x}{a} + c$

Integration involving exponential functions

The exponential function $f(x) = e^x$ is its own derivative and its own integral—this property makes it one of the most manageable functions in calculus. Exponential functions can be integrated using the following rules.

$$\int e^x dx = e^x + c$$

$$\int a^x dx = \frac{a^x}{\ln a} + c$$

Integration involving logarithmic functions

The integrals involving logarithmic functions follow very simple rules as given next.

$$\int \frac{f'(x)}{f(x)} = \ln|f(x)| + c$$

$$\int \ln x \, dx = x \ln x - x + c = x(\ln x - 1) + c$$

$$\int \log_a x \, dx = \frac{x}{\ln a}(\ln x - 1) + c$$

If $f(x) = x$,

$$\int \frac{f'(x)}{f(x)} = \frac{dx}{x} = \ln|x| + c$$

Integration by substitution

Sometimes a substitution of variables can make a complicated integral relatively simpler. This is illustrated in the following example.

Example 1.4.8: To evaluate the integral $\int \frac{2x\,dx}{(1+x^2)^2}$

We can make a substitution $u = 1 + x^2$, so that $du = 2x\,dx$, and

$$\int \frac{2xdx}{(1+x^2)^2} = \int \frac{du}{u^2} = -\frac{1}{u} + c$$

$$= -\frac{1}{1+x^2} + c \text{(on reverese substitution)}$$

Example 1.4.9: To evaluate

$$\int \frac{5}{x-10} dx$$

Substituting $u = x - 10$, $du = dx$. Therefore

$$\int \frac{5}{x-10} dx = 5\int \frac{du}{u} = 5\ln|u| + c = 5\ln|x-10| + c \quad x \neq 0$$

Integration by parts

When a function can be considered as a product of two separate functions or parts, the integration can be carried out based on the product rule of differentiation (1.70).

Integrating both sides of (1.70) with respect to x

$$uv = \int u\frac{dv}{dx}dx + \int v\frac{du}{dx}dx$$

$$= \int u\,dv + \int v\,du$$

So, we have the following theorem for integration.

Theorem 1.4.10 The integration-by-parts rule is

$$\int u\,dv = uv - \int v\,du \tag{1.80}$$

Example 1.4.10: To evaluate $\int x \sin x \, dx$

Let $u = x$ and $dv/dx = \sin x$ so that $du/dx = 1$ and $v = -\cos x$. Hence, the integral

$$\int x \sin x \, dx = \int u\,dv = x(-\cos x) - \int (1)(-\cos x)\,dx = -x\cos x + \sin x + c$$

Example 1.4.11: To evaluate $\int x \ln x \, dx$

Let $u = \ln x$ and $dv = x\,dx$
so that $du = \frac{1}{x}dx$ and $v = \frac{x^2}{2}$. Hence, the integral

$$\int x \ln x \, dx = \ln x \cdot \frac{x^2}{2} - \int \frac{x^2}{2}\left(\frac{1}{x}dx\right) = \frac{x^2}{2}\ln x - \frac{1}{2}\int x\,dx$$

$$= \frac{x^2}{2}\ln x - \frac{x^2}{4} + c$$

1.4.3 Multivariate function

In (1.68), we have defined the derivative of a function one variable. This definition needs to be amended in the case of a function of two variables $f(x, y)$. Here, we introduce the concept of a "partial derivative" and the corresponding notations $\partial/\partial x$ and $\partial/\partial y$.

Definition 1.6.5

The partial derivative of $f(x, y)$ with respect to x is formally defined as

$$\left(\frac{\partial f}{\partial x}\right)_y = \lim_{\Delta x \to 0} \frac{f(x+\Delta x, y) - f(x, y)}{\Delta x} \tag{1.81}$$

provided that the limit exists. It can be seen that y is treated as a constant.

Similarly, the partial derivative of $f(x, y)$ with respect to y (x being held constant) can be expressed as

$$\left(\frac{\partial f}{\partial y}\right)_x = \lim_{\Delta y \to 0} \frac{f(x, y+\Delta y) - f(x, y)}{\Delta y} \tag{1.82}$$

also contingent on the existence of a limit.

Further extending the definitions to the general n-variable function $f(x_1, x_2, \ldots, x_n)$

$$\frac{\partial f(x_1, x_2, \ldots, x_n)}{\partial x_i} = \lim_{\Delta x_i \to 0} \frac{[f(x_1, x_2, \ldots, x_i + \Delta x_i \ldots, x_n) - f(x_1, x_2, \ldots, x_i, \ldots, x_n)]}{\Delta x_i} \quad (1.83)$$

provided that the limit exists.

Just as we have seen the derivates of a function with multiple variables, the integral of a function with more than one variable can also be defined. Let us consider a function of two variables $f(x, y)$ defined over the domain $[a, b] \times [c, d]$. For the integral of a function of one variable, we had divided the interval $[a, b]$ into a number of subintervals along a straight line. Here, we consider the limits of integration to be represented by a closed curve in a (two-dimensional) region R in the xy-plane and subdivide the region into $m \times n$ subregions R_{ij} ($i = 1,2,\ldots, m$; $j = 1,2,\ldots, n$) each of area ΔA. Let $\left(x^*_{ij}, y^*_{ij}\right)$ be any arbitrary point in R_{ij}. A ***double Riemann sum*** for $f(x, y)$ over R is given by

$$\sum_{j=1}^{n} \sum_{i=1}^{m} f\left(x^*_{ij}, y^*_{ij}\right) \cdot \Delta A \quad (1.84)$$

which leads to the definition of the surface integral of $f(x, y)$ over the region R.

Definition 1.6.6

If a function $f(x, y)$ be defined over a domain $[a, b] \times [c, d]$, then the double integral of $f(x, y)$ over the domain $[a, b] \times [c, d]$ is given by

$$\int_c^d \int_a^b f(x, y)\, dx\, dy = \lim_{m,n \to \infty} \sum_{j=1}^{n} \sum_{i=1}^{m} f\left(x^*_{ij}, y^*_{ij}\right) \cdot \Delta A \quad (1.85)$$

provided the limit exists.

The form of the double integral in (1.85) implies that first integration is carried out with respect to x from a to b, treating y as constant, followed by integration with respect to y from c to d. However, if dx and dy, as well as the limits of integration, are interchanged, the double integral

$$\int_a^b \int_c^d f(x, y)\, dy\, dx$$

produces the same result. Moreover, the double integral can also be written in any of the following forms

$$\int_c^d dy \int_a^b f(x, y)\, dx \quad \text{or} \quad \int_a^b dx \int_c^d f(x, y)\, dy$$

as it is understood that each integral symbol acts on everything to its right.

Example 1.4.12: To evaluate

$$\int_1^2 \int_0^1 x^2 y^2 dx\, dy$$

Integrating first with respect to x and then with respect to y, we have

$$\int_1^2 dy \int_0^1 x^2 y^2 dx = \int_1^2 dy \left[\frac{x^3 y^2}{3}\right]_0^1 = \int_1^2 \left(\frac{y^2}{3}\right) dy = \left[\frac{y^3}{9}\right]_1^2 = \frac{7}{9}$$

If the order of integration is interchanged, we obtain

$$\int_0^1 dx \int_1^2 x^2 y^2 dy = \int_0^1 dx \left[\frac{x^2 y^3}{3}\right]_1^2 = \int_0^1 \left(\frac{7x^2}{3}\right) dx = \left[\frac{7x^3}{9}\right]_0^1 = \frac{7}{9}$$

1.5 Series and limits

There are several examples in physical sciences where the function describing a system is expressed as a series. This facilitates the solution of a relatively complicated function by means of approximation.

A series may contain a finite or infinite number of terms. If a series contains an infinite sequence of terms u_1, u_2, u_3,..., the ith partial sum is defined as

$$s_i = \sum_{n=1}^{i} u_n \tag{1.86}$$

If the partial sum s_i converges to a limit as $i \rightarrow \infty$, that is,

$$\lim_{i \rightarrow \infty} s_i = S$$

the infinite series

$$\sum_{n=1}^{\infty} u_n$$

is said to be convergent and the sum is given by the limit S.

In general, the terms of a series can be complex numbers. In such case the partial sum S_i will also be complex and can be expressed as

$$S_i = X_i + iY_i \tag{1.87}$$

where X_i and Y_i are the partial sums of the real and imaginary terms separately. X_i and Y_i are real.

In a certain type of series, each term depends on a variable, say x, and the series assumes the general form

$$P(x) = a_0 + a_1 x + a_2 x^2 + a_3 x^3 + \ldots \tag{1.88}$$

where a_0, a_1, a_2, etc. are constants. This type of series, occurring frequently in physical problems, is called a power series.

1.5.1 Taylor series

Using Taylor's theorem, a function can be expressed as a power series in x, known as a ***Taylor series***. The theorem, and hence the series, is applicable for functions which are continuous and differentiable within the domain of interest.

Theorem 1.5.1 If $f(x)$ has $n+1$ continuous derivatives on an interval containing a, then for any x in the interval

$$f(x) = \left[\sum_{k=0}^{n} \frac{f^{(k)}(a)}{k!}(x-a)^k\right] + R_{n+1}(x) \tag{1.89}$$

where the error term

$$R_{n+1}(x) = \frac{f^{(n+1)}(c)}{(n+1)!}(x-a)^{n+1}$$

for some c between a and x.

If

$$\lim_{n\to\infty} R_n(x) = 0$$

the infinite ***Taylor series*** converges to

$$\begin{aligned} f(x) &= \sum_{k=0}^{\infty} \frac{f^{(k)}(a)}{k!}(x-a)^k \\ &= f(a) + \frac{f'(a)}{1!}(x-a) + \frac{f''(a)}{2!}(x-a)^2 + \cdots \end{aligned} \tag{1.90}$$

Example 1.5.1: Let us obtain the Taylor series of

$$f(x) = \frac{1}{1-x}$$

about $x = 2$.

We calculate

$$f'(x) = \frac{1}{(1-x)^2},\ f''(x) = \frac{2}{(1-x)^3},\ f'''(x) = \frac{6}{(1-x)^4}, \ldots, f^{(k)}(x) = \frac{k!}{(1-x)^{k+1}}$$

so that

$$f'(2) = 1, \quad f''(2) = -2, \quad f'''(2) = 6, \ldots, \quad f^{(k)}(2) = -1^{k+1}k!$$

and

$$f(x) = \frac{1}{1-x} = -1 + (x-2) - (x-2)^2 + (x-2)^3 + \cdots + (-1)^{k+1}(x-2)^k + \cdots$$

If the reference point is set to be zero in (1.90) the expansion becomes a ***Maclaurin series***

$$f(x) = f(0) + x f'(0) + \frac{x^2}{2!} f''(0) + \cdots = \sum_{k=0}^{\infty} \frac{x^k}{k!} f^{(k)}(0) \tag{1.91}$$

Eq. (1.91) can be used to expand various functions such as exponential, logarithmic, and trigonometric into infinite (power) series. For an exponential function, we have

$$e^x = 1 + x + \frac{x^2}{2!} + \frac{x^3}{3!} + \cdots = \sum_{k=0}^{\infty} \frac{x^k}{k!} \tag{1.92}$$

The Maclaurin series for the trigonometric functions $\sin x$ and $\cos x$ are

$$\sin x = x - \frac{x^3}{3!} + \frac{x^5}{5!} \cdots + (-1)^n \frac{x^{2n+1}}{(2n+1)!} + \cdots \tag{1.93}$$

$$\cos x = 1 - \frac{x^2}{2!} + \frac{x^4}{4!} \cdots + (-1)^n \frac{x^{2n}}{2n!} + \cdots \tag{1.94}$$

If $x = i\phi$, where ϕ is real, then

$$\begin{aligned} e^{i\phi} &= 1 + i\phi - \frac{\phi^2}{2!} - \frac{i\phi^3}{3!} + \frac{\phi^4}{4!} + \frac{i\phi^5}{5!} - \cdots \\ &= \left(1 - \frac{\phi^2}{2!} + \frac{\phi^4}{4!} - \cdots\right) + i\left(\phi - \frac{\phi^3}{3!} + \frac{\phi^5}{5!} - \cdots\right) \end{aligned} \tag{1.95}$$

That is,

$$e^{i\phi} = \cos\phi + i\sin\phi \tag{1.96}$$

We cannot find a Maclaurin expansion for $\ln x$ since the function does not exist at $x = 0$. However, it is possible to obtain the Maclaurin expansion of $\ln(1 + x)$, as shown in the following example.

Example 1.5.2: To find the Maclaurin expansion of $f(x) = \ln(1 + x)$.

We have

$$f'(x) = \frac{1}{1+x},\ f''(x) = -\frac{1}{(1+x)^2},\ f'''(x) = \frac{2}{(1+x)^3},\ \ldots,\ f^k(x) = \frac{(-1)^{k+1}(k-1)!}{(1+x)^k}$$

so that

$$f(x) = \ln(1 + x) = x - \frac{x^2}{2} + \frac{x^3}{3} + \cdots + \frac{(-1)^{k+1}}{k} x^k + \cdots$$

1.5.2 Fourier series

Besides power series, a function may also be represented by a sum of sine and cosine terms. Such a representation is known as a ***Fourier series***. One of the conditions a function has to fulfill to be expanded as a Fourier series is that it must be periodic.

Let $f(x)$ be a $2L$-periodic function defined in the interval $[-L, L]$ and outside the interval by $f(x+2L) = f(x)$. The Fourier series or Fourier expansion is then given by

$$f(x) = \frac{a_0}{2} + \sum_{s=1}^{\infty}\left(a_s \cos\frac{s\pi x}{L} + b_s \sin\frac{s\pi x}{L}\right) \tag{1.97}$$

where the Fourier coefficients a_s and b_s are given by

$$a_s = \frac{1}{L}\int_{-L}^{L} f(x)\cos\frac{s\pi x}{L}dx \quad s = 0, 1, 2, \ldots \tag{1.98a}$$

$$b_s = \frac{1}{L}\int_{-L}^{L} f(x)\sin\frac{s\pi x}{L}dx \quad s = 0, 1, 2, \ldots \tag{1.98b}$$

The following orthogonality property of the sinusoidal functions makes calculation of Fourier expansion possible

$$\int_{-L}^{L}\cos\left(\frac{s\pi x}{L}\right)dx = 0 = \int_{-L}^{L}\sin\left(\frac{s\pi x}{L}\right)dx$$

$$\int_{-L}^{L}\cos\left(\frac{s\pi x}{L}\right)\cos\left(\frac{r\pi x}{L}\right)dx = \begin{cases} L & s = r \\ 0 & s \neq r \end{cases}$$

$$\int_{-L}^{L}\sin\left(\frac{s\pi x}{L}\right)\cos\left(\frac{r\pi x}{L}\right)dx = 0$$

$$\int_{-L}^{L}\sin\left(\frac{s\pi x}{L}\right)\sin\left(\frac{r\pi x}{L}\right)dx = \begin{cases} L & s = r \\ 0 & s \neq r \end{cases}$$

The Fourier series in (1.97) can also be written in a complex form as

$$f(x) = \sum_{s=-\infty}^{\infty} c_s\, e^{i\pi sx/L} \tag{1.99}$$

where

$$c_s = \frac{1}{2L}\int_{-L}^{L} f(x)\, e^{-i\pi sx/L}\, dx \tag{1.100}$$

Example 1.5.3: Let us expand $f(x) = x^2$, $0 < x < 2\pi$, in a Fourier series if the period is 2π.

We can calculate the Fourier coefficients from (1.98a,b) as follows.
Here the period is $2L = 2\pi$, hence $L = \pi$.
If s = 0,

$$a_0 = \frac{1}{\pi}\int_0^{2\pi} x^2 dx = \frac{8\pi^2}{3}$$

and further, using the orthogonality relations,

$$a_s = \frac{1}{L}\int_0^{2L} f(x)\cos\frac{s\pi x}{L}dx = \frac{1}{\pi}\int_0^{2\pi} x^2 \cos sx\, dx$$

$$= \frac{1}{\pi}\left[(x^2)\left(\frac{\sin sx}{s}\right) - (2x)\left(\frac{-\cos sx}{s^2}\right) + 2\left(\frac{-\sin sx}{s^3}\right)\right]_0^{2\pi} = \frac{4}{s^2}, \quad s \neq 0$$

Similarly, we can calculate

$$b_s = -\frac{4\pi}{s}$$

Thus we have the Fourier expansion

$$f(x) = x^2 = \frac{4\pi^2}{3} + \sum_{s=1}^{\infty}\left(\frac{4}{s^2}\cos sx - \frac{4\pi}{s}\sin sx\right)$$

Exercise 1

1. Find $\mathbf{u}\cdot\mathbf{v}$ for the following vectors.
   a. $\mathbf{u} = (2, -6, 0, 3)$, $\mathbf{v} = (-4, 1, 4, 3)$
   b. $\mathbf{u} = (-3, 6, 0)$, $\mathbf{v} = (2, -4, 1)$
   c. $\mathbf{u} = (5, 7)$, $\mathbf{v} = (-2, 4)$
   d. $\mathbf{u} = (2, -4, 2)$, $\mathbf{v} = (1, 1, 1)$
2. Calculate the cross product $\mathbf{w} = \mathbf{u} \times \mathbf{v}$ for the following vectors and show that for each case w is perpendicular to both $\mathbf{u}$ and $\mathbf{v}$.
   a. $\mathbf{u} = (1, 2, --2)$, $\mathbf{v} = (3, 0, 1)$
   b. $\mathbf{u} = (-6, 4, 2)$, $\mathbf{v} = (3, 1, 5)$
   c. $\mathbf{u} = (2, -1, 4)$, $\mathbf{v} = (1, 3, 7)$
3. Solve the following systems of equations using augmented matrices
   a. $3x + 4y + z = 1$
      $2x + 3y = 0$
      $4x + 3y - z = -2$

   b. $x + y + 2z = 9$
      $2x + 4y - 3z = 1$
      $3x + 6y - 5z = 0$
4. If $\mathbf{A} = \begin{bmatrix} 2 & 4 & -2 \\ 3 & 5 & 1 \end{bmatrix}$, show that $\mathbf{A}^T\mathbf{A}$ and $\mathbf{A}\mathbf{A}^T$ are both symmetric but $\mathbf{A}^T\mathbf{A} \neq \mathbf{A}\,\mathbf{A}^T$.
5. Find if the following matrices are invertible
   a. $\begin{bmatrix} 3 & -4 & -2 \\ 0 & 1 & 1 \\ -6 & 7 & 5 \end{bmatrix}$

b. $\begin{bmatrix} 1 & -2 & 3 \\ 5 & 6 & -1 \\ 3 & 2 & 1 \end{bmatrix}$

c. $\begin{bmatrix} 2 & -2 & 5 \\ -3 & 1 & 1 \\ 2 & 7 & -4 \end{bmatrix}$

6. Find the derivatives with respect to x of the following functions:
   a. $f(x) = 2^x$
   b. $f(x) = x^2 e^x$
   c. $f(x) = x^2 \ln x$
   d. $f(x) = \frac{\ln x}{x^3}$
   e. $f(x) = \frac{\sin x}{x}$
7. Evaluate the following integrals
   a. $\int_0^1 xe^x \, dx$
   b. $\int_1^2 x^2 \ln x \, dx$
   c. $\int \frac{\ln x}{x} dx$
   d. $\int 2xe^{x^2} dx$
   e. $\int x \sin x \, dx$
8. Evaluate the double integrals
   a. $\int_{-1}^3 \int_1^2 x^2 y \, dx \, dy$
   b. $\int_0^1 \int_{x^2}^x x \, y^2 dx \, dy$
   c. $\int_0^1 \int_0^1 x^2 e^y dx \, dy$
9. Obtain the Taylor series expansions of $f(x) = \frac{1}{x}$ about $x = 1$ and $x = 3$.
10. Find the Maclaurin expansion of $f(x) = \cos^2 x$ up to powers of x^4.
11. Compute the Fourier series for
    a. $f(x) = x, \quad -\pi \leq x \leq \pi$.
    b. $f(x) = \begin{cases} 0 & -2 < x < 0 \\ 2 & 0 < x < 2 \end{cases}$ period = 4

Further reading

Anton, H., Rorres, C., 2014. Elementary Linear Algebra, 11th (ed.) Wiley.

Arfken, G.B., Weber, H.J., Harris, F.E., 2013. Mathematical Methods for Physicists: A Comprehensive Guide, 7th (ed.) Elsevier Academic Press.

Bittinger, M.L., Ellenbogen, D.J., Surgent, S.A., 2012. Calculus and Its Applications, 10th (ed.) Addison-Wesley.

Nicholson, W.K., 2021. Linear Algebra with Applications, *Open Edition*. Creative Common License (CC BY-NC-SA).

Riley, K.F., Hobson, M.P., Bence, S.J., 2006. Mathematical methods for physics and engineering, 3rd (ed.) Cambridge University Press, Cambridge, U.K.

CHAPTER

Vector spaces and matrices

2

In Chapter 1, we have looked at vectors from three different perspectives. First, they were viewed as directed line segments (arrows). Next, vectors were considered as ordered pairs and ordered triple of real numbers, which could be visualized in respectively, two-dimensional and three-dimensional rectangular coordinate systems and extending the notion further by going beyond our visual experience, as n-tuples of real numbers. Besides, we have also characterized vectors as "special objects" that can be added together or multiplied by scalars to produce "objects of the same kind." In this chapter, the notion will be formalized by introducing vector space, a "structured space" where vectors reside. "Vector spaces" and "linear spaces" are considered to be synonymous.

2.1 Linear systems

In real world, there are very few physical elements that display truly linear characteristics. Nonetheless, as nonlinear dynamical equations are relatively difficult to solve, nonlinear systems are usually approximated by linear equations (linearization). With a large number of mathematical tools available for analysis, linear systems have found useful applications across a wide range of disciplines, including molecular structural biology.

"Linearity" is closely related to "proportionality"; it is the property of a mathematical relationship, say, a function $f(x)$ where $x \in \mathbb{R}$, that can be graphically represented by a straight line. Generalized for n dimension, linearity requires the property of a function being compatible with addition and scaling (superposition principle), that is,

$$f(\mathbf{x} + \mathbf{y}) = f(\mathbf{x}) + f(\mathbf{y})$$

and

$$f(\lambda \mathbf{x}) = \lambda f(\mathbf{x})$$

where λ is a real or complex number, $\mathbf{x}$ and $\mathbf{y}$ can be elements of an n-dimensional vector space.

Example 2.1.1: Let us consider a differential equation which has relevance in many oscillating systems.

$$L(x) = 0 \tag{2.1}$$

Mathematical Approaches to Molecular Structural Biology. DOI: https://doi.org/10.1016/B978-0-323-90397-4.00002-0
© 2023 Elsevier Inc. All rights reserved.

where

$$L = a\frac{d^2}{dt^2} + b\frac{d}{dt} + c \tag{2.2}$$

a, b, and c are constants. Looking at each term on the right-hand side, we find

$$c(x+y) = cx + cy,\ b\frac{d}{dt}(x+y) = b\frac{dx}{dt} + b\frac{dy}{dt} \text{ and } a\frac{d^2}{dt^2}(x+y) = a\frac{d^2x}{dt^2} + a\frac{d^2y}{dt^2}$$

so that,

$$L(x+y) = L(x) + L(y) \tag{2.3}$$

Also,

$$L(kx) = kL(x) \tag{2.4}$$

where k is a constant. Hence, (2.1) is a linear differential equation representing a linear oscillating system.

Exercises 2.1

1. State, giving reasons, if the following systems are linear/nonlinear.
   a. $f(x) = k^2x$
   b. $f(x) = kx^2$
2. In the following differential equations, x is the independent variable and y is the dependent variable. Identify which of the equations are nonlinear.
   a. $\frac{d^2y}{dt^2} + y = 0$
   b. $y\frac{d^2y}{dt^2} + y = 0$
   c. $x\frac{d^2y}{dt^2} + y = 0$
   d. $x^2\frac{d^2y}{dx^2} + x\frac{dy}{dx} = \sin x$
3. In the following partial differential equations, u is the dependent variable and x and y are independent variables. Comment on their linearity/nonlinearity.
   a. $\frac{\partial^2 u}{\partial x^2} + \sin y = 0$
   b. $x\frac{\partial u}{\partial x} + y\frac{\partial u}{\partial y} + u^2 = 0$
4. Are the following equations governing the motion of a pendulum linear/nonlinear?
   a. $\frac{d^2\theta}{dt^2} + \frac{g}{l}\sin\theta = 0$
   b. $\frac{d^2\theta}{dt^2} + \frac{g}{l}\theta = 0$

2.2 Sets and subsets

As we proceed to define a vector space, it will be useful to have some basic notion of "sets" and "subsets"—a notion which is considered to be somewhat "primitive" or "basic" in modern mathematics. Therefore neither a set nor its content is defined, they are to be understood intuitively. Yet, even for the sake of intuitive understanding, one needs at least an "informal" description of a set and its properties.

2.2.1 Set

Stating simply, a ***set*** is a collection of definite, distinct objects which are called members or elements of the set.

Some simple examples of a set would be a set of nucleotides, a set of acidic amino acids, etc. There is precisely one set with no members at all—it is called the empty set or ***null set***. A set with just one member is called a singleton or a ***singleton set***.

Some relevant notations

Usually sets are denoted by capital letters A, B, C, ... and the members of a set by small letters a, b, c, ...

If x is an element of set X, we write $x \in X$. If set X is a member of set Y, we write $X \in Y$.

If z is not a member of set X, we write $z \notin X$.

A set is generally specified by using curly braces. The same set can be described in three different ways:

a. by listing all its members, $S = \{2, 4, 6, 8\}$
b. by stating a property of its members
   $S = \{x \mid x \text{ is a natural number between 1 and 9 that is divisible by 2}\}$
   or, $S = \{y \mid y \text{ is an even number between 1 and 9}\}$
c. defining a set of recursive rules to identify its members
   (i) $2 \in S$
   (ii) if $x \in S$, then $x + 2 \in S$, $(x < 10)$
   (iii) nothing else belongs to S.

Example 2.2.1: Some frequently used sets:

1. $\mathbb{N} = \{0, 1, 2. 3, ...\}$ the set of all natural numbers
2. $\mathrm{N}^+ = \{n \in \mathrm{N}: n > 0\}$
3. $\mathbb{Z} = \{0, -1, +1, -2, +2, ...\}$ the set of all integers
4. $\mathbb{R}$ = the set of all real numbers
5. $\mathbb{R}^n$ = the set of all n-dimensional vectors of real numbers
6. $\mathbb{C}$ = the set of all complex numbers

Two sets S and T are equal if they have exactly the same members. For example,

$$S = \{5, 9, 3, 8\} \text{ and } T = \{3, 9, 5, 8\}$$

The order of listing of the members does not matter.

Venn diagrams. Sets and relationship between them are graphically represented by Venn diagrams. A set is shown as a circle or oval (Fig. 2.1) and its members are represented as objects included within the circle (or oval).

Examples 2.2.2: $N = \{0, 1, 2, 3, 4, 5, 6, 7, 8, 9\}$ $V = \{a, e, i, o, u\}$.

Union. There are three operations, each of which takes two sets and return a new set—(a) union, (b) intersection, and (c) set-theoretic difference.

The union of two sets S and T is the collection of all elements which belong to either S or T, or to both S and T (Fig. 2.2). Symbolically,

$$\begin{aligned} S \cup T &= \{x : (x \in \mathrm{S}) \text{ or } (x \in \mathrm{T})\} \\ &= \{x : (x \in \mathrm{S}) \vee (x \in \mathrm{T})\} \end{aligned}$$

Example 2.2.3: If $S = \{2, 3, 9, 5, 6, 1\}$ and $T = \{5, 4, 8, 3, 7\}$, then

$$S \cup T = \{1, 2, 3, 4, 5, 6, 7, 8, 9\}$$

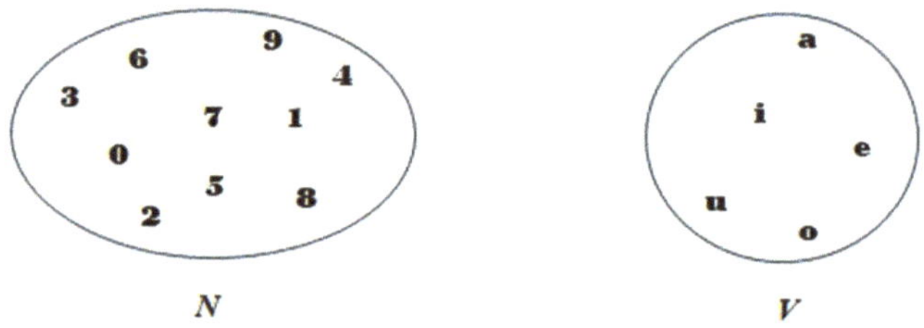

FIGURE 2.1

Sets N and V as described in the text.

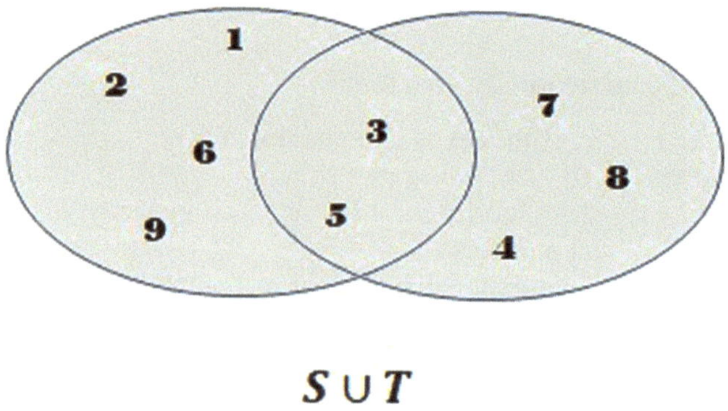

FIGURE 2.2

Union of two sets S and T.

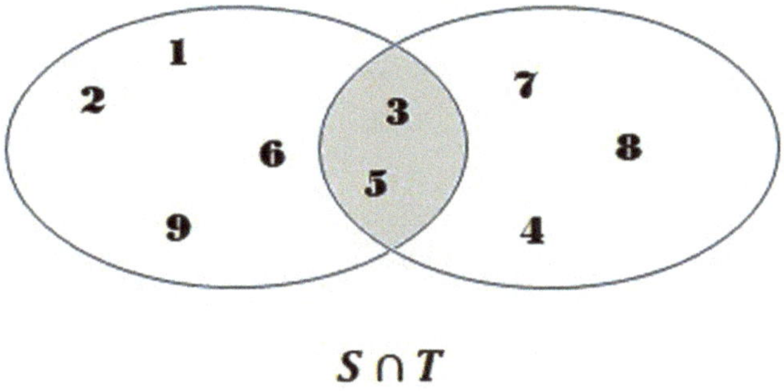

FIGURE 2.3

Intersection of two sets S and T (shaded).

Intersection. The intersection of two sets S and T is the collection of all elements which belong to both S and T (Fig. 2.3). Symbolically,

$$S \cap T = \{x : (x \in S) \text{ and } (x \in T)\}$$
$$= \{x : (x \in S) \wedge (x \in T)\}$$

Example 2.2.4: If $S = \{2, 3, 9, 5, 6, 1\}$ and $T = \{5, 4, 8, 3, 7\}$, as above, then

$$S \cap T = \{3, 5\}$$

Disjoint sets. If two sets have no elements in common, they are said to be disjoint. That is,

$$S \text{ and } T \text{ are disjoint if and only if } S \cap T = \varnothing$$

Set-theoretic difference. This operation subtracts the elements of one set from another. The difference between two sets S and T is the collection of elements in S that are not in T (Fig. 2.4). It is expressed as

$$S - T = \{x : (x \in S) \wedge (x \notin T)\}$$

Example 2.2.5: If $S = \{2, 3, 9, 5, 6, 1\}$ and $T = \{5, 4, 8, 3, 7\}$, as above, then

$$S - T = \{1, 2, 6, 9\}$$

2.2.2 Subset

For two sets A and B, it is said that A is a subset of B if and only if every element of A is also an element of B. The relation between two such sets is denoted by $A \subseteq B$. If $A \subseteq B$ and $A \neq B$, we call A a proper subset of B and denote the relation by $A \subset B$.

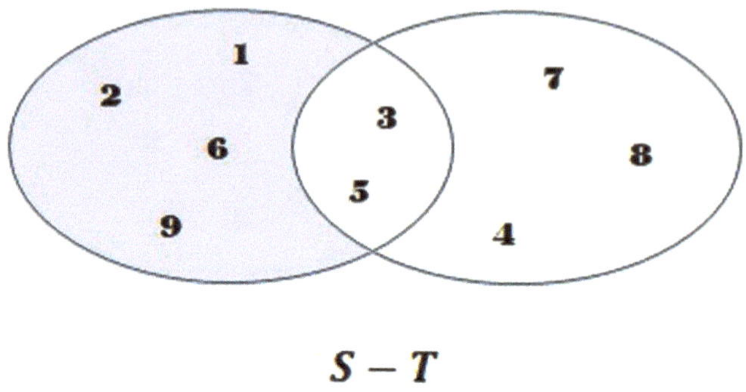

FIGURE 2.4

Set-theoretic difference between S and T (shaded).

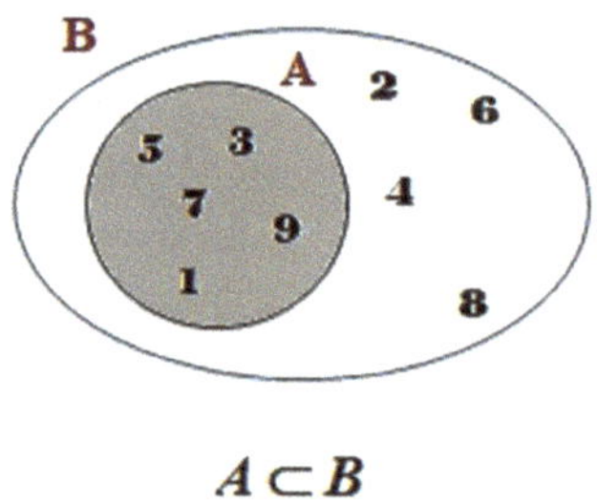

FIGURE 2.5

$A \subset B$ (shaded).

It can be proved that two sets S and T are equal if and only if each is a subset of the other. Symbolically,

$$S = T \Leftrightarrow (S \subseteq T) \wedge (T \subseteq S)$$

Example 2.2.6: If $A = \{1, 3, 5, 7, 9\}$ and $B = \{1, 2, 3, 4, 5, 6, 7, 8, 9\}$ then $A \subset B$ (Fig. 2.5).

Example 2.2.7: If $T = \{x, y, z\}$, then T has eight different subsets.

$$\begin{matrix} \varnothing & \{x\} & \{y\} & \{z\} \\ \{x, y\} & \{y, z\} & \{z, x\} & \{x, y, z\} \end{matrix}$$

The empty set $\varnothing$ is a subset of every set.

In fact, it can be shown that a set S with n elements has 2^n subsets.

Exercise 2.2

1. If S is a set of vectors $\in \mathbb{R}^4$, identify which of the following vectors are not members of S.

   $(3, 7, -2, 0, 1)$, $(0, 3, -2, 2)$, $(0, 0, 0, 0)$, $(7, -5, 2, -1)$, $(0, 0, 0)$, $(1, 2, -1, 0, 0)$, $(0, 0, 0, 1)$

2. S is a set of polynomials $P(x)$ having a root $x = 1$. Identify which of the following polynomials belong to the set.
   a. $-2x^2 - 3x + 5$
   b. $2x^2 - 7x + 1$
   c. $-2x^2 + 3x + 5$
   d. $x^6 + 4x^3 - 5x$
   e. $x^5 + 4x^3 - 5$
3. S is the set of all points (x, y) in $\mathbb{R}^2$ and T is a set of all points for which $x \geq 0$ and $y \leq 0$. Is S a subset of T? Is T a subset of $\mathbb{R}^2$?
4. $S = \{$all points in the xy-plane$\}$ $\quad T = \{$all points in the yz-plane$\}$

   What is $S \cap T$?
5. S is a set of k number of vectors in $\mathbb{R}^n$. How many subsets does S have?

2.3 Vector spaces and subspaces

2.3.1 Vector space

The formal definition of a vector space stipulates the relationship between the following:

(a) two sets—V, a nonempty set of vectors, and F, a scalar field (either a field $\mathbb{R}$ of real numbers or a field $\mathbb{C}$ of complex numbers), and
(b) two algebraic operations— addition and scalar multiplication.

In a general notion of a vector space, scalars can come from a mathematical structure known as a "field." They can be either real numbers or complex numbers—vector spaces with real scalars are called ***real vector spaces*** and those with complex scalars are called ***complex vector spaces***.

Definition 2.3.1 Axiomatic definition of a vector space.

V is called a vector space over F, if the following axioms are satisfied:

1. $\mathbf{x} + \mathbf{y} \in V$ for all $\mathbf{x}, \mathbf{y} \in V$ (closure property for vector addition).
2. $(\mathbf{x} + \mathbf{y}) + \mathbf{z} = \mathbf{x} + (\mathbf{y} + \mathbf{z})$ for every $\mathbf{x}, \mathbf{y}, \mathbf{z} \in V$.
3. $\mathbf{x} + \mathbf{y} = \mathbf{y} + \mathbf{x}$ for every $\mathbf{x}, \mathbf{y} \in V$.
4. There is an element $\mathbf{0} \in V$, called zero vector, such that $\mathbf{x} + \mathbf{0} = \mathbf{x}$ for every $\mathbf{x} \in V$.
5. For every $\mathbf{x} \in V$, there is an element $(-\mathbf{x}) \in V$ such that $\mathbf{x} + (-\mathbf{x}) = \mathbf{0}$

(Continued)

6. $k\,\mathbf{x} \in V$ for all $k \in F$ and $\mathbf{x} \in V$ (closure property for scalar multiplication).
7. $k(\mathbf{x}+\mathbf{y}) = k\,\mathbf{x}+k\,\mathbf{y}$ for every $k \in F$ and all $\mathbf{x}, \mathbf{y} \in V$.
8. $(k+m)\,\mathbf{x} = k\,\mathbf{x}+m\,\mathbf{x}$ for all $k, m \in F$ and every $\mathbf{x} \in V$.
9. $(k\,m)\,\mathbf{x} = k(m\,\mathbf{x})$ for all $k, m \in F$ and every $\mathbf{x} \in V$
10. $\mathbf{I}\,\mathbf{x} = \mathbf{x}$ for every $\mathbf{x} \in V$.

Here are some examples of vector spaces.

Example 2.3.1: If V contains a single element denoted by $\mathbf{0}$ and defined by

$$\mathbf{0}+\mathbf{0}=\mathbf{0} \text{ and } \lambda\mathbf{0}=\mathbf{0}$$

for all scalars λ, it is called the ***zero vector space***.

Example 2.3.2: **Real vector space $\mathbb{R}^n$**. For $\mathbf{u}, \mathbf{v} \in \mathbb{R}^n$ and $\lambda \in \mathbb{R}$

$$\mathbf{u}+\mathbf{v} = (u_1, u_2, \ldots\ldots, u_n) + (v_1, v_2, \ldots\ldots, v_n) = (u_1+v_1, u_2+v_2, \ldots\ldots, u_n+v_n) \in \mathbb{R}^n$$

and

$$\lambda\,\mathbf{u} = (\lambda u_1, \lambda u_2, \ldots\ldots, \lambda u_n) \in \mathbb{R}^n$$

that is, the set $V = \mathbb{R}^n$ is closed under addition and scalar multiplication. $\mathbb{R}^n$ is a finite-dimensional vector space.

Example 2.3.3: **Infinite-dimensional vector space**. Let $\mathbf{u}$ and $\mathbf{v}$ be infinite sequences of real numbers.

$$\mathbf{u} = (u_1, u_2, \ldots\ldots, u_n, \ldots\ldots) \text{ and } \mathbf{v} = (v_1, v_2, \ldots\ldots, v_n, \ldots\ldots)$$

For $\mathbf{u}, \mathbf{v} \in V$, and $\lambda \in \mathbb{R}$,

$$\mathbf{u}+\mathbf{v} = (u_1, u_2, \ldots\ldots, u_n, \ldots) + (v_1, v_2, \ldots\ldots, v_n, \ldots)$$

$$= (u_1+v_1, u_2+v_2, \ldots\ldots, u_n+v_n, \ldots) \in V$$

and

$$\lambda\,\mathbf{u} = (\lambda u_1, \lambda u_2, \ldots\ldots, \lambda u_n, \ldots\ldots) \in V$$

The set V containing objects which are infinite sequences of real numbers is denoted by $\mathbb{R}^{\propto}$.

Example 2.3.4: *Vector space of polynomials.*

Let us consider a polynomial

$$P(x) = a_0 + a_1 x + a_2 x^2 + \cdots\cdots + a_n x^n$$

where $a_0, a_1, a_2, \ldots\ldots, a_n$ are real numbers called the coefficients of the polynomial. If all the coefficients are zero, it is called a **zero polynomial**. Now, for $n \geq 1$, let P_n denote the set of all polynomials of degree $\leq n$. It can be seen that the sums and scalar multiple of all polynomials in P_n are again in P_n. Then, P_n, together with the zero polynomial, constitutes a vector space.

Vector space of $m \times n$ matrices

For matrices $\mathbf{A}, \mathbf{B} \in \mathbb{R}^{m\times n}$, $m, n \in \mathbb{N}$,

addition: $\mathbf{A} + \mathbf{B} =$

$$\begin{bmatrix} a_{11} & \cdots & a_{1n} \\ \vdots & \ddots & \vdots \\ a_{m1} & \cdots & a_{mn} \end{bmatrix} + \begin{bmatrix} b_{11} & \cdots & b_{1n} \\ \vdots & \ddots & \vdots \\ b_{m1} & \cdots & b_{mn} \end{bmatrix} = \begin{bmatrix} a_{11}+b_{11} & \cdots & a_{1n}+b_{1n} \\ \vdots & \ddots & \vdots \\ a_{m1}+b_{m1} & \cdots & a_{mn}+b_{mn} \end{bmatrix} \in \mathbb{R}^{m\times n}$$

and

multiplication by a scalar λ: $\lambda\mathbf{A} = \lambda \begin{bmatrix} a_{11} & \cdots & a_{1n} \\ \vdots & \ddots & \vdots \\ a_{m1} & \cdots & a_{mn} \end{bmatrix} = \begin{bmatrix} \lambda a_{11} & \cdots & \lambda a_{1n} \\ \vdots & \ddots & \vdots \\ \lambda a_{m1} & \cdots & \lambda a_{mn} \end{bmatrix} \in \mathbb{R}^{m\times n}$

that is, the vector space $V = \mathbb{R}^{m\times n}$ is closed under addition (axiom 1) and scalar multiplication (axiom 6). It can be shown that the other axioms 2–5 and 7–10 are also valid for $V = \mathbb{R}^{m\times n}$. The vector space of 2×2 matrices, which is easier to numerically illustrate, is a special case of $V = \mathbb{R}^{m\times n}$ where $m = n = 2$. It is to be noted that V must include

$$\mathbf{0}_{22} = \begin{bmatrix} 0 & 0 \\ 0 & 0 \end{bmatrix}$$

2.3.2 Vector subspaces

Vector subspaces are sets within the original vector space with the property that when vector space operations (addition and scalar multiplication) are performed within an individual subspace, we never leave it, that is, the "closure conditions" apply. Vector subspaces have been found to be useful for dimensionality reduction.

Definition 2.3.2 Formal definition of vector subspace.

Let us consider a nonempty subset S of a vector space V over scalar field F. If S is also a vector space over F under addition and multiplication defined on V, then S is said to be a subspace of V.

To rephrase, $S \subseteq V$, $S \neq \emptyset$, is a subspace of V if and only if

(a) $\mathbf{x}, \mathbf{y} \in S \Rightarrow \mathbf{x} + \mathbf{y} \in S$
(b) $\mathbf{x} \in S \Rightarrow \lambda\mathbf{x} \in S$ for all $\lambda \in F$.

Here are some examples of vector subspaces.

Example 2.3.5: Trivial subspace.

If in a vector space, $S = \{\mathbf{0}\}$ is the subset that contains the zero vector only, S is the zero subspace of V since

$\mathbf{0} + \mathbf{0} = \mathbf{0}$ and $\lambda\,\mathbf{0} = \mathbf{0}$ for any scalar λ
S is also called a trivial subspace of V.
Similarly, V is a trivial subspace of itself.

Example 2.3.6: Subspaces of $\mathbb{R}^2$.

If S is a straight line through the origin in $\mathbb{R}^2$, then adding two vectors (say, $\mathbf{x}$ and $\mathbf{y}$) on the line or multiplying a vector on the line by a scalar produces a vector on the same line (Fig. 2.6), that is S is closed under addition and scalar multiplication. Therefore straight lines passing through the origin are subspaces of $\mathbb{R}^2$. However, straight lines not passing through the origin cannot be subspaces as they do not contain the zero vector.

Example 2.3.7: Subspaces of $\mathbb{R}^3$.

As in $\mathbb{R}^2$, in $\mathbb{R}^3$ straight lines passing through the origin are subspaces. In addition, any plane P passing through the origin is also a subspace of $\mathbb{R}^3$ (Fig. 2.7).

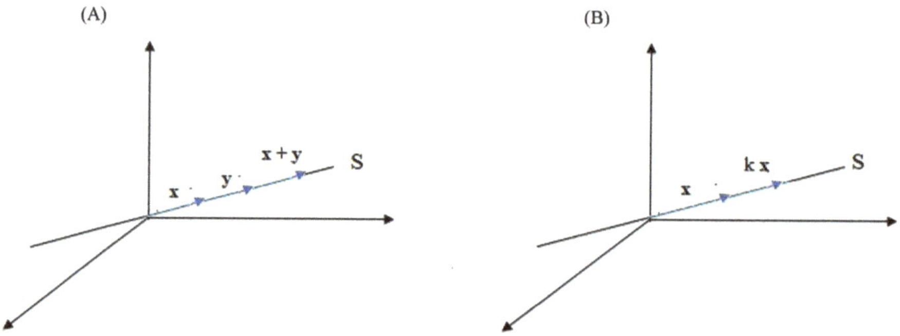

FIGURE 2.6

(A) S is closed under addition and (B) S is closed under scalar multiplication.

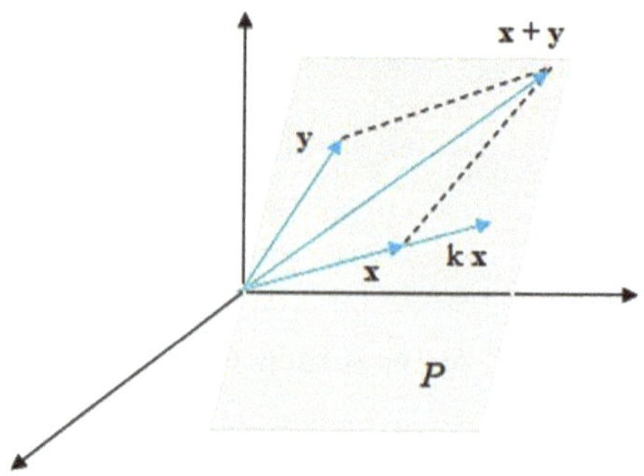

FIGURE 2.7

P is a subspace of $\mathbb{R}^3$.

Example 2.3.8: $\mathbb{R}^2$ not a subspace of $\mathbb{R}^3$.

The vectors in $\mathbb{R}^3$ have three entries whereas the vectors in $\mathbb{R}^2$ have only two. Hence $\mathbb{R}^2$ is not a subspace of $\mathbb{R}^3$. However, the set

$$W = \left\{ \begin{bmatrix} s \\ t \\ 0 \end{bmatrix} : s, t \in R \right\}$$

which appears to be equivalent to a plane passing through the origin is actually a subset of $\mathbb{R}^3$. It can be seen that the zero vector is in W. Besides, W is closed under both addition and scalar multiplication since these operations on vectors in W invariably produce vectors, the third entries of which are zero. Hence, W is a subspace of $\mathbb{R}^3$.

Example 2.3.9: A subset that is not a subspace.

In $\mathbb{R}^2$ consider the subset

$$\mathrm{W} = \left\{ \begin{bmatrix} \mathrm{s} \\ \mathrm{t} \end{bmatrix} : \mathrm{s} \geq 0, \mathrm{t} \geq 0 \right\}$$

The zero vector is in W, and W is closed under vector addition. However, for a vector (say) $\mathbf{v} = (1,\ 1) \in \mathrm{W}$, $(-1)\ \mathbf{v} = (-1,\ -1) \notin \mathrm{W}$. That is, W is not closed under scalar multiplication. Hence, W is not a subspace of $\mathbb{R}^2$.

2.3.3 Null space/row space/column space

Let us consider a matrix $\mathbf{A} \in \mathbb{R}^{m \times n}$

$$\mathbf{A} = \begin{bmatrix} a_{11} & a_{12} \cdots\cdots\cdots & a_{1n} \\ a_{21} & a_{22} \cdots\cdots\cdots & a_{2n} \\ & \cdots\cdots\cdots\cdots\cdots\cdots & \\ & \cdots\cdots\cdots\cdots\cdots\cdots & \\ a_{m1} & a_{m2} \ \cdots\cdots & a_{mn} \end{bmatrix}$$

If we write

$$\mathbf{r}_1 = [a_{11} \quad a_{12} \quad \cdots\cdots \quad a_{1n}]$$

$$\mathbf{r}_2 = [a_{21} \quad a_{22} \quad \cdots\cdots \quad a_{2n}]$$

$$\cdots\cdots\cdots\cdots\cdots$$

$$\cdots\cdots\cdots\cdots\cdots$$

$$\mathbf{r}_m = [a_{m1} \quad a_{m2} \quad \cdots\cdots \quad a_{mn}]$$

then $\mathbf{r}_1, \mathbf{r}_2, \ldots\ldots, \mathbf{r}_m \in \mathbb{R}^n$, formed by the rows of $\mathbf{A}$, are called the **row vectors** of $\mathbf{A}$.

Similarly, if we write

$$\mathbf{c}_1 = \begin{bmatrix} a_{11} \\ a_{21} \\ \cdot \\ \cdot \\ a_{m1} \end{bmatrix} \quad \mathbf{c}_2 = \begin{bmatrix} a_{12} \\ a_{22} \\ \cdot \\ \cdot \\ a_{m2} \end{bmatrix}, \ldots\ldots, \quad \mathbf{c}_n = \begin{bmatrix} a_{1n} \\ a_{2n} \\ \cdot \\ \cdot \\ a_{mn} \end{bmatrix}$$

then $\mathbf{c}_1, \mathbf{c}_2, \ldots\ldots, \mathbf{c}_n \in \mathbb{R}^m$, formed by the columns of **A**, are called the **column vectors** of A.

Example 2.3.10: .

Let us consider a 4×3 matrix

$$\mathbf{A} = \begin{bmatrix} 4 & 0 & 2 \\ -1 & 3 & 1 \\ 5 & -3 & 2 \\ 1 & 4 & 0 \end{bmatrix}$$

The row vectors of **A** are

$$\mathbf{r}_1 = [4 \ \ 0 \ \ 2], \ \mathbf{r}_2 = [-1 \ \ 3 \ \ 1], \ \mathbf{r}_3 = [5 - 3 \ \ 2], \text{and } \mathbf{r}_4 = [1 \ \ 4 \ \ 0]$$

and the column vectors of **A** are

$$\mathbf{c}_1 = \begin{bmatrix} 4 \\ -1 \\ 5 \\ 1 \end{bmatrix}, \mathbf{c}_2 = \begin{bmatrix} 0 \\ 3 \\ -3 \\ 4 \end{bmatrix}, \mathbf{c}_3 = \begin{bmatrix} 2 \\ 1 \\ 2 \\ 0 \end{bmatrix}$$

We can now define three important vector spaces associated with a matrix.

Definition 2.3.3. For a matrix $\mathbf{A} \in \mathbb{R}^{m \times n}$, the subspace of $\mathbb{R}^n$ spanned by the row vectors is called its ***row space***, while the subspace of $\mathbb{R}^m$ spanned by the column vectors is called its ***column space***. Further, the solution space of the homogeneous system of equations $\mathbf{Ax} = \mathbf{0}$ is called the ***null space*** of **A**. The null space contains all possible linear combinations of the elements in $\mathbb{R}^n$ that produce $\mathbf{0} \in \mathbb{R}^m$.

Evidently, the product **Ax** can be expressed as a linear combination of the column vectors

$$\mathbf{Ax} = x_1 \mathbf{c}_1 + x_2 \mathbf{c}_2 + \cdots\cdots + x_n \mathbf{c}_n \tag{2.5}$$

so that a linear system of m equations of n unknows can be written as

$$\mathbf{Ax} = x_1 \mathbf{c}_1 + x_2 \mathbf{c}_2 + \cdots\cdots + x_n \mathbf{c}_n = \mathbf{b} \tag{2.6}$$

Therefore, **b** is in the column space of **A**.

Let us illustrate this in the following example.

Example 2.3.11:

Let $\mathbf{Ax} = \mathbf{b}$ be the linear system

$$\begin{bmatrix} 2 & -3 & 5 \\ -1 & 2 & -2 \\ 3 & 1 & -1 \end{bmatrix} \begin{bmatrix} x_1 \\ x_2 \\ x_3 \end{bmatrix} = \begin{bmatrix} 16 \\ -6 \\ 4 \end{bmatrix}$$

and let us determine if $\mathbf{b}$ is in the column space of $\mathbf{A}$. It has to be shown that $\mathbf{b}$ can be expressed as a linear combination of the column vectors of $\mathbf{A}$.

Now, on solving the system it can be found that

$$x_1 = 2,\ x_2 = 1,\ x_3 = 3$$

With these values we have

$$\mathbf{A}\mathbf{x} = 2\begin{bmatrix} 2 \\ -1 \\ 3 \end{bmatrix} + 1\begin{bmatrix} -3 \\ 2 \\ 1 \end{bmatrix} + 3\begin{bmatrix} 5 \\ -2 \\ -1 \end{bmatrix} = \begin{bmatrix} 16 \\ -6 \\ 4 \end{bmatrix} = \mathbf{b}$$

The above example illustrates a specific case where a vector can be expressed as a linear combination of a set of three other vectors. In the following section the generalized concept of linear combination and linear independence of vectors will be discussed.

Exercise 2.3

1. State, giving reasons, if each of the following sets equipped with the given operations is a vector space.
   a. The set of all real numbers with standard operations of addition and multiplication.
   b. The set of all pairs of real numbers (m, n) where $m \geq 0$, with the standard operations on $\mathbb{R}^2$.
   c. The set of all 2×2 matrices of the form

$$\begin{bmatrix} \lambda_1 & 0 \\ 0 & \lambda_2 \end{bmatrix}$$

2. Are the following sets, along with given operations, vector spaces? Give reasons for your answer.
   a. The set of all polynomials of degree ≥ 4 together with 0; operations of P.
   b. The set of $1, y, y^2, \ldots\ldots\ldots$; operations of P.
3. Is the straight line $y = 5x + 3$ a subspace of $\mathbb{R}^2$?
4. State, giving reasons, if $U = \{x f(x) \mid f(x) \text{ in } P_3\}$ a subspace of P_3.
5. Are the following subspaces of $\mathbb{R}^3$?
   a. All vectors of the form (u_1, u_2, u_3), where $u_2 = u_1 + u_3$.
   b. All vectors of the form $(x_1, x_2, 0)$.
6. A set of 2×2 invertible matrices is not a subspace of $\mathbf{M}_{22}$—explain.

2.4 Liner combination/linear independence

Now that we have some preliminary idea about vector spaces, we can expect that it should be possible to find in a vector space a set of vectors with which every vector can be represented by adding them together and scaling them. In order to do so it is necessary to introduce the concepts of linear combination and linear independence.

In a rectangular xy-coordinate system $\mathbf{i} = (1, 0)$ and $\mathbf{j} = (0, 1)$ are the standard unit vectors. A vector $\mathbf{v}$ (v_1, v_2) in the plane can be expressed as

$$\mathbf{v}(v_1, v_2) = a_1\mathbf{i} + a_2\mathbf{j} = a_1(1, 0) + a_2(0, 1) \tag{2.7}$$

which can also be written as

$$a_1\mathbf{i} + a_2\mathbf{j} - \mathbf{v}(v_1, v_2) = \mathbf{0}$$

where $a_1 = v_1$ and $a_2 = v_2$. However, the equation

$$a_1\mathbf{i} + a_2\mathbf{j} = \mathbf{0}$$

has only the trivial solution, that is, $a_1 = a_2 = 0$, since $\mathbf{i}$ and $\mathbf{j}$ are independent of each other (one cannot be expressed in terms of terms of the other). On the other hand, $\mathbf{v}$ is dependent on $\mathbf{i}$ and $\mathbf{j}$ and, as (2.7) shows, it can be expressed as a linear combination of $\mathbf{i}$ and $\mathbf{j}$. It is to be noted that a pair of vectors is linearly dependent if and only if one is a scalar multiple of the other.

Similarly, a vector $\mathbf{v} \in \mathbb{R}^3$ can be expressed as a linear combination of $\mathbf{i} = (1, 0, 0)$, $\mathbf{j} = (0, 1, 0)$, and $\mathbf{k} = (0, 0, 1)$.

$$\mathbf{v}(v_1, v_2, v_3) = a_1\mathbf{i} + a_2\mathbf{j} + a_3\mathbf{k} = a_1(1, 0, 0) + a_2(0, 1, 0) + a_3(0, 0, 1) \tag{2.8}$$

and the equation

$$a_1\mathbf{i} + a_2\mathbf{j} + a_3\mathbf{k} = \mathbf{0}$$

has a trivial solution, $\mathrm{a}_1 = \mathrm{a}_2 = \mathrm{a}_3 = 0$

Generalized concept

A vector $\mathbf{w} \in V$ is said to be a linear combination of the vectors $\mathbf{w}_1, \mathbf{w}_2, \ldots\ldots, \mathbf{w}_r \in V$ if it can be expressed in the form

$$\mathbf{w} = \lambda_1\mathbf{w}_1 + \lambda_2\mathbf{w}_2 + \cdots\cdots\cdots + \lambda_r\mathbf{w}_r = \sum_{i=1}^{r} \lambda_i\mathbf{w}_i \tag{2.9}$$

where $\lambda_1, \lambda_2, \ldots\ldots\ldots\ldots, \lambda_r \in \mathbb{R}$. These scalars are called the coefficients of the linear combination.

If $S = \{\mathbf{w}_1, \mathbf{w}_2, \ldots\ldots, \mathbf{w}_r\}$ is a nonempty set of vectors in a vector space V, it can be proved that the set W of all possible linear combinations of the vectors in S is a subspace of V. Further, the subspace W of V, which contains all possible linear combinations of the vectors in S, is called the subspace of V generated by S and denoted as

$$W = \textbf{span}\ \{\mathbf{w}_1,\ \mathbf{w}_2, \cdots\cdots, \mathbf{w}_r\}\ \text{or}\ W = \textbf{span}\ (S)$$

Then,

$$\mathbf{w} \in \textbf{span}\ \{\mathbf{w}_1,\ \mathbf{w}_2, \cdots\cdots, \mathbf{w}_r\}$$

A set $S = \{\mathbf{w}_1, \mathbf{w}_2, \ldots, \mathbf{w}_r\}$ of two or more vectors in a vector space V is said to be a linearly independent set if no vector in S can be expressed as a linear combination of the others. On the other hand, the set S is linearly dependent if at least one member of the set can be expressed as a linear combination of the rest of the set.

The above definition of linear dependence/independence can be restated as follows:

Definition 2.4.1 A set of vectors $S = \{\mathbf{w}_1, \mathbf{w}_2, \ldots\ldots, \mathbf{w}_r\} \in V$ is said to be ***linearly dependent*** if there is a nontrivial linear combination, such that

$$\sum_{i=1}^{r} \lambda_i \mathbf{w}_i = 0 \tag{2.10}$$

with at least one $\lambda_i \neq 0$. If only trivial solution exists, that is, $\lambda_1 = \lambda_2 = \ldots\ldots = \lambda_r = 0$, the set $S = \{\mathbf{w}_1, \mathbf{w}_2, \ldots\ldots, \mathbf{w}_r\}$ is said to be ***linearly independent***.

In Chapter 1 (Section 1.5), we have seen for an $n \times n$ invertible matrix $\mathbf{A}$, the homogenous system $\mathbf{Ax} = \mathbf{0}$ has only the trivial solution $\mathbf{x} = \mathbf{0}$. Now, if $\mathbf{c}_1, \mathbf{c}_2, \ldots\ldots, \mathbf{c}_n$ are the columns of $\mathbf{A}$ and we write

$$\mathbf{x} = \begin{bmatrix} x_1 \\ x_2 \\ \cdot \\ \cdot \\ \cdot \\ x_n \end{bmatrix}$$

so that from (2.10)

$$\mathbf{Ax} = x_1\,\mathbf{c}_1 + x_2\,\mathbf{c}_2 + \cdots\cdots + x_n\,\mathbf{c}_n = \sum_{i=1}^{n} x_i\,\mathbf{c}_i = \mathbf{0}$$

and we have the following theorem.

Theorem 2.4.1 The columns $\{\ \mathbf{c}_1, \mathbf{c}_2, \ldots\ldots, \mathbf{c}_n\ \}$ of a matrix $\mathbf{A}$ is a linearly independent set if and only if $\mathbf{Ax} = \mathbf{0}$ implies $\mathbf{x} = \mathbf{0}$.

Hence, it can be said that the columns of an invertible matrix $\mathbf{A}$ are linearly independent. Similarly, it can be shown that the rows of an invertible matrix

A are linearly independent. It is to be noted that while invertibility relates to only square matrices, Theorem 2.4.1 holds for any $m \times n$ matrix.

Example 2.4.1: Consider vectors $\mathbf{u} = (2, -1, 2, 1)$ and $\mathbf{v} = (3, 4, -1, 1) \in \mathbb{R}^4$. Let us determine if each of the vectors $\mathbf{w} = (0, -11, 8, 1)$ and $\mathbf{w}' = (2, 3, 1, 2) \in \mathbb{R}^4$ can be expressed as a linear combination of **u** and **v**.

Solution. For w to be a linear combination of **u** and **v**, there must be scalars k_1 and k_2 such that

$$\mathbf{w} = k_1\mathbf{u} + k_2\mathbf{v}$$

that is,

$$\begin{bmatrix} 0 \\ -11 \\ 8 \\ 1 \end{bmatrix} = k_1 \begin{bmatrix} 2 \\ -1 \\ 2 \\ 1 \end{bmatrix} + k_2 \begin{bmatrix} 3 \\ 4 \\ -1 \\ 1 \end{bmatrix}$$

Equating corresponding components gives equations

$$2k_1 + 3k_2 = 0$$

$$-k_1 + 4k_2 = -11$$

$$2k_1 - k_2 = 8$$

$$k_1 + k_2 = 1$$

This linear system has a solution: $k_1 = 3$, $k_2 = -2$, so that

$$\mathbf{w} = 3\mathbf{u} - 2\mathbf{v}$$

Similarly, for $\mathbf{w}'$ to be a linear combination of **u** and **v**, there must be scalars k_1 and k_2 such that

$$\mathbf{w}' = k_1\mathbf{u} + k_2\mathbf{v}$$

that is,

$$\begin{bmatrix} 2 \\ 3 \\ 1 \\ 2 \end{bmatrix} = k_1 \begin{bmatrix} 2 \\ -1 \\ 2 \\ 1 \end{bmatrix} + k_2 \begin{bmatrix} 3 \\ 4 \\ -1 \\ 1 \end{bmatrix}$$

Equating corresponding components gives equations

$$2k_1 + 3k_2 = 2$$

$$-k_1 + 4k_2 = 3$$

$$2k_1 - k_2 = 1$$

$$k_1 + k_2 = 2$$

It can be verified that this system of equations is "inconsistent"—it has no solution. Consequently, $\mathbf{w}'$ cannot be expressed as a linear combination of $\mathbf{u}$ and $\mathbf{v}$.

Example 2.4.2: To show if the vectors $\mathbf{v}_1 = (3, 0, -3, 6)$, $\mathbf{v}_2 = (0, 2, 3, 1)$, $\mathbf{v}_3 = (0, -2, -2, 0)$, and $\mathbf{v}_4 = (-2, 1, 2, 1) \in \mathbb{R}^4$ are linearly independent (or dependent).

The set of vectors $S = \{\mathbf{v}_1, \mathbf{v}_2, \mathbf{v}_3, \mathbf{v}_4\} \in \mathbb{R}^4$ is linearly independent if and only if coefficients satisfy the vector equation

$$\lambda_1 \mathbf{v}_1 + \lambda_2 \mathbf{v}_2 + \lambda_3 \mathbf{v}_3 + \lambda_4 \mathbf{v}_4 = \mathbf{0}$$

are $\lambda_1 = \lambda_2 = \lambda_3 = \lambda_4 = 0$. The above equation can be rewritten in the component form as

$$\lambda_1 (3, 0, -3, 6) + \lambda_2 (0, 2, 3, 1) + \lambda_3 (0, -2, -2, 0) + \lambda_4 (-2, 1, 2, 1) = (0, 0, 0, 0)$$

which yields the homogenous linear system

$$\begin{array}{ccccccccc} 3\lambda_1 & & & + & -3\lambda_3 & + & 6\lambda_4 & = & 0 \\ & & 2\lambda_2 & + & 3\lambda_3 & + & \lambda_4 & = & 0 \\ & & -2\lambda_2 & - & 2\lambda_3 & & & = & 0 \\ -2\lambda_1 & + & \lambda_2 & + & 2\lambda_3 & + & \lambda_4 & = & 0 \end{array}$$

It can be verified that the system has only trivial solution, that is

$$\lambda_1 = 0,\ \lambda_2 = 0,\ \lambda_3 = 0,\ \lambda_4 = 0$$

so that it can be concluded that $\mathbf{v}_1$, $\mathbf{v}_2$, $\mathbf{v}_3$, and $\mathbf{v}_4$ are linearly independent.

Exercise 2.4

1. Let $\mathbf{x} = (2, 1, -1)$, $\mathbf{y} = (1, 0, 1)$ and $\mathbf{z} = (1, 1, -2)$. Determine if each of the vectors $\mathbf{u} = (4, 3, 4)$ and $\mathbf{v} = (3, 0, 3)$ can be expressed as a linear combination of $\mathbf{x}$, $\mathbf{y}$ and $\mathbf{z}$.
2. Let $\mathbf{u} = (2, 1, 0, 3)$ and $\mathbf{v} = (-1, 2, 1, 0) \in \mathbb{R}^4$. Determine if each of the vectors $\mathbf{w} = (3, 0, -1, 6)$ and $\mathbf{w}' = (2, 1, -1, 4) \in \mathbb{R}^4$ can be expressed as a linear combination of $\mathbf{u}$ and $\mathbf{v}$.
3. $\mathbf{A} = \begin{bmatrix} 2 & -1 & 3 \\ 0 & -2 & 1 \\ 1 & 4 & 0 \end{bmatrix}, \mathbf{x} = \begin{bmatrix} 2 \\ 3 \\ 4 \end{bmatrix}$
   Write the product $\mathbf{Ax}$ as a linear combination of the column vectors of $\mathbf{A}$.
4. $\mathbf{x} = (2, 1, -1, 2)$ and $\mathbf{y} = (1, 2, -1, 3) \in \mathbb{R}^4$. Determine if each of the vectors $\mathbf{u} = (4, 5, -3, 8)$ and $\mathbf{v} = (3, 2, 5, -4) \in \mathbb{R}^4$ lies in $W = \text{span}\ \{\mathbf{x}, \mathbf{y}\}$.

5. Determine if each of the following subsets is independent.
   a. $\{(1, 1, 1), (1, 0, 0), (1, -1, 1)\} \in \mathbb{R}^3$
   b. $\{(1, 0, 1, 0), (1, 1, 0, 0), (0, 1, 0, 1), (0, 0, 1, 1)\} \in \mathbb{R}^4$
6. Do the columns of the matrix

$$\mathbf{A} = \begin{bmatrix} 1 & -2 & 2 \\ 2 & 1 & 3 \\ 1 & 2 & -1 \end{bmatrix}$$

   form a linearly independent set in $\mathbb{R}^3$?

2.5 Basis vectors

We have seen above that a set of unit vectors **i** and **j**, which are "linearly independent," "span the 2-space," that is, any vector in 2-space can be expressed as a linear combination of **i** and **j**. Similarly, a set of unit vectors **i**, **j**, and **k**, which are also linearly independent, "spans" the 3-space. It can also be said that the vectors **i** and **j** form a "***basis***" for the 2-space while the vectors **i**, **j**, and **k** form a "***basis***" for the 3-space.

Now, we are in a position to extend the concept of "***basis vectors***" to general vector spaces. A vector space V can be either ***finite dimensional*** or ***infinite dimensional***. In a finite-dimensional vector space, there is a finite set of vectors that spans the space. In an infinite-dimensional vector space, no such set exists.

Definition 2.5.1 If $S = \{\mathbf{v}_1, \mathbf{v}_2, \ldots\ldots, \mathbf{v}_r\}$ is a linearly independent set of vectors that spans a finite-dimensional vector space V, S is called a **basis** for V.

It is to be noted that if any vector is dropped from the basis, it will no longer span the space. A **basis** for a space, therefore, is a **"minimal spanning set."** The basis is, at the same time, a "**maximal linearly independent set**"—an additional vector to the set will make it linearly dependent.

The following theorem states the uniqueness of basis representation.

Theorem 2.5.1 If $S = \{\mathbf{v}_1, \mathbf{v}_2, \ldots\ldots, \mathbf{v}_n\}$ is a **basis** for a vector space V, every vector $\mathbf{v} \in V$ can be expressed in the form

$$\mathbf{v} = c_1\mathbf{v}_1 + c_2\mathbf{v}_2 + \cdots\cdots + c_n\mathbf{v}_n$$

in exactly one way.

In such case, the scalars $c_1, c_2, \ldots\ldots, c_n$ are called the ***coordinates*** of **v** relative to the basis S. The vector $(c_1, c_2, \ldots\ldots, c_n) \in \mathbb{R}^n$ constructed from the coordinates and denoted by

$$(\mathbf{v})_S = (c_1, c_2, \ldots\ldots, c_n) \in \mathbb{R}^n$$

is called the ***coordinate vector*** of **v** relative to S.

Theorem 2.5.1 establishes a one-to-one correlation between vectors in $\mathbb{R}^n$ and vectors in V.

Example 2.5.1: The vectors $\mathbf{u}_1 = (1, -1)$ and $\mathbf{u}_2 = (2, 3)$ form a basis for $\mathbb{R}^2$. Let us find out the coordinate vector of $\mathbf{v} = (-1, -9)$ relative to the basis $S = \{\mathbf{u}_1, \mathbf{u}_2\}$.

Solution. **v** can be expressed as a linear combination of the vectors in S

$$\mathbf{v} = c_1\,\mathbf{u}_1 + c_2\,\mathbf{u}_2$$

$$= c_1\begin{bmatrix} 1 \\ -1 \end{bmatrix} + c_2\begin{bmatrix} 2 \\ 3 \end{bmatrix} = \begin{bmatrix} -1 \\ -9 \end{bmatrix}$$

From above, we have a system of linear equations

$$c_1 + 2c_2 = -1$$

$$-c_1 + 3c_2 = -9$$

The system can be solved to obtain $c_1 = 3$, $c_2 = -2$, so the coordinate vector

$$(\mathbf{v})_S = (3, -2)$$

Example 2.5.2: Let us consider a basis $S = \{\mathbf{u}_1, \mathbf{u}_2, \mathbf{u}_3\}$ for $\mathbb{R}^3$ where $\mathbf{u}_1 = (1, 0, 0)$, $\mathbf{u}_2 = (2, 2, 0)$, and $\mathbf{u}_3 = (3, 3, 3)$. In a manner similar to the earlier example, it is possible to find out the vector $\mathbf{v} \in \mathbb{R}^3$, the coordinate vector of which relative to S is $(\mathbf{v})_S = (3, -2, 1)$.

Solution. **v** can be expressed as a linear combination of the vectors in S

$$\mathbf{v} = c_1\,\mathbf{u}_1 + c_2\,\mathbf{u}_2 + c_3\,\mathbf{u}_3$$

where c_1, c_2, c_3 are the coordinates of the vector **v** relative to S. So, we have

$$\mathbf{v} = 3\begin{bmatrix} 1 \\ 0 \\ 0 \end{bmatrix} + (-2)\begin{bmatrix} 2 \\ 2 \\ 0 \end{bmatrix} + 1\begin{bmatrix} 3 \\ 3 \\ 3 \end{bmatrix} = \begin{bmatrix} 2 \\ -1 \\ 3 \end{bmatrix}$$

which can be also written as $\mathbf{v} = (2, -1, 3)$

The standard basis for $\mathbb{R}^n$ is constituted by the following **standard unit vectors**

$$\mathbf{e}_1 = (1, 0, 0, \ldots\ldots, 0),\ \mathbf{e}_1 = (1, 0, 0, \ldots\ldots, 0), \ldots\ldots\ldots\ldots \mathbf{e}_n = (0, 0, 0, \ldots\ldots, 1),$$

and every vector $\mathbf{v} = (v_1, v_2, \ldots\ldots\ldots\ldots, v_n) \in \mathbb{R}^n$ can be expressed as

$$\mathbf{v} = v_1\,\mathbf{e}_1 + v_2\,\mathbf{e}_2 + \cdots\cdots + v_n\,\mathbf{e}_n$$

In the particular case where $n = 3$, the unit vectors that form the standard basis for $\mathbb{R}^3$ are

$$\mathbf{i} = (1, 0, 0),\ \mathbf{j} = (0, 1, 0),\ \mathbf{k} = (0, 0, 1)$$

which we have seen earlier. Evidently, any vector $\mathbf{v} = (v_1, v_2, v_3) \in \mathbb{R}^3$ can be expressed as

$$\mathbf{v} = (v_1, v_2, v_3) = v_1(1, 0, 0) + v_2(0, 1, 0) + v_3(0, 0, 1).$$

Example 2.5.3: To show if the vectors $\mathbf{v}_1 = (3, 0, -6)$, $\mathbf{v}_2 = (-4, 1, 7)$, and $\mathbf{v}_3 = (-2, 1, 5)$ form a basis for $\mathbb{R}^3$.

We have to show that $S = \{\mathbf{v}_1, \mathbf{v}_2, \mathbf{v}_3\}$ is linearly independent and spans $\mathbb{R}^3$.
To prove linear independence, it must be shown that the vector equation

$$a_1\mathbf{v}_1 + a_2\mathbf{v}_2 + a_3\mathbf{v}_3 = \mathbf{0}$$

has only the trivial solution, that is, $a_1 = a_2 = a_3 = 0$. The above equation can be expressed as a homogeneous linear system

$$\mathbf{Ma} = \mathbf{0}$$

where the coefficient matrix

$$\mathbf{M} = \begin{bmatrix} 3 & -4 & -2 \\ 0 & 1 & 1 \\ -6 & 7 & 5 \end{bmatrix}$$

and $\mathbf{a} = (a_1, a_2, a_3)$. The homogeneous system has trivial solution if $\det(\mathbf{M}) \neq 0$. We can calculate $\det(\mathbf{M}) = 6 \neq 0$.

Further, to prove that S spans $\mathbb{R}^3$, it must be shown that for every vector $\mathbf{b} = (b_1, b_2, b_3) \in \mathbb{R}^3$ can be expressed as a linear combination of $\mathbf{v}_1$, $\mathbf{v}_2$, $\mathbf{v}_3$, that is

$$c_1\mathbf{v}_1 + c_2\mathbf{v}_2 + c_3\mathbf{v}_3 = \mathbf{b}$$

and the nonhomogeneous system

$$\mathbf{Mc} = \mathbf{b}$$

where $\mathbf{c} = (c_1, c_2, c_3)$ and $\mathbf{M}$ is the coefficient matrix. The nonhomogeneous system will be consistent for all values of b_1, b_2, and b_3 only if $\det(\mathbf{M}) \neq 0$ which has already been shown above.

So, we can conclude that the vectors $\mathbf{v}_1$, $\mathbf{v}_2$, and $\mathbf{v}_3$ form a basis for $\mathbb{R}^3$.

The standard basis for $m \times n$ matrices

Earlier we have introduced the vector space $V = \mathbb{R}^{m \times n}$ of $m \times n$ matrices. Let us now consider the vector space M_{22} of 2×2 matrices and determine if the matrices

$$\mathbf{M}_1 = \begin{bmatrix} 1 & 0 \\ 0 & 0 \end{bmatrix}, \mathbf{M}_2 = \begin{bmatrix} 0 & 1 \\ 0 & 0 \end{bmatrix}, \mathbf{M}_3 = \begin{bmatrix} 0 & 0 \\ 1 & 0 \end{bmatrix}, \mathbf{M}_4 = \begin{bmatrix} 0 & 0 \\ 0 & 1 \end{bmatrix}$$

form a basis for M_{22}.

Here, it has to be shown that the matrices are linearly independent and span M_{22}. Linear independence is proved if the equation

$$a_1\mathbf{M}_1 + a_2\mathbf{M}_2 + a_3\mathbf{M}_3 + a_4\mathbf{M}_4 = \mathbf{0} \tag{2.11}$$

where $\mathbf{0}$ is the 2×2 zero matrix and has only the trivial solution. The matrix form of (2.11) is

$$a_1\begin{bmatrix}1 & 0\\0 & 0\end{bmatrix} + a_2\begin{bmatrix}0 & 1\\0 & 0\end{bmatrix} + \mathrm{a}_3\begin{bmatrix}0 & 0\\1 & 0\end{bmatrix} + a_4\begin{bmatrix}0 & 0\\0 & 1\end{bmatrix} = \begin{bmatrix}0 & 0\\0 & 0\end{bmatrix}$$

or

$$\begin{bmatrix}a_1 & a_2\\a_3 & a_4\end{bmatrix} = \begin{bmatrix}0 & 0\\0 & 0\end{bmatrix}$$

which shows $a_1 = a_2 = a_3 = a_4 = 0$, that is, (2.11) has only the trivial solution.

Further, to prove that the matrices span M_{22}, it must be shown that every 2×2 matrix

$$\mathbf{P} = \begin{bmatrix}a & b\\c & d\end{bmatrix}$$

can be expressed as

$$c_1\mathbf{M}_1 + c_2\mathbf{M}_2 + c_3\mathbf{M}_3 + c_4\mathbf{M}_4 = \mathbf{P}$$

which, in the matrix form, can be written as

$$c_1\begin{bmatrix}1 & 0\\0 & 0\end{bmatrix} + \mathrm{c}_2\begin{bmatrix}0 & 1\\0 & 0\end{bmatrix} + c_3\begin{bmatrix}0 & 0\\1 & 0\end{bmatrix} + \mathrm{c}_4\begin{bmatrix}0 & 0\\0 & 1\end{bmatrix} = \begin{bmatrix}a & b\\c & d\end{bmatrix}$$

or

$$\begin{bmatrix}c_1 & c_2\\c_3 & c_4\end{bmatrix} = \begin{bmatrix}a & b\\c & d\end{bmatrix}$$

This equation has a nontrivial solution

$$c_1 = a,\ c_2 = b,\ c_3 = c,\ c_4 = d$$

indicating that the matrices span M_{22}.

Thus we can conclude that the matrices $\mathbf{M}_1$, $\mathbf{M}_2$, $\mathbf{M}_3$, and $\mathbf{M}_4$ form a basis for M_{22}.

In general, the mn different matrices whose entries are all zero except for a single entry of 1 form the ***standard basis*** for M_{mn}.

Exercise 2.5

1. $\mathbf{u}_1 = (2, -1)$ and $\mathbf{u}_2 = (1, 1)$ form a basis for $\mathbb{R}^2$. Find out the coordinate vector of $\mathrm{v} = (3, 1)$ relative to the basis $S = \{\mathbf{u}_1, \mathbf{u}_2\}$.
2. Determine if the vectors $\mathbf{v}_1 = (1, 2, 1)$, $\mathbf{v}_2 = (-2, 1, 2)$ and $\mathbf{v}_3 = (2, 3, -1)$ form a basis for $\mathbb{R}^3$.

3. Consider a basis $S = \{\mathbf{u}_1, \mathbf{u}_2, \mathbf{u}_3\}$ for $\mathbb{R}^3$ where $\mathbf{u}_1 = (1, 1, 0)$, $\mathbf{u}_2 = (0, 1, 1)$, and $\mathbf{u}_3 = (1, 0, 1)$. Find out the vector $\mathbf{v} \in \mathbb{R}^3$ the coordinate vector of which relative to S is $(\mathbf{v})_S = (3, -2, 1)$.
4. $\mathbf{A}_1 = \begin{bmatrix} 1 & 1 \\ 0 & 1 \end{bmatrix}, \mathbf{A}_2 = \begin{bmatrix} 0 & 1 \\ 1 & 0 \end{bmatrix}, \mathbf{A}_3 = \begin{bmatrix} 0 & 0 \\ 1 & 1 \end{bmatrix}, \mathbf{A}_4 = \begin{bmatrix} 0 & 0 \\ 0 & 1 \end{bmatrix}$

   Show that $S = \{\mathbf{A}_1, \mathbf{A}_2, \mathbf{A}_3, \mathbf{A}_4\}$ is a basis for M_{22}. Express

$$\mathbf{A} = \begin{bmatrix} 1 & 1 \\ 1 & 1 \end{bmatrix}$$

   as a linear combination of $\mathbf{A}_1$, $\mathbf{A}_2$, $\mathbf{A}_3$, and $\mathbf{A}_4$ and find the coordinate vector of $\mathbf{A}$ relative to S.

2.6 Dimension and rank

In geometry, $\mathbb{R}^3$ is considered to be three-dimensional, planes are said to be two-dimensional, and lines one-dimensional. However, the concept of "dimension" can be formalized based on the following theorem.

Theorem 2.6.1 Let $\{\mathbf{v}_1, \mathbf{v}_2, \ldots\ldots, \mathbf{v}_n\}$ be any basis for an n-dimensional vector space V, then,

a. if a set in V has more than n vectors, it is linearly dependent, and
b. if a set in V has fewer than n vectors, it does not span V

Evidently, the criteria for linear independence and spanning imply that unless a set in V has exactly n vectors it cannot be a basis. Further, it can also be proved that all bases for a finite-dimensional vector space have the same number of vectors.

We have seen that the standard basis for $\mathbb{R}^n$ has n vectors (two vectors for $\mathbb{R}^2$, three vectors for $\mathbb{R}^3$, and so on). There seems to be a link between the number of vectors in a basis and the dimension of a vector space. Therefore the following can be considered as a definition of dimension.

Definition 2.6.1 The ***dimension*** of a finite-dimensional vector space V is formally defined to be the number of vectors in a basis for V and denoted by

$$\dim(V)$$

Example 2.6.1: The dimension of the space spanned by a linearly independent set of vectors $S = \{\mathbf{v}_1, \mathbf{v}_2, \ldots\ldots, \mathbf{v}_r\}$ is equal to the number of vectors in that set, that is,

$$\dim\ [\text{span}\ \{\mathbf{v}_1,\ \mathbf{v}_2, \ldots\ldots, \mathbf{v}_r\}] = r$$

The zero vector space is zero-dimensional.

Example 2.6.2: Let us consider the homogeneous system

$$\begin{array}{rcrcrcrcrcl} x_1 & + & 3x_2 & - & 2x_3 & & & + & 2x_5 & = & 0 \\ 2x_1 & + & 6x_2 & - & 5x_3 & - & 2x_4 & + & 4x_5 & = & 0 \\ & & & & 5x_3 & + & 10x_4 & & & = & 0 \\ 2x_1 & + & 6x_2 & & & + & 8x_4 & + & 4x_5 & = & 0 \end{array}$$

and try to determine the dimension of its solution space.

The augmented matrix for the given system is

$$\begin{bmatrix} 1 & 3 & -2 & 0 & 2 & 0 \\ 2 & 6 & -5 & -2 & 4 & 0 \\ 0 & 0 & 5 & 10 & 0 & 0 \\ 2 & 6 & 0 & 8 & 4 & 0 \end{bmatrix}$$

the reduced row-echelon form of which generated by elementary row operations is

$$\begin{bmatrix} 1 & 3 & 0 & 4 & 2 & 0 \\ 0 & 0 & 1 & 2 & 0 & 0 \\ 0 & 0 & 0 & 0 & 0 & 0 \\ 0 & 0 & 0 & 0 & 0 & 0 \end{bmatrix}$$

The solution to the system in vector form can be calculated to be

$$(x_1, x_2, x_3, x_4, x_5) = (-3r - 4s - 2t, r, -2s, s, t)$$

which can be alternatively written as

$$\begin{bmatrix} x_1 \\ x_2 \\ x_3 \\ x_4 \\ x_5 \end{bmatrix} = r\begin{bmatrix} -3 \\ 1 \\ 0 \\ 0 \\ 0 \end{bmatrix} + s\begin{bmatrix} -4 \\ 0 \\ -2 \\ 1 \\ 0 \end{bmatrix} + t\begin{bmatrix} -2 \\ 0 \\ 0 \\ 0 \\ 1 \end{bmatrix}$$

where r, s, and t are arbitrary parameters.

This shows that the vectors

$$\mathbf{v}_1 = (-3, 1, 0, 0, 0), \ \mathbf{v}_2 = (-4, 0, -2, 1, 0), \ \text{and} \ \mathbf{v}_3 = (-2, 0, 0, 0, 1)$$

span the solution space. It can be shown that the equation

$$k_1 \mathbf{v}_1 + k_2 \mathbf{v}_2 + k_3 \mathbf{v}_3 = \mathbf{0}$$

is satisfied only when $k_1 = k_2 = k_3 = 0$. Hence, the vectors $\mathbf{v}_1$, $\mathbf{v}_2$, and $\mathbf{v}_3$ are linearly independent and the solution space has dimension = 3.

Example 2.6.3: Dimension of row space and column space.

Let us consider a matrix in its row-echelon form

$$\mathbf{A}_{re} = \begin{bmatrix} 1 & 3 & -1 & 0 \\ 0 & 1 & 2 & 0 \\ 0 & 0 & 0 & 1 \end{bmatrix}$$

It can be clearly seen that the row vectors with leading 1s (nonzero row vectors) are

$$\mathbf{r}_1 = [1 \;\; 3 - 1 \;\; 0], \quad \mathbf{r}_2 = [0 \;\; 1 \;\; 2 \;\; 0] \text{ and } \mathbf{r}_3 = [0 \;\; 0 \;\; 0 \;\; 1]$$

Let us determine the condition for

$$k_1\,\mathbf{r}_1 + k_2\,\mathbf{r}_2 + k_3\,\mathbf{r}_3 = \mathbf{0} \tag{2.12}$$

or, in the component form

$$k_1[1 \;\; 3 - 1 \;\; 0] + k_2[0 \;\; 1 \;\; 2 \;\; 0] + k_3[0 \;\; 0 \;\; 0 \;\; 1] = [0 \;\; 0 \;\; 0 \;\; 0]$$

Equating the fourth entries, we find, $k_3 = 0$. so that we have from (2.12)

$$k_1\mathbf{r}_1 + k_2\,\mathbf{r}_2 = \mathbf{0}$$

Then, equating the first entries, we have $k_1 = 0$, so finally we get $k_2 = 0$. That is, (2.12) is satisfied if $k_1 = k_2 = k_3 = 0$. Hence, the row vectors $\mathbf{r}_1$, $\mathbf{r}_2$, and $\mathbf{r}_3$ are linearly independent and form a basis for the row space of $\mathbf{A}_{re}$. As defined earlier, the dimension of a vector space V is equal to the number of vectors in a basis for V. Hence,

$$\dim \text{ (row space of } \mathbf{A}_{re}) = 3$$

The column vectors of $\mathbf{A}_{re}$ are

$$\mathbf{c}_1 = \begin{bmatrix} 1 \\ 0 \\ 0 \end{bmatrix}, \mathbf{c}_2 = \begin{bmatrix} 3 \\ 1 \\ 0 \end{bmatrix}, \mathbf{c}_3 = \begin{bmatrix} -1 \\ 2 \\ 0 \end{bmatrix}, \mathbf{c}_4 = \begin{bmatrix} 0 \\ 0 \\ 1 \end{bmatrix}$$

of which $\mathbf{c}_1$, $\mathbf{c}_2$, and $\mathbf{c}_4$ contain the leading 1s of the row vectors while $\mathbf{c}_3$ does not. If we choose $\mathbf{c}_1$, $\mathbf{c}_2$, and $\mathbf{c}_3$ we find that for the equation

$$l_1\,\mathbf{c}_1 + l_2\,\mathbf{c}_2 + l_3\,\mathbf{c}_3 = \mathbf{0}$$

which can be written as

$$\begin{bmatrix} 1 & 3 & -1 \\ 0 & 1 & 2 \\ 0 & 0 & 0 \end{bmatrix} \begin{bmatrix} l_1 \\ l_2 \\ l_3 \end{bmatrix} = \mathbf{0}$$

the determinant

$$\begin{vmatrix} 1 & 3 & -1 \\ 0 & 1 & 2 \\ 0 & 0 & 0 \end{vmatrix} = 0$$

indicating that vectors $\mathbf{c}_1$, $\mathbf{c}_2$, and $\mathbf{c}_3$ are not linearly independent.

However, if we choose $\mathbf{c}_1$, $\mathbf{c}_2$, and $\mathbf{c}_4$, we have for the equation

$$l_1\,\mathbf{c}_1 + l_2\,\mathbf{c}_2 + l_4\,\mathbf{c}_4 = \mathbf{0}$$

or,

$$\begin{bmatrix} 1 & 3 & 0 \\ 0 & 1 & 0 \\ 0 & 0 & 1 \end{bmatrix} \begin{bmatrix} l_1 \\ l_2 \\ l_4 \end{bmatrix} = \mathbf{0}$$

the determinant

$$\begin{bmatrix} 1 & 3 & 0 \\ 0 & 1 & 0 \\ 0 & 0 & 1 \end{bmatrix} = 1 \neq 0$$

So, the vectors $\mathbf{c}_1$, $\mathbf{c}_2$, and $\mathbf{c}_4$, which contain the leading 1s of the row vectors, are linearly independent and form a basis for the column space of $\mathbf{A}_{re}$. Hence,

$$\dim \text{ (column space of } \mathbf{A}_{re}) = 3$$

The dimension of the null space of $\mathbf{A}_{re}$ is given by the solution space of $\mathbf{A}_{re}\,\mathbf{x} = \mathbf{0}$. The augmented matrix for the given homogenous system

$$\begin{bmatrix} 1 & 3 & -1 & 2 & 0 \\ 0 & 1 & 2 & -1 & 0 \\ 0 & 0 & 0 & 1 & 0 \end{bmatrix}$$

can be reduced to

$$\begin{bmatrix} 1 & 0 & -7 & 0 & 0 \\ 0 & 1 & 2 & 0 & 0 \\ 0 & 0 & 0 & 1 & 0 \end{bmatrix}$$

from which we obtain $(x_1, x_2, x_3, x_4) = r(7, -2, 1, 0)$ where r is an arbitrary parameter.

Thus we find that the solution space has a single basis vector and, therefore, the dimension of the null pace of $\mathbf{A}_{re}$ is equal to 1.

It is to be noted that for the given $\mathbf{A}_{re}$, which have four columns,

$$\dim \left(\text{row/column space}\right) + \dim \text{ (null space)} = 4$$

In general, if a matrix $\mathbf{A}_{re}$ is in a row-echelon form, then

a. The nonzero row vectors form a basis for the row space of $\mathbf{A}_{re}$.
b. The column vectors with the leading 1s of the row vectors form a basis for the column space of $\mathbf{A}_{re}$.
c. dim (column space of $\mathbf{A}_{re}$) = dim (row space of $\mathbf{A}_{re}$).

Further, it can be proved that elementary row operations, as defined in Chapter 1, do not affect either the row space or the null space of a matrix. Therefore if $\mathbf{A}_{re}$ is the row-echelon form of the matrix $\mathbf{A}$,

$$\dim \text{ (row space of } \mathbf{A}) = \dim \text{ (row space of } \mathbf{A}_{re})$$

$$\dim \text{ (column space of } \mathbf{A}) = \dim \text{ (column space of } \mathbf{A}_{re})$$

and we have the following theorem.

Theorem 2.6.2 For a matrix **A**, the row space and the column space have the same dimension.

Now we can formalize the definition of the rank that has been preliminarily given in Chapter 1.

Definition 2.6.2 The common dimension of the row space and the column space of a matrix A is called its ***rank*** and denoted by rank (**A**).

Besides, the dimension of the null space of a matrix A is called its ***nullity*** and denoted by nullity (**A**).

The result of Example 2.6.3 has been generalized in the following theorem.

Theorem 2.6.3 For a matrix **A** with n columns,

$$\text{rank}\ (\mathbf{A}) + \text{nullity}\ (\mathbf{A}) = n$$

Exercise 2.6

1. Consider the following homogeneous system

$$\begin{array}{rcrcrcl} x_1 & - & 3x_2 & + & x_3 & = & 0 \\ 2x_1 & - & 6x_2 & + & 2x_3 & = & 0 \\ 3x_1 & - & 9x_2 & + & 3x_3 & = & 0 \end{array}$$

and determine the dimension of its solution space.

2. Determine the dimensions of row space and column spaces of a matrix, the row-echelon form of which is

$$\mathbf{A}_{re} \begin{bmatrix} 1 & 2 & 1 & -1 \\ 0 & 1 & 3 & 0 \\ 0 & 0 & 0 & 1 \end{bmatrix}$$

3. Row reduce each of the following matrices and determine its rank

$$\mathbf{A} = \begin{bmatrix} 1 & -2 & 1 & 0 \\ 2 & -1 & 0 & 2 \\ 0 & -2 & 3 & 1 \end{bmatrix}$$

$$\mathbf{B} = \begin{bmatrix} 2 & -3 & 5 & 16 \\ -1 & 2 & -2 & -6 \\ 3 & 1 & -1 & 4 \end{bmatrix}$$

4. Show for each of the following matrices that the sum of rank and nullity is equal to the number of columns.

a. $\mathbf{A} = \begin{bmatrix} 2 & 1 & 0 & 4 \\ 3 & -2 & 0 & 1 \\ 1 & 4 & -1 & 5 \end{bmatrix}$

b. $\mathbf{B} = \begin{bmatrix} 1 & 1 & 2 & 9 \\ 2 & 4 & -3 & 1 \\ 3 & 6 & -5 & 0 \end{bmatrix}$

2.7 Inner product space

In Chapter 1 we have seen that a vector can be represented by an arrow in 2-space or 3-space—mathematicians call this a "geometric vector." For geometric interpretation of the concepts in $\mathbb{R}^2$ and $\mathbb{R}^3$, it is necessary to define the length of a vector, and the distance and angle between two vectors. These definitions are then extended to nonvisual higher dimension spaces.

Norm

It has been earlier said that the length of a vector, represented by an arrow drawn to scale, specifies its magnitude. A common mathematical synonym for length is the term "norm" (also called Euclidean norm). As indicated in Fig. 2.8, for a

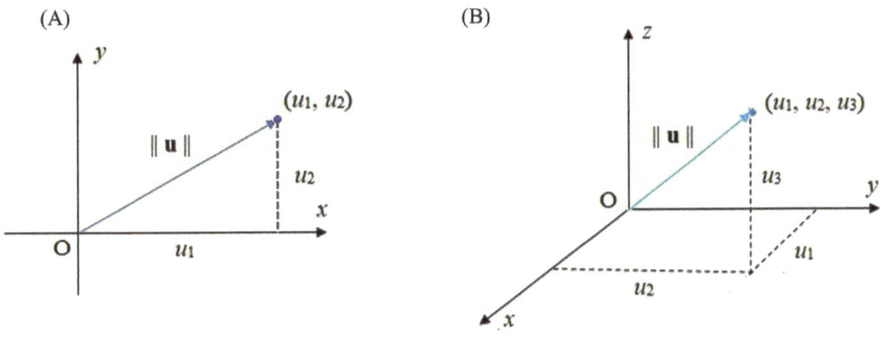

FIGURE 2.8

Norm of a vector in $\mathbb{R}^2$ (A) and $\mathbb{R}^3$ (B).

vector $\mathbf{u} = (u_1, u_2) \in \mathbb{R}^2$, the norm (or length or magnitude) is obtained from Pythagorean theorem as

$$\|\mathbf{u}\| = \sqrt{u_1^2 + u_2^2} \tag{2.13}$$

Similarly, for a vector $\mathbf{u} = (u_1, u_2, u_3) \in \mathbb{R}^3$,

$$\|\mathbf{u}\| = \sqrt{u_1^2 + u_2^2 + u_3^2} \tag{2.14}$$

The definition can be extended to n-space so that for a vector $\mathbf{u} = (u_1, u_2, \ldots \ldots\ldots, u_n) \in \mathbb{R}^n$, the Euclidean vector norm can be written as

$$||\mathbf{u}|| = (\mathbf{u}^T\mathbf{u})^{1/2} = \sqrt{u_1^2 + u_2^2 + \cdots\cdots\cdots + u_n^2} \tag{2.15}$$

It follows from the above equation that

(a) $\|\mathbf{u}\| \geq 0$
(b) $\|\mathbf{u}\| = 0$ if and only if $\mathbf{u} = 0$

Further, if k is a scalar, then

$$||k\mathbf{u}|| = |k|||\mathbf{u}||$$

Definition 2.7.1 A vector having norm 1 is called a ***unit vector***. If $\mathbf{u} \in \mathbb{R}^n$ be a nonzero vector, a unit vector that is in the same direction as $\mathbf{u}$ is defined as

$$\hat{\mathbf{u}} = \frac{1}{||\mathbf{u}||}\mathbf{u} \tag{2.16}$$

It can be shown that $\|\hat{\mathbf{u}}\| = 1$. Hence, it appears that multiplication of a nonzero vector by the reciprocal of its magnitude produces a unit vector. The process is called ***normalizing*** a vector.

Example 2.7.1: To find a unit vector $\hat{\mathbf{u}}$ that has same direction as $\mathbf{u} = (-4, 2, 4)$.

Solution. The vector $\mathbf{u}$ has length

$$\|\mathbf{u}\| = \sqrt{(-4)^2 + 2^2 + 4^2 = 6}$$

From (2.16),

$$\hat{\mathbf{u}} = \frac{1}{||\mathbf{u}||}\mathbf{u} = \frac{1}{6}(-4, 2, 4) = \left(-\frac{2}{3}, \frac{1}{3}, \frac{2}{3}\right)$$

It can be verified that $\|\hat{\mathbf{u}}\| = 1$

Distance

From the basic concept of a vector (as introduced in Chapter 1), one can say that the terms "vector" and "point" are interchangeable. As such, the point in three-dimensional space that is defined as the origin is a vector $\mathbf{0} = (0, 0, 0)$. Hence, for the vector $\mathbf{u}$ which has the point $P = (u_1, u_2, u_3)$, the length $\|\mathbf{u}\|$ given by (2.14) is also defined to be the "distance" from the origin to P.

Now, let us consider two vectors $\mathbf{u}$ and $\mathbf{v}$ in $\mathbb{R}^2$ denoted respectively by points $P_1 = (u_1, u_2)$ and $P_2 = (v_1, v_2)$. Application of vector subtraction gives the distance between the two vectors.

Theorem 2.7.1 If $\mathbf{u}$ and $\mathbf{v}$ have a common tail, then $\mathbf{u} - \mathbf{v}$ is the vector from the tip of $\mathbf{v}$ to the tip of $\mathbf{u}$.

The distance d between the two vectors is given by the

$$d = \| \overrightarrow{P_1P_2} \| = \|\mathbf{u} - \mathbf{v}\| = \sqrt{(u_1 - v_1)^2 + (u_2 - v_2)^2} \tag{2.17}$$

Similarly, the distance between vectors $\mathbf{u} = (u_1, u_2, u_3)$ and $\mathbf{v} = (v_1, v_2, v_3) \in \mathbb{R}^3$ is defined to be

$$d(\mathbf{u}, \mathbf{v}) = \|\mathbf{u} - \mathbf{v}\| = \sqrt{(u_1 - v_1)^2 + (u_2 - v_2)^2 + (u_3 - v_3)^2} \tag{2.18}$$

Example 2.7.2: For two points P_1 and P_3 in $\mathbb{R}^3$ represented by vectors $\mathbf{u} = (2, -1, 3)$ and $\mathbf{v} = (1, 1, 4)$ respectively, the vector $\overrightarrow{P_1P_2} = \mathbf{u} - \mathbf{v} = (1, -2, -1)$ and

$$d = \sqrt{(1)^2 + (-2)^2 + (-1)^2} = \sqrt{6}$$

Definition 2.7.2 Considering $\mathbf{u} = (u_1, u_2, \ldots\ldots\ldots, u_n)$ and $\mathbf{v} = (v_1, v_2, \ldots\ldots\ldots, v_n)$ as points in $\mathbb{R}^n$, the distance (Fig. 2.9) between $\mathbf{u}$ and $\mathbf{v}$ is expressed as

$$d(\mathbf{u}, \mathbf{v}) = \|\mathbf{u} - \mathbf{v}\| = \sqrt{(u_1 - v_1)^2 + (u_2 - v_2)^2 + \cdots\cdots + (u_n - v_n)^2}$$

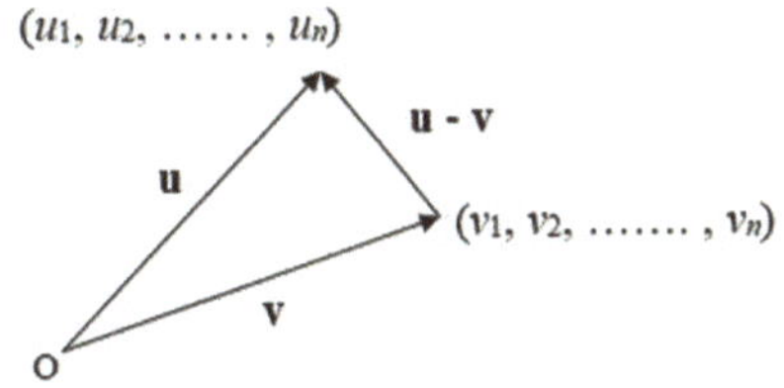

FIGURE 2.9

Distance between two vectors.

Dot product

For the purpose of defining the dot product between two vectors **u** and **v** in $\mathbb{R}^2$ or $\mathbb{R}^3$, the two vectors are positioned in such a way that their initial points coincide. Let θ be the angle between them. The dot product (also called the Euclidean inner product) between **u** and **v** is then defined as

$$\mathbf{u}\cdot\mathbf{v} = \|\mathbf{u}\|\,\|\mathbf{v}\|\cos\theta$$

If either $\mathbf{u} = \mathbf{0}$ or $\mathbf{v} = \mathbf{0}$, $\mathbf{u}\cdot\mathbf{v} = 0$.

When necessary, the angle between the vectors u and v can be written as

$$\cos\theta = \frac{\mathbf{u}\cdot\mathbf{v}}{\|\mathbf{u}\|\,\|\mathbf{v}\|}$$

In component form, the dot product between two nonzero vectors $\mathbf{u} = (u_1, u_2, u_3)$ and $\mathbf{v} = (v_1, v_2, v_3) \in \mathbb{R}^3$ can be written as

$$\mathbf{u}\cdot\mathbf{v} = u_1 v_1 + u_2 v_2 + u_3 v_3$$

Example 2.7.3: For $\mathbf{u} = (2, -3, 5, 4)$ and $\mathbf{v} = (1, -4, 1, 0) \in \mathbb{R}^4$, the dot product

$$\mathbf{u}\cdot\mathbf{v} = (2)(1) + (-3)(-4) + (5)(1) + (4)(0) = 19$$

Extending the operation to $\mathbb{R}^n$, the dot product (also called the Euclidean inner product) between two vectors $\mathbf{u} = (u_1, u_2, \ldots\ldots\ldots, u_n)$ and $\mathbf{v} = (v_1, v_2, \ldots\ldots\ldots, v_n)$ is defined as

$$\mathbf{u}\cdot\mathbf{v} = u_1 v_1 + u_2 v_2 + \cdots\cdots u_n v_n$$

This can also be written in the form

$$\mathbf{u}^T\mathbf{v} = \sum_{i=1}^{n} u_i v_i \in \mathbb{R}$$

which is a scalar called the **standard inner product** for $\mathbb{R}^n$.

Multiplication of a vector by a matrix can be depicted as dot product of vectors. Considering a 4-vector

$$\mathbf{u} = \begin{bmatrix} u_1 \\ u_2 \\ u_3 \\ u_4 \end{bmatrix}$$

and a 3×4 matrix

$$\mathbf{A} = \begin{bmatrix} \mathbf{r}_1 \\ \mathbf{r}_2 \\ \mathbf{r}_3 \end{bmatrix}$$

where

$$\mathbf{r}_1 = (r_{11}, r_{12}, r_{13}, r_{14})$$

$$\mathbf{r}_2 = (r_{21}, r_{22}, r_{23}, r_{24})$$

$$\mathbf{r}_3 = (r_{31}, r_{32}, r_{33}, r_{34})$$

the product of **A** and **u** can be computed as

$$\mathbf{Au} = \begin{bmatrix} \mathbf{r}_1 \\ \mathbf{r}_2 \\ \mathbf{r}_3 \end{bmatrix} \mathbf{u} = \begin{bmatrix} \mathbf{r}_1 \cdot \mathbf{u} \\ \mathbf{r}_2 \cdot \mathbf{u} \\ \mathbf{r}_3 \cdot \mathbf{u} \end{bmatrix} = \begin{bmatrix} r_{11}u_1 + r_{12}u_2 + r_{13}u_3 + r_{14}u_4 \\ r_{21}u_1 + r_{22}u_2 + r_{23}u_3 + r_{24}u_4 \\ r_{31}u_1 + r_{32}u_2 + r_{33}u_3 + r_{34}u_4 \end{bmatrix}$$

Thus we see that each entry of **Au** is the dot product of the corresponding row (vector) with **u**.

So far, the definition of dot product or inner product has been coordinate-dependent. However, it needs to be appreciated that dot products are more general concepts with specific properties. Therefore, to formulate a more general coordinate-free definition, the notion of a dot product/inner product can be extended to general real vector spaces based on its essential properties.

Definition 2.7.3 General inner product. An inner product on a vector space V is a function that maps each ordered pair of vectors $\mathbf{u}, \mathbf{v} \in V$ to a scalar $\langle \mathbf{u}, \mathbf{v} \rangle$ such that the following axioms are satisfied for all vectors $\mathbf{u}$, $\mathbf{v}$ and $\mathbf{w} \in \mathrm{V}$ and all scalars λ.

a. $\langle \mathbf{u}, \mathbf{v} \rangle = \langle \mathbf{v}, \mathbf{u} \rangle$ (symmetry)
b. $\langle \mathbf{u} + \mathbf{v}, \mathbf{w} \rangle = \langle \mathbf{u}, \mathbf{w} \rangle + \langle \mathbf{v}, \mathbf{w} \rangle$ (additivity)
c. $\langle \lambda\, \mathbf{u}, \mathbf{v} \rangle = \lambda\, \langle \mathbf{u}, \mathbf{v} \rangle$ (homogeneity)
d. $\langle \mathbf{u}, \mathbf{u} \rangle \geq 0$ and $\langle \mathbf{u}, \mathbf{u} \rangle = 0$ if and only if $\mathbf{u} = 0$ (positivity)

A real vector space with an inner product $\langle , \rangle$ is called a ***real inner product space***.

Similarly, the earlier coordinate-dependent definitions of the norm of a vector and the distance between two vectors can be generalized for real inner product spaces as follows.

Definition 2.7.4 If **u** and **v** are vectors in a real inner product space V and λ is a scalar, the norm (or length) of **u** is defined as

$$\|\mathbf{u}\| = \sqrt{\langle \mathbf{u}, \mathbf{u} \rangle}$$

and the distance between **u** and **v** is defined as

$$d(\mathbf{u}, \mathbf{v}) = \|\mathbf{u} - \mathbf{v}\| = \sqrt{\langle \mathbf{u} - \mathbf{v}, \mathbf{u} - \mathbf{v} \rangle}$$

provided that the following conditions are satisfied

a. $\|\mathbf{u}\| \geq 0$ with equality if and only if $\mathbf{u} = \mathbf{0}$
b. $\|\lambda\, \mathbf{u}\| = |\lambda|\, \|\mathbf{u}\|$
c. $d\,(\mathbf{u}, \mathbf{v}) = d\,(\mathbf{v}, \mathbf{u})$
d. $d\,(\mathbf{u}, \mathbf{v}) \geq 0$ with equality if and only if $\mathbf{u} = \mathbf{v}$

With the help of inner products, we can depict the geometry of a vector space by defining the angle between two vectors. In a real inner product space V, the angle between two nonzero vectors $\mathbf{u}$ and $\mathbf{v} \in V$ is defined to be the number $\theta \in [0, \pi]$ such that

$$\cos\theta = \frac{\langle \mathbf{u}, \mathbf{v} \rangle}{\|\mathbf{u}\|\,\|\mathbf{v}\|} \tag{2.19}$$

Exercise 2.7

1. Determine the norm of $\mathbf{u}$, and a unit vector that has the same direction as $\mathbf{u}$.
   a. $\mathbf{u} = (2, 3, 4)$
   b. $\mathbf{u} = (1, 2, 3, 1, 4)$
   c. $\mathbf{u} = (2, -2, -1, 4)$
2. Determine the Euclidean distance between $\mathbf{u}$ and $\mathbf{v}$ and the cosine of the angle between the two vectors.
   a. $\mathbf{u} = (2, -1, 2)$, $\mathbf{v} = (2, 0, 1)$
   b. $\mathbf{u} = (4, 0, -2)$, $\mathbf{v} = (3, 2. 0)$
   c. $\mathbf{u} = (1, 0, -2, 1, 2)$, $\mathbf{v} = (2, 1, -1, 0, 2)$
3. $\mathbf{A} = \begin{bmatrix} 2 & -1 & 3 \\ 4 & 0 & 2 \\ -1 & 2 & 1 \end{bmatrix}, \mathbf{u} = \begin{bmatrix} 1 \\ -2 \\ -1 \end{bmatrix}, \mathbf{v} = \begin{bmatrix} 2 \\ -1 \\ 0 \end{bmatrix}$

   Show, using dot products, $\mathbf{Au} \cdot \mathbf{v} = \mathbf{u} \cdot \mathbf{A}^T\mathbf{v}$

2.8 Orthogonality

Two vectors are perpendicular if the angle between them is $\pi/2$, that is, they are at right angles to each other. A right angle is visually perceivable in $\mathbb{R}^2$ and $\mathbb{R}^3$, but not in higher dimensions. Orthogonality is the generalization of the concept of perpendicularity. Geometrically, we can consider orthogonal vectors with respect to a specific inner product.

In an inner product space V, two vectors $\mathbf{u}$ and $\mathbf{v}$ are said to be orthogonal (to each other) if and only if $\langle \mathbf{u}, \mathbf{v} \rangle = 0$; this is denoted by writing $\mathbf{u} \perp \mathbf{v}$

For $\mathbb{R}^n$. with standard inner product, $\mathbf{u} \perp \mathbf{v} = \mathbf{u}^T\, \mathbf{v} = 0$

Further, if $\|\mathbf{u}\| = 1 = \|\mathbf{v}\|$, that is, $\mathbf{u}$ and $\mathbf{v}$ are unit vectors, the vectors are said to be orthonormal.

Example 2.8.1: Two vectors $\mathbf{u} = (3, 1, -2, 4)$ and $\mathbf{v} = (2, 0, 1, -1) \in \mathbb{R}^4$ are orthogonal since

$$\mathbf{u} \cdot \mathbf{v} = (3)(2) + (1)(0) + (-2)(1) + (4)(-1) = 0$$

Orthogonal and orthonormal set

A set $S = \{\mathbf{u}_1, \mathbf{u}_2, \ldots, \mathbf{u}_n\}$ of two or more vectors in a real inner product space V is said to be an orthogonal set if all pairs of distinct vectors in the set are orthogonal. That is,

$$\mathbf{u}_i \perp \mathbf{u}_j \text{ for all } i \neq j$$

In addition, if $\|\mathbf{u}_i\| = 1$ for each i so that

$$\langle \mathbf{u}_i, \mathbf{u}_j \rangle = \begin{cases} 1 & \text{when } i = j \\ 0 & \text{when } i \neq j \end{cases} \tag{2.20}$$

the set is said to be orthonormal.

Example 2.8.2: Let us consider a set of vectors.

$$S' = \left\{ \mathbf{u}_1 = \begin{bmatrix} 1 \\ 1 \\ 1 \end{bmatrix}, \mathbf{u}_2 = \begin{bmatrix} 4 \\ 1 \\ -5 \end{bmatrix}, \mathbf{u}_3 = \begin{bmatrix} 2 \\ -3 \\ 1 \end{bmatrix} \right\}$$

We have

$$\mathbf{u}_1{}^T \mathbf{u}_2 = 4 + 1 - 5 = 0$$

$$\mathbf{u}_2{}^T \mathbf{u}_3 = 8 - 3 - 5 = 0$$

$$\mathbf{u}_3{}^T \mathbf{u}_1 = 2 - 3 + 1 = 0$$

Therefore S' is an orthogonal set of vectors in $\mathbb{R}^3$.

However,

$$||\mathbf{u}_1|| = \sqrt{3}, \ ||\mathbf{u}_2|| = \sqrt{42}, \ ||\mathbf{u}_3|| = \sqrt{14}$$

So, S' is not an orthonormal set since each vector does not have a unit length.

An orthogonal set (not containing a zero vector) can be converted into an orthonormal set by normalizing each vector. It follows that

$$S = \left\{ \mathbf{u}_1 = \frac{1}{\sqrt{3}} \begin{bmatrix} 1 \\ 1 \\ 1 \end{bmatrix}, \mathbf{u}_2 = \frac{1}{\sqrt{42}} \begin{bmatrix} 4 \\ 1 \\ -5 \end{bmatrix}, \mathbf{u}_3 = \frac{1}{\sqrt{14}} \begin{bmatrix} 2 \\ -3 \\ 1 \end{bmatrix} \right\}$$

is an orthonormal set. This result leads us to the following definition.

Definition 2.8.1 The result of normalizing an orthogonal set $\{\mathbf{u}_1, \mathbf{u}_2, \ldots\ldots\ldots, \mathbf{u}_k\}$ is an orthonormal set $\left\{\frac{1}{\|\mathbf{u}_1\|}\mathbf{u}_1, \frac{1}{\|\mathbf{u}_2\|}\mathbf{u}_2. \cdots, \frac{1}{\|\mathbf{u}_k\|}\mathbf{u}_k\right\}$.

Example 2.8.2: Evidently, the set of standard unit vectors in $\mathbb{R}^3$

$$S = \left\{ \mathbf{i} = \begin{bmatrix} 1 \\ 0 \\ 0 \end{bmatrix}, \ \mathbf{j} = \begin{bmatrix} 0 \\ 1 \\ 0 \end{bmatrix}, \mathbf{k} = \begin{bmatrix} 0 \\ 0 \\ 1 \end{bmatrix} \right\}$$

is an orthonormal set since

$$\mathbf{i}\cdot\mathbf{j} = (1,\ 0,\ 0)\cdot(0,1,0) = 0$$
$$\mathbf{j}\cdot\mathbf{k} = (0,\ 1,\ 0)\cdot(0,\ 0,\ 1) = 0$$
$$\mathbf{k}\cdot\mathbf{i} = (0,\ 0,\ 1)\cdot(1,\ 0,\ 0) = 0.$$

Let us consider an orthogonal set $S = \{\mathbf{u}_1,\ \mathbf{u}_2, \ldots,\ \mathbf{u}_k\} \in V$ and a linear combination

$$\sum_{i=1}^{k} \lambda_i \mathbf{u}_i = \mathit{0} \tag{2.21}$$

Using (2.20), we can write

$$\sum_{i=1}^{k} \lambda_i \langle \mathbf{u}_i, \mathbf{u}_j \rangle = 0$$

that is,

$$\lambda_j \|\mathbf{u}_j\|^2 = 0 \tag{2.22}$$

Since $\|\mathbf{u}_j\|^2 \neq 0$, (2.21) implies $\lambda_j = 0$ for $j = 1,\ 2, \ldots\ldots\ldots,\ k$. Hence, we have the following theorem.

Theorem 2.8.1 Every orthogonal set in V is linearly independent.

Example 2.8.3: Orthogonal matrix.

Looking at a 2×2 matrix

$$\mathbf{M}_{22} = \begin{bmatrix} \frac{1}{\sqrt{2}} & \frac{1}{\sqrt{2}} \\ \frac{1}{\sqrt{2}} & -\frac{1}{\sqrt{2}} \end{bmatrix}$$

it can be seen that the columns as well as the rows of the matrix are orthonormal. Hence, $\mathbf{M}_{22}$ is an orthogonal matrix.

The determinant of $\mathbf{M}_{22}$ is

$$\det\ (\mathbf{M}_{22}) = \begin{vmatrix} \frac{1}{\sqrt{2}} & \frac{1}{\sqrt{2}} \\ \frac{1}{\sqrt{2}} & -\frac{1}{\sqrt{2}} \end{vmatrix} = -1$$

If the rows of M_{22} are interchanged so that

$$\mathbf{M}'_{22} = \begin{bmatrix} \frac{1}{\sqrt{2}} & -\frac{1}{\sqrt{2}} \\ \frac{1}{\sqrt{2}} & \frac{1}{\sqrt{2}} \end{bmatrix}$$

we have

$$\det\left(\mathbf{M}'_{22}\right) = \begin{vmatrix} \frac{1}{\sqrt{2}} & -\frac{1}{\sqrt{2}} \\ \frac{1}{\sqrt{2}} & \frac{1}{\sqrt{2}} \end{vmatrix} = 1$$

Example 2.8.4: Let us take a 3×3 matrix

$$\mathbf{M}_{33} = \begin{bmatrix} 2 & 1 & 1 \\ -1 & 1 & 1 \\ 0 & -1 & 1 \end{bmatrix}$$

The rows of $\mathbf{M}_{33}$ are

$$\mathbf{r}_1 = [2 \;\; 1 \;\; 1],\; \mathbf{r}_2 = [-1 \;\; 1 \;\; 1],\; \mathbf{r}_3 = [0 - 1 \;\; 1]$$

and the columns of $\mathbf{M}_{22}$ are

$$\mathbf{c}_1 = \begin{bmatrix} 2 \\ -1 \\ 0 \end{bmatrix}, \mathbf{c}_2 = \begin{bmatrix} 1 \\ 1 \\ -1 \end{bmatrix}, \mathbf{c}_3 = \begin{bmatrix} 1 \\ 1 \\ 1 \end{bmatrix}$$

It can be seen that

$$\mathbf{r}_1 \cdot \mathbf{r}_2 = 0$$

$$\mathbf{r}_2 \cdot \mathbf{r}_3 = 0$$

$$\mathbf{r}_3 \cdot \mathbf{r}_1 = 0$$

that is, the rows of the matrix are orthogonal. However, it is to be noted that orthogonality of the rows alone does not make a matrix orthogonal. In case of the columns, it is found that

$$\mathbf{c}_1 \cdot \mathbf{c}_2 = 1$$

$$\mathbf{c}_2 \cdot \mathbf{c}_3 = 1$$

$$\mathbf{c}_3 \cdot \mathbf{c}_1 = 1$$

which means that the columns are not orthogonal. If the rows are normalized, also the columns of the resulting matrix

$$\mathbf{M}_{33}' = \begin{bmatrix} 2/\sqrt{6} & 1/\sqrt{6} & 1/\sqrt{6} \\ -1/\sqrt{3} & 1/\sqrt{3} & 1/\sqrt{3} \\ 0 & -1/\sqrt{2} & 1/\sqrt{2} \end{bmatrix}$$

are orthogonal (can be verified). Hence, $\mathbf{M}_{33}'$ is an orthogonal matrix.

Now, let us consider a square matrix $\mathbf{A} \in \mathbb{R}^{n \times n}$ and write it as a row of orthogonal column vectors

$$\mathbf{A} = [\mathbf{v}_1 \ \ \mathbf{v}_2 \cdots \mathbf{v}_j \cdots \mathbf{v}_n]$$

where

$$\mathbf{v}_j = \begin{bmatrix} v_{1j} \\ v_{2j} \\ \cdot \\ \cdot \\ v_{ij} \\ \cdot \\ \cdot \\ \cdot \\ v_{nj} \end{bmatrix}$$

The transpose $\mathbf{A}^T \in \mathbb{R}^{n \times n}$ can be written as a column of n row vectors

$$\mathbf{A}^T = \begin{bmatrix} \mathbf{v}_1 \\ \mathbf{v}_2 \\ \cdot \\ \cdot \\ \mathbf{v}_i \\ \cdot \\ \cdot \\ \cdot \\ \mathbf{v}_n \end{bmatrix}$$

where

$$\mathbf{v}_i = \begin{bmatrix} v_{i1} \ \ v_{i2} \cdots v_{ij} \cdots v_{in} \end{bmatrix}$$

so that

$$\mathbf{A}^T\mathbf{A} = \begin{bmatrix} \mathbf{v}_1 \\ \mathbf{v}_2 \\ \cdot \\ \cdot \\ \mathbf{v}_i \\ \cdot \\ \cdot \\ \cdot \\ \mathbf{v}_n \end{bmatrix} [\mathbf{v}_1 \mathbf{v}_2 \cdots \mathbf{v}_j \cdots \mathbf{v}_n]$$

$$= \begin{bmatrix} v_1 \cdot v_1 & v_1 \cdot v_2 & \cdots & v_1 \cdot v_j & \cdots & v_1 \cdot v_n \\ v_2 \cdot v_1 & v_2 \cdot v_2 & \cdots & v_2 \cdot v_j & \cdots & v_2 \cdot v_n \\ \vdots & \vdots & \cdots & \vdots & \cdots & \vdots \\ v_i \cdot v_1 & v_i \cdot v_2 & \cdots & v_i \cdot v_j & \cdots & v_i \cdot v_n \\ \vdots & \vdots & \cdots & \vdots & \cdots & \vdots \\ v_n \cdot v_1 & v_n \cdot v_2 & \cdots & v_n \cdot v_j & \cdots & v_n \cdot v_n \end{bmatrix}$$

Since $\{\mathbf{v}_1 \ \mathbf{v}_2 \cdots \mathbf{v}_j \cdots \mathbf{v}_n\}$ is an orthogonal set, $\mathbf{v}_i \cdot \mathbf{v}_j = 0$, when $i \neq j$. Therefore

$$\mathbf{A}^T\mathbf{A} = \begin{bmatrix} v_1 \cdot v_1 & 0 & \cdots & 0 & \cdots & 0 \\ 0 & v_2 \cdot v_2 & \cdots & 0 & \cdots & 0 \\ \vdots & \vdots & \cdots & \vdots & \cdots & \vdots \\ 0 & 0 & \cdots & v_i \cdot v_i & \cdots & 0 \\ \vdots & \vdots & \cdots & \vdots & \cdots & \vdots \\ 0 & 0 & \cdots & 0 & \cdots & v_n \cdot v_n \end{bmatrix}$$

which is a diagonal matrix. Further, if $\|\mathbf{v}_i\| = 1$ for each i. we have

$$\mathbf{A}^T\mathbf{A} = \mathbf{I} = \mathbf{A}\mathbf{A}^T$$

which implies

$$\mathbf{A}^{-1} = \mathbf{A}^T$$

Thus, we are led to the following definition of an orthogonal matrix:

Definition 2.8.2 A square matrix is said to be ***orthogonal*** if its transpose is same as its inverse.

Theorem 2.8.2 For a square matrix $\mathbf{A} \in \mathbb{R}^{n \times n}$, the following properties are equivalent:

a. The inverse of **A** is its transpose.
b. The row vectors of **A** form an orthonormal set in $\mathbb{R}^n$.
c. The column vectors of **A** form an orthonormal set in $\mathbb{R}^n$.

Therefore a real, square matrix satisfying any of the properties listed in the above theorem is said to be an orthogonal matrix.

Here are a few more fundamental properties of an orthogonal matrix:

a. The transpose of an orthogonal matrix is orthogonal.
b. The inverse of an orthogonal matrix is orthogonal.
c. A product of orthogonal matrices is orthogonal.
d. For an orthogonal matrix **A**, det (**A**) $\pm$ 1.

It is interesting to note that transformations by orthogonal matrices do not change the length of a vector. If a vector $\mathbf{v} \in \mathbb{R}^n$ is transformed with an orthogonal matrix $\mathbf{A} \in \mathbb{R}^{n \times n}$, then

$$\begin{aligned}||\mathbf{Av}|| &= (\mathbf{Av})^T(\mathbf{Av}) \\ &= \mathbf{v}^T\mathbf{A}^T\mathbf{Av} \text{ (using reverse order law for transposition)} \\ &= \mathbf{v}^T\mathbf{Iv} \\ &= \mathbf{v}^T\mathbf{v} = ||\mathbf{v}||\end{aligned}$$

Similarly, if each of two vectors $\mathbf{u}$ and $\mathbf{v} \in \mathbb{R}^n$ is transformed by an orthogonal matrix $\mathbf{A}$, the distance between the vectors also remains unchanged.

$$(\mathbf{Au})^T(\mathbf{Av}) = \mathbf{u}^T\mathbf{A}^T\mathbf{Av} = \mathbf{u}^T\mathbf{Iv} = \mathbf{u}^T\mathbf{v}$$

The transformation T_A: $\mathbb{R}^n \rightarrow \mathbb{R}^n$ is essentially multiplication by the orthogonal matrix $\mathbf{A}$. T_A is called an **orthogonal operator** on $\mathbb{R}^n$. It follows from (2.19) that the angle between the two transformed vectors T_A ($\mathbf{u}$) and T_A ($\mathbf{v}$) $\in \mathbb{R}^n$ is

$$\frac{(\mathbf{Au})^T(\mathbf{Av})}{||\mathbf{Au}||\,||\mathbf{Av}||} = \frac{\mathbf{u}^T\mathbf{v}}{||\mathbf{u}||\,||\mathbf{v}||} = \cos\theta$$

That is, the orthogonal operator T_A preserves the angle between the vectors $\mathbf{u}$ and $\mathbf{v}$ in $\mathbb{R}^n$ (Fig. 2.10).

Coordinates relative to orthogonal basis

We have seen earlier that every vector in a vector space V can be expressed as a linear combination of basis vectors $S = \{\mathbf{v}_1, \mathbf{v}_2, \ldots\ldots, \mathbf{v}_n\}$

$$\mathbf{u} = c_1\mathbf{v}_1 + c_2\mathbf{v}_2 + \cdots\cdots + c_n\mathbf{v}_n \tag{2.23}$$

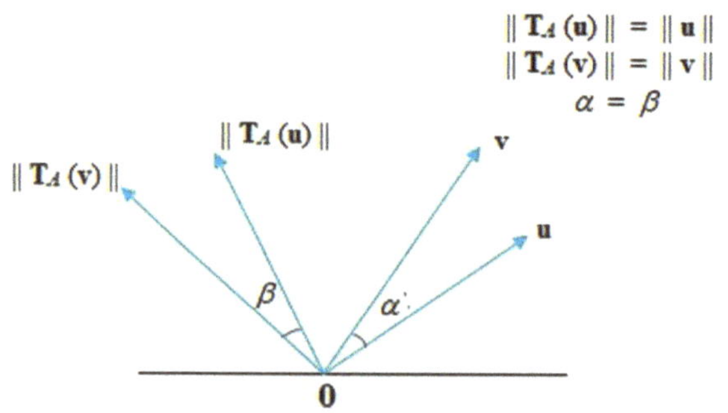

FIGURE 2.10

Orthogonal matrix preserves the length of a vector and the angle between two vectors.

Now, the inner product

$$\begin{aligned}\langle \mathbf{u}, \mathbf{v}_i \rangle &= \langle c_1 \mathbf{v}_1 + c_2 \mathbf{v}_2 + \cdots\cdots + c_n \mathbf{v}_n, \mathbf{v}_i \rangle \\ &= c_1 \langle \mathbf{v}_1, \mathbf{v}_i \rangle + c_2 \langle \mathbf{v}_2, \mathbf{v}_i \rangle + \cdots\cdots + c_n \langle \mathbf{v}_n, \mathbf{v}_i \rangle\end{aligned} \tag{2.24}$$

Since S is an orthogonal set,

$$\langle \mathbf{v}_i, \mathbf{v}_j \rangle = \begin{cases} 1 & \text{when } i = j \\ 0 & \text{when } i \neq j \end{cases}$$

so that we have from (2.24)

$$\langle \mathbf{u}, \mathbf{v}_i \rangle = c_i \langle \mathbf{v}_i, \mathbf{v}_i \rangle = c_i \|\mathbf{v}_i\|^2$$

that is,

$$c_i = \frac{\langle \mathbf{u}, \mathbf{v}_i \rangle}{\|\mathbf{v}_i\|^2}$$

Hence, we have the following theorem which substitutes the c_i's in (2.23) with the corresponding inner products.

Theorem 2.8.3 If $S = \{\mathbf{v}_1, \mathbf{v}_2, \ldots\ldots, \mathbf{v}_n\}$ is an orthogonal basis for an inner product space V, any vector $\mathbf{u} \in V$ can be expressed as

$$\mathbf{u} = \frac{\langle \mathbf{u}, \mathbf{v}_1 \rangle}{\|\mathbf{v}_1\|^2} \mathbf{v}_1 + \frac{\langle \mathbf{u}, \mathbf{v}_2 \rangle}{\|\mathbf{v}_2\|^2} \mathbf{v}_2 + \cdots\cdots + \frac{\langle \mathbf{u}, \mathbf{v}_n \rangle}{\|\mathbf{v}_n\|^2} \mathbf{v}_n \tag{2.25}$$

If $S = \{\mathbf{v}_1, \mathbf{v}_2, \ldots\ldots, \mathbf{v}_n\}$ is an orthonormal basis for an inner product space V, then $\mathbf{u} \in V$ can be expressed as

$$\mathbf{u} = \langle \mathbf{u}, \mathbf{v}_1 \rangle \mathbf{v}_1 + \langle \mathbf{u}, \mathbf{v}_2 \rangle \mathbf{v}_2 + \cdots\cdots + \langle \mathbf{u}, \mathbf{v}_n \rangle \mathbf{v}_n \tag{2.26}$$

Example 2.8.5: Let us consider a set S of vectors $\mathbf{w}_1, \mathbf{w}_2, \mathbf{w}_3 \in \mathbb{R}^3$

$$\mathbf{w}_1 = (0, 1, 0), \quad \mathbf{w}_2 = (-2, 0, 2), \quad \text{and} \quad \mathbf{w}_3 = (1, 0, 1)$$

$\mathbf{w}_1, \mathbf{w}_2, \mathbf{w}_3$ form an orthogonal set since

$$\langle \mathbf{w}_1, \mathbf{w}_2 \rangle = 0, \langle \mathbf{w}_2, \mathbf{w}_3 \rangle = 0, \langle \mathbf{w}_3, \mathbf{w}_1 \rangle = 0$$

Hence, these vectors are linearly independent and form a basis for $\mathbb{R}^3$. However, only $\mathbf{w}_1$ is a unit vector, $\mathbf{w}_2$ and $\mathbf{w}_3$ are not. An orthonormal basis can be derived by normalizing each vector.

$$\mathbf{v}_1 = \frac{\mathbf{w}_1}{\|\mathbf{w}_1\|} = (0, 1, 0), \quad \mathbf{v}_2 = \frac{\mathbf{w}_2}{\|\mathbf{w}_2\|} = \left(-\frac{1}{\sqrt{2}}, 0, \frac{1}{\sqrt{2}}\right), \quad \mathbf{v}_3 = \frac{\mathbf{w}_3}{\|\mathbf{w}_3\|} = \left(\frac{1}{\sqrt{2}}, 0, \frac{1}{\sqrt{2}}\right)$$

Given a vector $\mathbf{u} = (2, 1, 3)$, it can be expressed as a linear combination of the above orthonormal basis using (2.26).

We calculate

$$\langle \mathbf{u}, \mathbf{v}_1 \rangle = (2,\ 1,\ 3)\cdot(0,\ 1,\ 0) = 1$$

$$\langle \mathbf{u}, \mathbf{v}_2 \rangle = (2,\ 1,\ 3)\cdot\left(-\frac{1}{\sqrt{2}},\ 0,\ \frac{1}{\sqrt{2}}\right) = \frac{1}{\sqrt{2}}$$

$$\langle \mathbf{u}, \mathbf{v}_3 \rangle = (2,\ 1,\ 3)\cdot\left(\frac{1}{\sqrt{2}},\ 0,\ \frac{1}{\sqrt{2}}\right) = \frac{5}{\sqrt{2}}$$

So, we express **u** as

$$(2, 1, 3) = (0,\ 1,\ 0) + \frac{1}{\sqrt{2}}\left(-\frac{1}{\sqrt{2}},\ 0,\ \frac{1}{\sqrt{2}}\right) + \frac{5}{\sqrt{2}}\left(\frac{1}{\sqrt{2}},\ 0,\ \frac{1}{\sqrt{2}}\right)$$

Orthogonal projection

In certain applications such as principal component analysis based on dimensionality reduction, it becomes necessary to take the "orthogonal projection" of a vector **u** on a specified nonzero vector **a**. In other words, with the initial points of **u** and **a** coinciding, **u** is "decomposed" into a sum of two terms—one term $\mathbf{w}_1$ is a scalar multiple of **a**, while the other term $\mathbf{w}_2$ is orthogonal to **a**.

We can write

$$\mathbf{w}_1 = k\mathbf{a}$$

where k is a scalar
and

$$\mathbf{u} = \mathbf{w}_1 + \mathbf{w}_2 \tag{2.27}$$

We can rewrite (2.27) as

$$\mathbf{u} = k\mathbf{a} + \mathbf{w}_2$$

and, by applying the properties of dot products

$$\mathbf{u}\cdot\mathbf{a} = (k\mathbf{a} + \mathbf{w}_2)\cdot\mathbf{a} = k||\mathbf{a}||^2 + \mathbf{w}_2\cdot\mathbf{a}$$

Since $\mathbf{w}_2$ is orthogonal to **a**, $\mathbf{w}_2\cdot\mathbf{a} = 0$. Hence

$$\mathbf{u}\cdot\mathbf{a} = k||\mathbf{a}||^2$$

which gives

$$k = \frac{\mathbf{u}\cdot\mathbf{a}}{||\mathbf{a}||^2}$$

So, we can rewrite

$$\mathbf{w}_2 = \mathbf{u} - \frac{\mathbf{u}\cdot\mathbf{a}}{||\mathbf{a}||^2}\mathbf{a}$$

and verify

$$\mathbf{w}_2 \cdot \mathbf{a} = 0$$

The vector $\mathbf{w}_1$ is called the **orthogonal projection of u on a** and can be expressed as

$$\text{proj}_a \mathbf{u} = \frac{\mathbf{u} \cdot \mathbf{a}}{\|\mathbf{a}\|^2} \mathbf{a} \tag{2.28}$$

while the vector component orthogonal to **a** is

$$\mathbf{u} - \text{proj}_a \mathbf{u} = \mathbf{u} - \frac{\mathbf{u} \cdot \mathbf{a}}{\|\mathbf{a}\|^2} \mathbf{a} \tag{2.29}$$

Example 2.8.5: Considering $\mathbf{u} = (1, 1, -3)$ and $\mathbf{a} = (2. 4, -1)$ let us determine the vector components of **u** in the direction of **a** and orthogonal to **a**.

We have

$$\mathbf{u} \cdot \mathbf{a} = (1)(2) + (1)(4) + (-3)(-1) = 9$$

and

$$\|\mathbf{a}\|^2 = 2^2 + 4^2 + (-1)^2 = 21$$

Therefore the vector component of **u** in the direction of **a** is

$$\text{proj}_a \mathbf{u} = \frac{\mathbf{u} \cdot \mathbf{a}}{\|\mathbf{a}\|^2} \mathbf{a} = \frac{9}{21}(2, 4, -1) = \left(\frac{6}{7}, \frac{12}{7}, -\frac{3}{7}\right)$$

and the component of **u** orthogonal to **a** is

$$\mathbf{u} - \text{proj}_a \mathbf{u} = \mathbf{u} - \frac{\mathbf{u} \cdot \mathbf{a}}{\|\mathbf{a}\|^2} \mathbf{a} = (1, 1, -3) - \left(\frac{6}{7}, \frac{12}{7}, -\frac{3}{7}\right) = \left(\frac{1}{7}, -\frac{5}{7}, -\frac{18}{7}\right)$$

It can be seen that

$$(\text{proj}_a \mathbf{u}) \cdot (\mathbf{u} - \text{proj}_a \mathbf{u}) = \frac{(6 - 60 + 54)}{49} = 0$$

which verifies that $\text{proj}_a \mathbf{u}$ and $\mathbf{u} - \text{proj}_a \mathbf{u}$ are perpendicular.

The "decomposition" of **u** is generalized in the following theorem known as the ***Projection Theorem***.

> **Theorem 2.8.4** If W is a finite-dimensional subspace of an inner product space V, and $W^{\perp}$ is a finite-dimensional orthogonal subspace of V, then every vector $\mathbf{u} \in V$ can be expressed in exactly one way as
>
> $$\mathbf{u} = \mathbf{w}_1 + \mathbf{w}_2$$
>
> where $\mathbf{w}_1$ is in W and $\mathbf{w}_2$ is in $W^{\perp}$.

Let us consider $\{\mathbf{v}_1, \mathbf{v}_2, \ldots\ldots, \mathbf{v}_r\}$ as an orthogonal basis for W. Then, following Theorem 2.8.3, the component $\mathbf{w}_1$ can be expressed in terms of the basis vectors for W as

$$\text{proj}_W\mathbf{u} = \mathbf{w}_1 = \frac{\langle \mathbf{w}_1, \mathbf{v}_1 \rangle}{\|\mathbf{v}_1\|^2}\mathbf{v}_1 + \frac{\langle \mathbf{w}_1, \mathbf{v}_2 \rangle}{\|\mathbf{v}_2\|^2}\mathbf{v}_2 + \cdots\cdots + \frac{\langle \mathbf{w}_1, \mathbf{v}_r \rangle}{\|\mathbf{v}_r\|^2}\mathbf{v}_r \tag{2.30}$$

Since $\mathbf{w}_2$ is orthogonal to W,

$$\langle \mathbf{w}_2, \mathbf{v}_i \rangle = 0 \text{ for } i = 1, 2, \ldots\ldots, r$$

and (2.30) can be rewritten as

$$\text{proj}_W\mathbf{u} = \frac{\langle \mathbf{w}_1 + \mathbf{w}_2, \mathbf{v}_1 \rangle}{\|\mathbf{v}_1\|^2}\mathbf{v}_1 + \frac{\langle \mathbf{w}_1 + \mathbf{w}_2, \mathbf{v}_2 \rangle}{\|\mathbf{v}_2\|^2}\mathbf{v}_2 + \cdots\cdots + \frac{\langle \mathbf{w}_1 + \mathbf{w}_2, \mathbf{v}_r \rangle}{\|\mathbf{v}_r\|^2}\mathbf{v}_r$$

$$= \frac{\langle \mathbf{u}, \mathbf{v}_1 \rangle}{\|\mathbf{v}_1\|^2}\mathbf{v}_1 + \frac{\langle \mathbf{u}, \mathbf{v}_2 \rangle}{\|\mathbf{v}_2\|^2}\mathbf{v}_2 + \cdots\cdots + \frac{\langle \mathbf{u}, \mathbf{v}_r \rangle}{\|\mathbf{v}_r\|^2}\mathbf{v}_r$$

So, we have the following theorem for calculating orthogonal projections.

Theorem 2.8.5 Let W be a finite-dimensional subspace of an inner product space V with orthogonal basis $\{\mathbf{v}_1, \mathbf{v}_2, \ldots\ldots, \mathbf{v}_r\}$. If a vector $\mathbf{u} \in V$, then

$$\text{proj}_W\mathbf{u} = \frac{\langle \mathbf{u}, \mathbf{v}_1 \rangle}{\|\mathbf{v}_1\|^2}\mathbf{v}_1 + \frac{\langle \mathbf{u}, \mathbf{v}_2 \rangle}{\|\mathbf{v}_2\|^2}\mathbf{v}_2 + \cdots\cdots + \frac{\langle \mathbf{u}, \mathbf{v}_r \rangle}{\|\mathbf{v}_r\|^2}\mathbf{v}_r \tag{2.31}$$

is the orthogonal projection of u on W.

If $\{\mathbf{v}_1, \mathbf{v}_2, \ldots\ldots, \mathbf{v}_r\}$ is an orthonormal basis for W and $\mathbf{u} \in V$, then

$$\text{proj}_W\mathbf{u} = \langle \mathbf{u}, \mathbf{v}_1 \rangle\, \mathbf{v}_1 + \langle \mathbf{u}, \mathbf{v}_2 \rangle\, \mathbf{v}_2 + \cdots\cdots + \langle \mathbf{u}, \mathbf{v}_r \rangle\, \mathbf{v}_r \tag{2.32}$$

Having a variety of useful properties, orthonormal bases are very convenient to work with. The following theorem provides that opportunity.

Theorem 2.8.6 Every nonzero finite-dimensional inner product space has an orthonormal basis.

So, for any such space, it is possible to convert any arbitrary basis into an orthonormal basis. The conversion can be carried out following a sequence of steps described here.

Let us start with an arbitrary basis $\{\mathbf{u}_1, \mathbf{u}_2, \ldots\ldots\ldots, \mathbf{u}_r\}$ for a nonzero finite-dimensional subspace W of an inner product space. The objective is to derive an orthogonal basis $\{\mathbf{v}_1, \mathbf{v}_2, \ldots\ldots, \mathbf{v}_r\}$ for W. Subsequently, the vectors in the orthogonal basis can be normalized to produce an orthonormal basis.

As the first step, let $\mathbf{u}_1 = \mathbf{v}_1$. Then, $\mathbf{u}_2$ is not in span $\{\mathbf{v}_1\}$ since $\{\mathbf{u}_1, \mathbf{u}_2\}$ is independent. We can obtain a vector $\mathbf{v}_2$ that is orthogonal to $\mathbf{v}_1$ by calculating the component of $\mathbf{u}_2$ orthogonal to the space W_1 spanned by $\mathbf{v}_1$. So, we obtain using (2.29)

$$\mathbf{v}_2 = \mathbf{u}_2 - \text{proj}_{W_1}\, \mathbf{u}_2 = \mathbf{u}_2 - \frac{\langle \mathbf{u}_2, \mathbf{v}_1 \rangle}{\|\mathbf{v}_1\|^2}\mathbf{v}_1$$

It can be seen from the above equation that $\mathbf{v}_2 \neq \mathbf{0}$, otherwise the linear independence of the basis $\{\mathbf{u}_1, \mathbf{u}_2, \ldots\ldots\ldots, \mathbf{u}_r\}$ will be violated.

The next one to be constructed is a vector $\mathbf{v}_3$ that is orthogonal to both $\mathbf{v}_1$ and $\mathbf{v}_2$. For this, we compute the component of $\mathbf{u}_3$ that is orthogonal to the space W_2 spanned by $\mathbf{v}_1$ and $\mathbf{v}_2$.

$$\mathbf{v}_3 = \mathbf{u}_3 - \text{proj}_{W_2}\, \mathbf{u}_3 = \mathbf{u}_3 - \frac{\langle \mathbf{u}_3, \mathbf{v}_1 \rangle}{\|\mathbf{v}_1\|^2}\mathbf{v}_1 - \frac{\langle \mathbf{u}_3, \mathbf{v}_2 \rangle}{\|\mathbf{v}_2\|^2}\mathbf{v}_2$$

Here again, the independence of $\{\mathbf{u}_1, \mathbf{u}_2, \ldots\ldots\ldots, \mathbf{u}_r\}$ requires that $\mathbf{v}_3 \neq \mathbf{0}$. Proceeding further,

$$\mathbf{v}_4 = \mathbf{u}_4 - \text{proj}_{W_3}\, \mathbf{u}_4 = \mathbf{u}_4 - \frac{\langle \mathbf{u}_4, \mathbf{v}_1 \rangle}{\|\mathbf{v}_1\|^2}\mathbf{v}_1 - \frac{\langle \mathbf{u}_4, \mathbf{v}_2 \rangle}{\|\mathbf{v}_2\|^2}\mathbf{v}_2 - \frac{\langle \mathbf{u}_4, \mathbf{v}_3 \rangle}{\|\mathbf{v}_3\|^2}\mathbf{v}_3$$

Evidently, in r such steps an orthogonal set of nonzero vectors $\{\mathbf{v}_1, \mathbf{v}_2, \ldots\ldots, \mathbf{v}_r\}$ will be obtained. Being linearly independent, the set will form an orthogonal basis for the r-dimensional space W. Normalization of the basis vectors will lead to an orthonormal basis. This systematic procedure is known as the **Gram–Schmidt algorithm**.

Exercise 2.8

1. Determine if **u** and **v** are orthogonal vectors.
   a. $\mathbf{u} = (4, 0)$, $\mathbf{v} = (0, 4)$
   b. $\mathbf{u} = (0, 2, 2)$, $\mathbf{v} = (1, 1, -1)$
   c. $\mathbf{u} = (1, 4, 0, 2)$, $\mathbf{v} = (-2, 0, 1, 2)$
   d. $\mathbf{u} = (2, -1, 0, 3)$, $\mathbf{v} = (3, 0, 1, -2)$
2. Obtain orthonormal bases of $\mathbb{R}^3$ from the following orthogonal sets.
   a. $\mathbf{u} = (1, 2, -1)$, $\mathbf{v} = (-1, 1, 1)$, $\mathbf{w} = (2, 0, 2)$
   b. $\mathbf{u} = (2, 2, 2)$, $\mathbf{v} = (4, 1, -5)$, $\mathbf{w} = (2, -3, 1)$
3. $\mathbf{u} = (1, 1, 1)$, $\mathbf{v} = (2, -3, 1)$, $\mathbf{w} = (a, b, c)$

   Find all (a, b, c) in $\mathbb{R}^3$ so that the set S = (u, v, w) is orthogonal.
4. Normalize the rows to make each of the following matrices orthogonal.

   a. $\mathbf{A} = \begin{bmatrix} 8 & -6 \\ 6 & 8 \end{bmatrix}$

   b. $\mathrm{B} = \begin{bmatrix} 2 & 1 & 1 \\ -2 & 2 & 2 \\ 0 & -3 & 3 \end{bmatrix}$

5. Taking $\mathbf{u} = (1, \ -2, \ 2)$ and $\mathbf{a} = (2, \ 1, \ -1)$, determine the projection of $\mathbf{u}$ in the direction and orthogonal to $\mathbf{a}$.
6. Apply Gram–Schmidt algorithm to transform the basis $\{\mathbf{u}_1, \mathbf{u}_2\}$ into an orthonormal basis

$$\mathbf{u}_1 = (1, \ -2), \ \mathbf{u}_2 = (2, \ 4)$$

7. Use the Gram–Schmidt process to transform the basis $\{\mathbf{u}_1, \mathbf{u}_2, \mathbf{u}_3\}$ into an orthonormal basis in $\mathbb{R}^3$.

$$\mathbf{u}_1 = (1, \ 2, \ 2), \ \ \mathbf{u}_2 = (2, \ 1, \ 1), \ \ \mathbf{u}_3 = (1, \ 0, \ 1)$$

2.9 Mapping and transformation

If S and T are two nonempty sets, a function that maps S to T is a rule that assigns to each element $s \in S$ a unique element $t \in T$. Functions are often called mappings to emphasize the association between the two sets. S is called the domain and T the codomain of the mapping. A mapping from S to T can be denoted as

$$f : S \rightarrow T$$

and the association of the element t with the element s is written as

$$t = f(s)$$

t, said to be the image of s under $f(s)$, is the value of f at s. It is to be noted that every member s in the domain S is associated to a unique member t of the codomain T. However, every element in the codomain needs not be associated with an element in the domain.

Let us consider two sets X and Y. For $x \in X$, and $y \in Y$, both functions $y = f_1(x) = ax + b$ and $y = f_2(x) = cx^2$ (where a, b, and c are constants) map X to Y. However, as it can be seen, f_1 is a linear function (graph is a straight line), while f_2 is a nonlinear function.

Linear mappings on vector spaces preserve their structure. We have seen that vectors are objects that can be added together and multiplied by a scalar to produce a resulting object which is also a vector. Considering two vector spaces V and W, a mapping $\psi : V \rightarrow W$ preserves the structure of the vector space V if

$$\psi(\mathbf{x} + \mathbf{y}) = \psi(\mathbf{x}) + \psi(\mathbf{y})$$

and

$$\psi(\lambda \mathbf{x}) = \lambda \psi(\mathbf{x})$$

for all $\mathbf{x}, \mathbf{y} \in V$ and $\lambda = \mathbb{R}$.

For vector spaces V and W, and $\forall$ $\mathbf{x}$, $\mathbf{y} \in V$ and $\forall$ λ, δ $\forall$ $\mathbb{R}$, $\psi: V \rightarrow W$ is a linear mapping if

$$\psi(\lambda\mathbf{x} + \delta\mathbf{y}) = \lambda\psi(\mathbf{x}) + \delta\psi(\mathbf{y})$$

A linear transformation from V into W is defined to be a linear function T that maps V into W. If, for a function T, $\mathbb{R}^n$ is the domain and $\mathbb{R}^m$ is the codomain, it is said that T is a transformation from $\mathbb{R}^n$ to $\mathbb{R}^m$. The transformation is expressed as

$$T\colon \mathbb{R}^n \rightarrow \mathbb{R}^m$$

In the special case, when $m = n$, the transformation is often called an **operator** on $\mathbb{R}^n$. In other words, a linear operator on V is defined to be a liner transformation from V into itself.

Definition 2.9.1 A transformation T: $\mathbb{R}^n \rightarrow \mathbb{R}^m$ is called a ***linear transformation*** if it satisfies the following conditions for vectors $\mathbf{x}$, $\mathbf{y} \in \mathbb{R}^n$ and all scalars λ:

(a) Additivity: $T(\mathbf{x} + \mathbf{y}) = T(\mathbf{x}) + T(\mathbf{y})$
(b) Homogeneity: $T(\lambda\,\mathbf{x}) = \lambda\,T(\mathbf{x})$

Let us try to understand the nature of T. We have seen in Chapter 1 that the generalized form of a linear system can be written as

$$a_{11}x_1 + a_{12}x_2 + \ldots\ldots\ldots + a_{1n}x_n = b_1$$

$$a_{21}x_1 + a_{22}x_2 + \ldots\ldots\ldots + a_{2n}x_n = b_2$$

$$\ldots\ldots\ldots\ldots\ldots\ldots\ldots\ldots\ldots\ldots\ldots$$

$$\ldots\ldots\ldots\ldots\ldots\ldots\ldots\ldots\ldots\ldots\ldots$$

$$a_{m1}x_1 + a_{m2}x_2 + \ldots\ldots\ldots + a_{mn}x_n = b_m$$

which can be compactly represented as

$$\begin{bmatrix} a_{11} & a_{12} & \cdots & a_{1n} \\ a_{21} & a_{22} & \cdots & a_{2n} \\ \vdots & \vdots & \vdots & \vdots \\ a_{m1} & a_{m2} & \cdots & a_{mn} \end{bmatrix} \begin{bmatrix} x_1 \\ x_2 \\ \vdots \\ x_n \end{bmatrix} = \begin{bmatrix} b_1 \\ b_2 \\ \vdots \\ b_m \end{bmatrix} \tag{2.33}$$

or even more compactly as a single matrix equation

$$\mathbf{Ax} = \mathbf{b} \tag{2.34}$$

where

$$\mathbf{A} = \begin{bmatrix} a_{11} & a_{12} & \cdots & a_{1n} \\ a_{21} & a_{22} & \cdots & a_{2n} \\ \vdots & \vdots & \cdots & \vdots \\ a_{m1} & a_{m2} & \cdots & a_{mn} \end{bmatrix} \text{ is an } m \times n \text{ coefficient matrix}$$

$$\mathbf{x} = \begin{bmatrix} x_1 \\ x_2 \\ \cdot \\ \cdot \\ x_n \end{bmatrix} \text{ is the variable vector and } \mathbf{b} = \begin{bmatrix} b_1 \\ b_2 \\ \cdot \\ \cdot \\ b_m \end{bmatrix} \text{ is the constant vector.}$$

(2.34) can be seen as a transformation that maps a vector **x** in $\mathbb{R}^n$ into a vector **b** in $\mathbb{R}^m$ by multiplying **x** on the left by an m x n matrix **A**. It is called a **matrix transformation** denoted by

$$T_A: \mathbb{R}^n \rightarrow \mathbb{R}^m$$

In cases where specification of the domain and codomain is not essential, the transformation can be expressed as

$$\mathbf{b} = T_A(\mathbf{x})$$

or alternatively as

$$\mathrm{x} \xrightarrow{T_A} \mathrm{b}$$

It is sometimes useful to consider an $m \times n$ matrix as a row of n column vectors. Thus we can write

$$\mathrm{A} = [\mathbf{a}_1\ \mathbf{a}_2 .. \mathbf{a}_j .. \mathbf{a}_n]$$

where

$$\mathbf{a}_j = \begin{bmatrix} a_{1j} \\ a_{2j} \\ \cdot \\ \cdot \\ \cdot \\ a_{mj} \end{bmatrix}$$

then, the system of linear equations can be written as a single vector equation

$$x_1\,\mathbf{a}_1 + x_2\,\mathbf{a}_2 + \cdots\cdots + x_n\,\mathbf{a}_n = \mathbf{b}$$

that is, **b** is expressed as a linear combination of the columns of A.

We have seen in (2.34) that **Ax** is a product of **matrix-vector multiplication**, that is multiplication of an n-vector **x** by an $m \times n$ matrix **A**. Therefore it is worthwhile to consider some algebraic properties of matrix-vector multiplication that have been found to be useful in many applications of linear algebra.

Theorem 2.9.1

For matrices **A, B** $\in \mathbb{R}^{m \times n}$ and vectors **x, y** $\in \mathbb{R}^n$

a. $\mathbf{A}(\mathbf{x}+\mathbf{y}) = \mathbf{A}\,\mathbf{x} + \mathbf{A}\,\mathbf{y}$
b. $\mathbf{A}(\lambda\,\mathbf{x}) = \lambda\,\mathbf{A}(\mathbf{x}) = (\lambda\,\mathbf{A})\,\mathbf{x}$ for all scalars λ
c. $(\mathbf{A}+\mathbf{B})\,\mathbf{x} = \mathbf{A}\,\mathbf{x} + \mathbf{B}\,\mathbf{x}$

Example 2.9.1: One of the ways to transform the vector spaces $\mathbb{R}^2$ and $\mathbb{R}^3$ is by the use of matrix operators, for example, an operator that maps each point into its symmetric image about a fixed line or a fixed plane that contains the origin.

$$\begin{bmatrix} a_1 \\ -a_2 \end{bmatrix} = \mathbf{T}\begin{bmatrix} a_1 \\ a_2 \end{bmatrix}$$

where the matrix

$$\mathbf{T} = \begin{bmatrix} 1 & 0 \\ 0 & -1 \end{bmatrix}$$

carries the vector $\begin{bmatrix} a_1 \\ a_2 \end{bmatrix}$ to its reflection $\begin{bmatrix} a_1 \\ -a_2 \end{bmatrix}$. This is a case of matrix transformation $\mathbb{R}^2 \rightarrow \mathbb{R}^2$.

Example 2.9.2: An example of a transformation T_A: $\mathbb{R}^4 \rightarrow \mathbb{R}^3$ is

$$\begin{bmatrix} 1 & 1 & 0 & 0 \\ 0 & 1 & 1 & 0 \\ 0 & 0 & 1 & 1 \end{bmatrix}\begin{bmatrix} \mathrm{x}_1 \\ \mathrm{x}_2 \\ \mathrm{x}_1 \\ \mathrm{x}_4 \end{bmatrix} = \begin{bmatrix} \mathrm{x}_1 + \mathrm{x}_2 \\ \mathrm{x}_2 + \mathrm{x}_3 \\ \mathrm{x}_3 + \mathrm{x}_4 \end{bmatrix}$$

Now, as we have seen above, if matrix $\mathbf{A} \in \mathbb{R}^{m \times n}$, the matrix transformation T_A: $\mathbb{R}^n \rightarrow \mathbb{R}^m$ can be expressed as

$$T_A(\mathbf{x}) = \mathbf{Ax}$$

for all $\mathbf{x} \in \mathbb{R}^n$, so Theorem 2.9.1 gives

$$T_A(\mathbf{x} + \mathbf{y}) = \mathbf{A}(\mathbf{x} + \mathbf{y}) = \mathbf{Ax} + \mathbf{Ay} = T_A(\mathbf{x}) + T_A(\mathbf{y})$$

and

$$T_A(\lambda\mathbf{x}) = \mathbf{A}(\lambda\mathbf{x}) = \lambda(\mathbf{Ax}) = \lambda T_A(\mathbf{x})$$

showing that T_A satisfies both additivity and homogeneity conditions. Hence, a matrix transformation is a linear transformation.

Let us find out if the converse is also true. It has been already seen that the standard basis of $\mathbb{R}^n$ is a set of column vectors

$$(\mathbf{e}_1, \mathbf{e}_2, \ldots\ldots, \mathbf{e}_n)$$

of the identity matrix $\mathbf{I}_n$. Each $\mathbf{e}_i \in \mathbb{R}^n$ and every vector $\mathbf{x} \in \mathbb{R}^n$ is a linear combination of the $\mathbf{e}_i$'s

$$\mathbf{x} = x_1\mathbf{e}_1 + x_2\mathbf{e}_2 + \cdots\cdots + x_n\mathbf{e}_n$$

so that

$$T(\mathbf{x}) = T(x_1\mathbf{e}_1 + x_2\mathbf{e}_2 + \cdots\cdots + x_n\mathbf{e}_n) = x_1 T(\mathbf{e}_1) + x_2 T(\mathbf{e}_2) + \cdots\cdots + x_n T(\mathbf{e}_n)$$

that is, $T(\mathbf{x})$ is now expressed as a linear combination of the column vectors of the $m \times n$ matrix

$$\mathbf{A} = [T(\mathbf{e}_1) \quad T(\mathbf{e}_2) \cdots \cdots T(\mathbf{e}_n)]$$

and we can write

$$T(\mathbf{x}) = [T(\mathbf{e}_1) \quad T(\mathbf{e}_2) \cdots \cdots T(\mathbf{e}_n)] \begin{bmatrix} x_1 \\ x_2 \\ \cdot \\ \cdot \\ \cdot \\ x_n \end{bmatrix} = \mathbf{A}\mathbf{x} \tag{2.35}$$

Since the above equation is valid for every $\mathbf{x} \in \mathbb{R}^n$, it can be concluded that every linear transformation T_A: $\mathbb{R}^n \to \mathbb{R}^m$ is a matrix transformation.

Theorem 2.9.2 A transformation T: $\mathbb{R}^n \to \mathbb{R}^m$ is linear only if it is a matrix transformation.

Basic matrix transformations

We have already seen an example of an operator that carries a vector to its reflection about the x-axis. Reflections about lines and planes through the origin, projections onto lines, and planes through the origin and rotation about the origin constitute some of the basic matrix transformations. They have important applications in the study of macromolecular structure and dynamics.

Reflection. Basic matrix operators on $\mathbb{R}^2$ and $\mathbb{R}^3$ that map each point into its symmetric image about a fixed line or a fixed plane containing the origin are called **reflection operators**. These are 2×2 or 3×3 matrices. The operator we have seen above

$$\mathbf{T} = \begin{bmatrix} 1 & 0 \\ 0 & -1 \end{bmatrix}$$

denotes reflection about the x-axis (Fig. 2.11A)

$$\mathbf{T}(x, y) = (x, -y)$$

Similarly,

$$\mathbf{T} = \begin{bmatrix} 0 & 1 \\ 1 & 0 \end{bmatrix}$$

denotes reflection about the line $y = x$ (Fig. 2.11B)

$$\mathbf{T}(x, y) = (y, x)$$

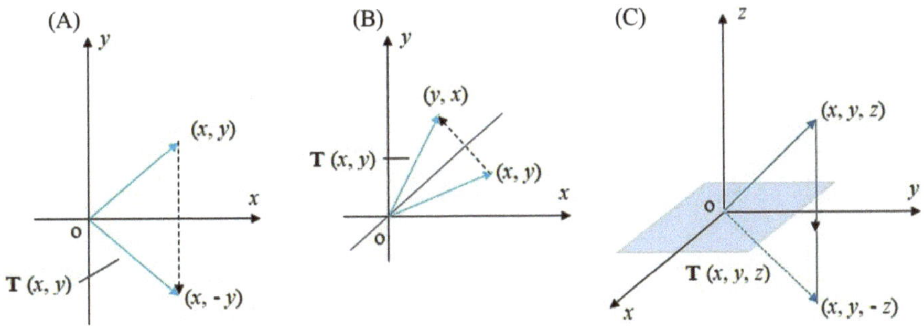

FIGURE 2.11

(A) Reflection about the *x*-axis, (B) reflection about the line *y* = *x*, and (C) reflection about the *xy*-plane.

and

$$\mathbf{T} = \begin{bmatrix} 1 & 0 & 0 \\ 0 & 1 & 0 \\ 0 & 0 & -1 \end{bmatrix}$$

denotes reflection about the *xy*-plane (Fig. 2.11C)

$$\mathbf{T}(x,\ y,\ z) = (x,\ y,\ -z)$$

Projection operators. Specific matrix operators on $\mathbb{R}^2$ and $\mathbb{R}^3$ map each point to its orthogonal projection on a fixed line or plane through the origin. These matrices are called ***projection operators*** or, more precisely, ***orthogonal projection operators***.

The standard matrix for orthogonal projection of a point onto the *x*-axis is

$$\mathbf{T}_{\mathrm{x}} = \begin{bmatrix} 1 & 0 \\ 0 & 0 \end{bmatrix}$$

We can see,

$$\mathbf{T}_x(x,\ y) = \begin{bmatrix} 1 & 0 \\ 0 & 0 \end{bmatrix} \begin{bmatrix} x \\ y \end{bmatrix} = \begin{bmatrix} x \\ 0 \end{bmatrix}$$

Similarly, a matrix

$$\mathbf{T}_{\mathrm{y}} = \begin{bmatrix} 0 & 0 \\ 0 & 1 \end{bmatrix}$$

orthogonally projects a point onto the *y*-axis, that is, $\mathbf{T}_y\ (x,\ y) = (0,\ y)$

Further, the matrix

$$\mathbf{T}_{xy} = \begin{bmatrix} 1 & 0 & 0 \\ 0 & 1 & 0 \\ 0 & 0 & 0 \end{bmatrix}$$

orthogonally projects a point onto the xy-plane

$$\mathbf{T}_{xy}(x, y, z) = \begin{bmatrix} 1 & 0 & 0 \\ 0 & 1 & 0 \\ 0 & 0 & 0 \end{bmatrix} \begin{bmatrix} x \\ y \\ z \end{bmatrix} = \begin{bmatrix} x \\ y \\ 0 \end{bmatrix}$$

The standard matrix operators for orthogonal projection of a point onto the xz- and yz-planes are respectively

$$T_{xz} = \begin{bmatrix} 1 & 0 & 0 \\ 0 & 0 & 0 \\ 0 & 0 & 1 \end{bmatrix} \quad \text{and} \quad T_{yz} = \begin{bmatrix} 0 & 0 & 0 \\ 0 & 1 & 0 \\ 0 & 0 & 1 \end{bmatrix}$$

Rotation operators. Comparison of proteins structures is one very important aspect of computational structural biology. Protein structures are represented by vector sets. A common metric to determine the similarity between two vector sets is the root-mean-square distance (RMSD). An optimal translation vector and rotation matrix are found to minimize the RMSD between corresponding atoms in two coordinate sets.

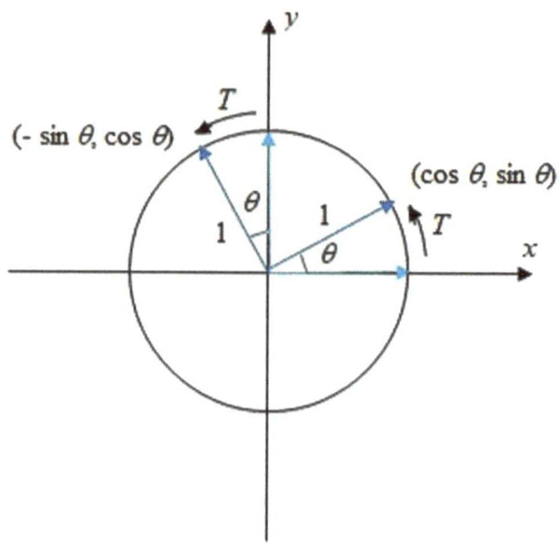

FIGURE 2.12

Rotation of the standard basis vectors.

Rotation operators are matrices $\in \mathbb{R}^2$ or $\mathbb{R}^3$ that move points along arcs of circles centered at the origin. Let us consider the rotation operator T: $\mathbb{R}^2 \to \mathbb{R}^2$ that moves points counterclockwise about the origin through an angle θ (Fig. 2.12). As shown in the figure, the images of the standard basis vectors under T are

$$T(1,\ 0) = (\cos\theta,\ \sin\theta) \text{ and } T(0,\ 1) = (-\sin\theta,\ \cos\theta)$$

Hence, the standard matrix for T is

$$\mathbf{R}_\theta = [T(1,\ 0) \quad T(0,1)] = \begin{bmatrix} \cos\theta & -\sin\theta \\ \sin\theta & \cos\theta \end{bmatrix}$$

which is called the ***rotation matrix*** for $\mathbb{R}^2$.

If $\mathbf{u} = (x_1, y_1) \in \mathbb{R}^2$ and $\mathbf{v} = (x_2, y_2) \in \mathbb{R}^2$ is its image under $\mathbf{R}_\theta$, then

$$\mathbf{v} = \begin{bmatrix} x_2 \\ y_2 \end{bmatrix} = \mathbf{R}_\theta \mathbf{u} = \begin{bmatrix} \cos\theta & -\sin\theta \\ \sin\theta & \cos\theta \end{bmatrix} \begin{bmatrix} x_1 \\ y_1 \end{bmatrix} = \begin{bmatrix} x_1\cos\theta - y_1\sin\theta \\ x_1\sin\theta + y_1\cos\vartheta \end{bmatrix} \tag{2.36}$$

Let us find the standard matrix for the rotation of a three-dimensional vector through an angle θ around the z-axis. (2.36) shows how the coordinates in the xy-plane change as the vector is rotated through an angle θ. Here, in this three-dimensional rotation, the z-coordinate remains unchanged, and the standard rotation matrix is

$$\mathbf{R}_{z,\theta} = \begin{bmatrix} \cos\theta & -\sin\theta & 0 \\ \sin\theta & \cos\theta & 0 \\ 0 & 0 & 1 \end{bmatrix}$$

Similarly, for rotations about the x-axis and y-axis, the three-dimensional rotation matrices are respectively

$$\mathbf{R}_{x,\theta} = \begin{bmatrix} 1 & 0 & 0 \\ 0 & \cos\theta & -\sin\theta \\ 0 & \sin\theta & \cos\theta \end{bmatrix} \quad \text{and} \quad \mathbf{R}_{y,\theta} = \begin{bmatrix} \cos\theta & 0 & \sin\theta \\ 0 & 1 & 0 \\ -\sin\theta & 0 & \cos\theta \end{bmatrix}$$

Exercise 2.9

1. A transformation T: $\mathbb{R}^3 \to \mathbb{R}^2$ is defined by

$$T\begin{bmatrix} x_1 \\ x_2 \\ x_3 \end{bmatrix} = \begin{bmatrix} x_1 \\ x_2 \end{bmatrix} \text{ for all } \begin{bmatrix} x_1 \\ x_2 \\ x_3 \end{bmatrix} \text{ in } \mathbb{R}^3$$

Is it a linear transformation? If so. find its matrix.

2. Given a vector (x, y), find a matrix that will
   a. expand the vector by a factor h in the x-direction,
   b. compress the vector by a factor k in the y-direction,
   c. simultaneously compress the vector by a factor of h in the x-direction and expand the vector by a factor k in the y-direction.
3. Given a vector (x, y), find a matrix that will
   a. only stretch the vector by a factor of λ,
   b. only rotate the vector by $\pi/2$ radians about the origin, and
   c. both rotate the vector by $\pi/2$ radians about the origin and stretch the vector by a factor of λ.
4. Show, by using matrix operators, that the result of projection of a vector (x, y, z) first onto the xy-plane followed by projection onto the x-axis is the same as projection of the vector first onto the xz-plane followed by projection onto the x-axis.

2.10 Change of basis

Earlier, in Section 2.5, we have familiarized ourselves with the concepts of basis and coordinate vector in a vector space. We have also seen that for many problems in either $\mathbb{R}^3$ or $\mathbb{R}^n$, the standard basis is very useful to work with. However, it is a fact that a basis suitable for one problem may not be so for another. Sometimes, it may be necessary to change from one basis to another.

Let us consider the change of basis from G to H of a vector $\mathbf{v} \in V$. It is important to understand how are the coordinate vectors $(\mathbf{v})_G$ and $(\mathbf{v})_H$ related. It is convenient to solve the problem in two-dimensional space, then the solution in n-dimensional space will be just an extension.

Let the basis vectors in the two systems be, respectively,

$$G = \{\hat{\mathrm{g}}_1, \hat{\mathrm{g}}_2\} \text{ and } H = \{\hat{\mathrm{h}}_1, \hat{\mathrm{h}}_2\}$$

and let the coordinate vectors for H relative to G be

$$[\hat{\mathbf{h}}_1]_G = \begin{bmatrix} a \\ b \end{bmatrix} \text{ and } [\hat{\mathbf{h}}_2]_G = \begin{bmatrix} c \\ d \end{bmatrix}$$

so that we can write

$$\hat{\mathbf{h}}_1 = a\hat{\mathbf{g}}_1 + b\hat{\mathbf{g}}_2$$

$$\hat{\mathbf{h}}_2 = c\hat{\mathbf{g}}_1 + d\hat{\mathbf{g}}_2$$

For any vector $\mathbf{v} \in V$, let

$$[\mathbf{v}]_H = \begin{bmatrix} k_1 \\ k_2 \end{bmatrix}$$

be the coordinate vector relative to H, so that

$$\mathbf{v} = k_1 \hat{\mathbf{h}}_1 + k_2 \hat{\mathbf{h}}_2 \tag{2.37}$$

Expressed in terms of the basis G

$$\begin{aligned} \mathbf{v} &= k_1 \left(a \hat{\mathbf{g}}_1 + b \hat{\mathbf{g}}_2\right) + k_2 \left(c \hat{\mathbf{g}}_1 + d \hat{\mathbf{g}}_2\right) \\ &= (k_1 a + k_2 c) \hat{\mathbf{g}}_1 + (k_1 b + k_2 d) \hat{\mathbf{g}}_2 \end{aligned} \tag{2.38}$$

Thus the coordinate vector for v relative to G is

$$[\mathbf{v}]_G = \begin{bmatrix} k_1 a + k_2 c \\ k_1 b + k_2 d \end{bmatrix} \tag{2.39}$$

which can also be written as

$$[\mathbf{v}]_G = \begin{bmatrix} a & c \\ b & d \end{bmatrix} \begin{bmatrix} k_1 \\ k_2 \end{bmatrix} = \begin{bmatrix} a & c \\ b & d \end{bmatrix} [\mathbf{v}]_H \tag{2.40}$$

(2.40) shows that coordinate vector $[\mathbf{v}]_G$ results when coordinate vector $[\mathbf{v}]_H$ is multiplied on the left by the matrix

$$\mathbf{P} = \begin{bmatrix} a & c \\ b & d \end{bmatrix}$$

the columns of which are the coordinates of the basis vectors for H relative to the basis G. The change of basis problem that has been solved for a two-dimensional space can now be extended to an n-dimensional vector space as follows:

If the basis of a vector space V is changed from $G = \{\hat{\mathbf{g}}_1, \hat{\mathbf{g}}_2, \ldots\ldots, \hat{\mathbf{g}}_n\}$ to $H = \{\hat{\mathbf{h}}_1, \hat{\mathbf{h}}_2, \ldots, \hat{\mathbf{h}}_n\}$, then for each vector $\mathbf{v} \in V$, the coordinate vectors $[\mathbf{v}]_G$ and $[\mathbf{v}]_H$ are related by the equation

$$[\mathbf{v}]_G = \mathbf{P}[\mathbf{v}]_H \tag{2.41}$$

where columns of $\mathbf{P}$ are the coordinate vectors of basis vectors for H relative to the basis G, that is

$$\mathbf{P}_{H \to G} = \left[[\hat{\mathbf{h}}_1]_G \; [\hat{\mathbf{h}}_2]_G \cdots [\hat{\mathbf{h}}_n]_G \right]$$

The matrix $\mathbf{P}$ in (2.41) is called the **transition matrix** from H to G. If $\mathbf{P}_{G \to H}$ is the transition matrix from G to H, it can be shown that

$$(\mathbf{P}_{G \to H}) \; (\mathbf{P}_{H \to G}) = \mathbf{I}$$

that is,

$$(\mathbf{P}_{G \to H}) = (\mathbf{P}_{H \to G})^{-1}$$

This is stated in the following theorem.

Theorem 2.10.1 The transition matrix **P** from a basis H to a basis G for a finite-dimensional vector space V is invertible, and $\mathbf{P}^{-1}$ is the transition matrix from G to H.

Example 2.10.1: Let us consider bases G and H where

$$\hat{\mathbf{g}}_1 = \begin{bmatrix} 1 \\ 0 \end{bmatrix}, \quad \hat{\mathbf{g}}_2 = \begin{bmatrix} 0 \\ 1 \end{bmatrix}$$

is the standard basis in $\mathbb{R}^2$, and

$$\hat{\mathbf{h}}_1 = \begin{bmatrix} 2 \\ 1 \end{bmatrix}, \quad \hat{\mathbf{h}}_2 = \begin{bmatrix} -1 \\ 1 \end{bmatrix}$$

are the coordinate vectors of H relative to G.

The transition matrix from H to G is then

$$\mathbf{P}_{H\to G} = \begin{bmatrix} 2 & -1 \\ 1 & 1 \end{bmatrix}$$

and the transition matrix from G to H is

$$\mathbf{P}_{G\to H} = \begin{bmatrix} 2 & -1 \\ 1 & 1 \end{bmatrix}^{-1} = \begin{bmatrix} 1/3 & 1/3 \\ -1/3 & 2/3 \end{bmatrix}$$

For a fixed vector **v**, the coordinate vector of which relative to G is given as

$$\mathbf{v}_{\mathrm{G}} = \begin{bmatrix} 3 \\ 2 \end{bmatrix}$$

the coordinate vector in H is

$$\mathbf{v}_{\mathrm{H}} = \begin{bmatrix} 1/3 & 1/3 \\ -1/3 & 2/3 \end{bmatrix} \begin{bmatrix} 3 \\ 2 \end{bmatrix} = \begin{bmatrix} 5/3 \\ 1/3 \end{bmatrix}$$

Exercise 2.10

1. Consider the bases $G = \{\mathbf{g}_1, \mathbf{g}_2\}$ and $H = \{\mathbf{h}_1, \mathbf{h}_{2|}\}$ for $\mathbb{R}^2$ where

$$\mathbf{g}_1 = \begin{bmatrix} 2 \\ 1 \end{bmatrix}, \quad \mathbf{g}_2 = \begin{bmatrix} -1 \\ 4 \end{bmatrix}, \quad \mathbf{h}_1 = \begin{bmatrix} 1 \\ 1 \end{bmatrix}, \quad \mathbf{h}_2 = \begin{bmatrix} -2 \\ 3 \end{bmatrix}$$

Calculate the transition matrices $\mathbf{P}_{G\to H}$ and $\mathbf{P}_{H\to G}$.

2. Consider the bases $G = \{\mathbf{g}_1, \mathbf{g}_2, \mathbf{g}_3\}$ and $H = \{\mathbf{h}_1, \mathbf{h}_2, \mathbf{h}_3\}$ for $\mathbb{R}^3$ where

$$\mathbf{g}_1 = \begin{bmatrix} 2 \\ 1 \\ -2 \end{bmatrix}, \quad \mathbf{g}_2 = \begin{bmatrix} 0 \\ 2 \\ -1 \end{bmatrix}, \quad \mathbf{g}_3 = \begin{bmatrix} 1 \\ -1 \\ 2 \end{bmatrix},$$

$$\mathbf{h}_1 = \begin{bmatrix} 2 \\ 1 \\ -1 \end{bmatrix}, \quad \mathbf{h}_2 = \begin{bmatrix} 1 \\ 2 \\ -3 \end{bmatrix}, \quad \mathbf{h}_3 = \begin{bmatrix} -1 \\ 0 \\ 2 \end{bmatrix}$$

Calculate the transition matrices $\mathbf{P}_{G \to H}$ and $\mathbf{P}_{H \to G.}$

3. Consider the bases $U = \{\mathbf{u}_1, \mathbf{u}_2\}$, $V = \{\mathbf{v}_1, \mathbf{v}_2\}$ and $W = \{\mathbf{w}_1, \mathbf{w}_2\}$ for $\mathbb{R}^3$ where

$$\mathbf{u}_1 = \begin{bmatrix} 1 \\ 0 \end{bmatrix}, \quad \mathbf{u}_2 = \begin{bmatrix} 0 \\ 1 \end{bmatrix},$$

$$\mathbf{v}_1 = \begin{bmatrix} 2 \\ 1 \end{bmatrix}, \quad \mathbf{v}_2 = \begin{bmatrix} -1 \\ 4 \end{bmatrix},$$

$$\mathbf{w}_1 = \begin{bmatrix} 1 \\ 1 \end{bmatrix}, \quad \mathbf{w}_2 = \begin{bmatrix} -2 \\ 3 \end{bmatrix}$$

and the vectors whose coordinate vectors relative to U are given as

a. $\mathbf{x}_U = \begin{bmatrix} 1 \\ 1 \end{bmatrix}$ and b. $\mathbf{y}_U = \begin{bmatrix} 2 \\ 3 \end{bmatrix}$

Find out the results of sequential changes of basis, $U \to V$, $V \to W$, and $W \to U$, on each of these vectors.

Further reading

Anton, H., Rorres, C., 2014. Elementary Linear Algebra. Wiley.

Deisenroth, M.P., Faisal, A.A., Ong, C.S., 2020. Mathematics for Machine Learning. Cambridge University Press.

Nicholson, W.K., 2021. Linear Algebra with Applications, Open Edition. Creative Common License (CC BY-NC-SA).

CHAPTER

Matrix decomposition

3

3.1 Eigensystems from different perspectives

We already know that the general effect of matrix transformation of vectors is a combination of rotation and stretching (in both positive and negative senses). In other words, multiplication of a vector $\mathbf{x}$ in a vector space V by a matrix $\mathbf{A}$ changes the direction (rotates) and magnitude (stretches or compresses) of $\mathbf{x}$. However, for a specific vector $\mathbf{u}$ and a scalar λ, multiplication of $\mathbf{u}$ by $\mathbf{A}$ is equivalent to multiplication of $\mathbf{u}$ by λ,

$$\mathbf{A}\mathbf{u} = \lambda\mathbf{u} \tag{3.1}$$

that is, there is a change only in magnitude but not in direction of the vector under transformation. For each matrix $\mathbf{A}$, only some of the vectors in a vector space V have this property—these "special" vectors are called the ***eigenvectors*** and their corresponding scalars are called the ***eigenvalues*** of $\mathbf{A}$.

Eigenvalues and eigenvectors are associated with square matrices and, together, they facilitate ***eigendecomposition*** of a matrix, which analyzes its structure. Eigenvalues and eigenvectors can be looked at from different perspectives. Here, we shall consider two of them and move on to discuss some of their basic properties.

3.1.1 A stable distribution vector

Let us suppose that in a living cell a protein molecule can exist in two different states, S_1 and S_2, and the population distribution of the states at a certain time is denoted by the vector $\mathbf{x} = (x_1, x_2)$. The transition occurs between the states according to the following rule: in 1 second, 20% of the molecules in state S_1 switch over to S_2 while 10% of the molecules in S_2 revert back to S_1.

Let us suppose that at time $t = 0$, the population distribution between the two states is given by the vector

$$(x_{01}, x_{02}) = (300, 300)$$

We find that at the end of one second, the number of molecules in state S_1 will be given by the dot product

$$(0.8, \ 0.1) \cdot (300, \ 300) = 270$$

Mathematical Approaches to Molecular Structural Biology. DOI: https://doi.org/10.1016/B978-0-323-90397-4.00006-8
© 2023 Elsevier Inc. All rights reserved.

and the number of molecules in state S_2 will be

$$(0.2,\ 0.9) \cdot (300,\ 300) = 330$$

We can put the transition rates, which remain constant over time, into a matrix

$$\mathbf{A} = \begin{bmatrix} 0.8 & 0.1 \\ 0.2 & 0.9 \end{bmatrix}$$

Multiplying the initial distribution vector (300, 300) by the matrix **A**, we obtain the distribution vector at the end of 1 second

$$\begin{bmatrix} 0.8 & 0.1 \\ 0.2 & 0.9 \end{bmatrix}\begin{bmatrix} x_{01} \\ x_{02} \end{bmatrix} = \begin{bmatrix} 0.8 & 0.1 \\ 0.2 & 0.9 \end{bmatrix}\begin{bmatrix} 300 \\ 300 \end{bmatrix} = \begin{bmatrix} 270 \\ 330 \end{bmatrix} = \begin{bmatrix} x_{11} \\ x_{12} \end{bmatrix}$$

Geometrically, multiplication by **A** has rotated the vector (x_{01}, x_{02}) into the vector (x_{11}, x_{12}) (Fig. 3.1). In other words, although the total number of molecules remain the same, the distribution (vector) has changed.

Similarly, at the end of 2 seconds, the distribution between the two states will be given by

$$\begin{bmatrix} 0.8 & 0.1 \\ 0.2 & 0.9 \end{bmatrix}\begin{bmatrix} x_{11} \\ x_{12} \end{bmatrix} = \begin{bmatrix} 0.8 & 0.1 \\ 0.2 & 0.9 \end{bmatrix}\begin{bmatrix} 270 \\ 330 \end{bmatrix} = \begin{bmatrix} 249 \\ 351 \end{bmatrix} = \begin{bmatrix} x_{21} \\ x_{22} \end{bmatrix}$$

It can be seen that in every second, the number of molecules in state S_1 is decreasing while that in S_2 is increasing.

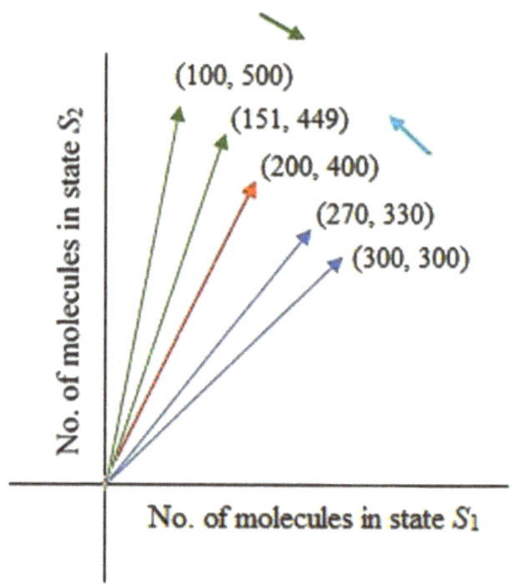

FIGURE 3.1

Transformation of the vectors showing changes in distribution of the molecules with time.

The population distribution will reach a steady state when

$$A\begin{bmatrix} x_{s1} \\ x_{s2} \end{bmatrix} = \lambda \begin{bmatrix} x_{s1} \\ x_{s2} \end{bmatrix}$$

That is

$$\mathbf{A}\,\mathbf{x}_s = \lambda \mathbf{x}_s \tag{3.2}$$

$\mathbf{x}_s$ is called the ***eigenvector*** and λ is called the ***eigenvalue*** of the matrix **A**. In the following section, we are going to discuss extensively about eigenvectors and eigenvalues of a matrix and the methods to calculate them. At this point, we can satisfy ourselves by taking $\lambda = 1$ and $\mathbf{x}_s = (200, 400)$ and verifying that

$$\begin{bmatrix} 0.8 & 0.1 \\ 0.2 & 0.9 \end{bmatrix}\begin{bmatrix} 200 \\ 400 \end{bmatrix} = 1\begin{vmatrix} 200 \\ 400 \end{vmatrix}$$

In other words, the distribution will reach a steady state represented by the vector (colored red in Fig. 3.1), when there are 200 molecules in S_1 and 400 molecules in S_2. The distribution vector (colored blue in Fig. 3.1) that started at (300, 300) moves toward the equilibrium with the passage of time.

If the population distribution starts with $(x_{01}, x_{02}) = (100, 500)$, then at the end of 1 second

$$\begin{bmatrix} 0.8 & 0.1 \\ 0.2 & 0.9 \end{bmatrix}\begin{bmatrix} x_{01} \\ x_{02} \end{bmatrix} = \begin{bmatrix} 0.8 & 0.1 \\ 0.2 & 0.9 \end{bmatrix}\begin{bmatrix} 100 \\ 500 \end{bmatrix} = \begin{bmatrix} 130 \\ 470 \end{bmatrix} = \begin{bmatrix} x_{11} \\ x_{12} \end{bmatrix}$$

and at the end of 2 seconds

$$\begin{bmatrix} 0.8 & 0.1 \\ 0.2 & 0.9 \end{bmatrix}\begin{bmatrix} x_{11} \\ x_{12} \end{bmatrix} = \begin{bmatrix} 0.8 & 0.1 \\ 0.2 & 0.9 \end{bmatrix}\begin{bmatrix} 130 \\ 470 \end{bmatrix} = \begin{bmatrix} 151 \\ 449 \end{bmatrix} = \begin{bmatrix} x_{21} \\ x_{22} \end{bmatrix}$$

In this case, it can be seen geometrically (Fig. 3.1) that the distribution vector (colored green) moves towards the steady state vector (colored red) from the left.

3.1.2 System of linear differential equations

Let us look at eigensystems in the perspective of linear differential equations. A simple first-order linear differential equation

$$u' = \lambda u \tag{3.3}$$

will have a solution in the form

$$u = \alpha e^{\lambda t} \tag{3.4}$$

Now, let us consider a system of two first-order linear differential equations

$$u_1' = u_1 + 2u_2 \tag{3.5a}$$

$$u_2' = 2u_1 + u_2 \tag{3.5b}$$

which can be written in the form

$$\mathbf{u}' = \mathbf{A}\mathbf{u} \tag{3.6}$$

where

$$\mathbf{u}' = \begin{bmatrix} u_1' \\ u_2' \end{bmatrix}, \quad \mathbf{A} = \begin{bmatrix} 1 & 2 \\ 2 & 1 \end{bmatrix}, \quad \mathbf{u} = \begin{bmatrix} u_1 \\ u_2 \end{bmatrix}$$

Based on (3.4) we can seek the solution of (3.6) as

$$u_1 = \alpha_1 e^{\lambda t} \text{ and } u_2 = \alpha_2 e^{\lambda t} \tag{3.7}$$

From (3.6) and (3.7), we can derive

$$\begin{bmatrix} 1 & 2 \\ 2 & 1 \end{bmatrix}\begin{bmatrix} \alpha_1 \\ \alpha_2 \end{bmatrix} = \lambda \begin{bmatrix} \alpha_1 \\ \alpha_2 \end{bmatrix}$$

which can be written as

$$\mathbf{A}\,\mathbf{x} = \lambda\,\mathbf{x} \tag{3.8}$$

where

$$\mathbf{A} = \begin{bmatrix} 1 & 2 \\ 2 & 1 \end{bmatrix}, \quad \mathbf{x} = \begin{bmatrix} \alpha_1 \\ \alpha_2 \end{bmatrix}$$

Hence, solutions to (3.6) in the form (3.7) can be constructed if solutions for λ and $\mathbf{x}$ in matrix equation (3.8) can be found. $\mathbf{x} = \mathbf{0}$ is a trivial solution to (3.8) and it provides no useful information. Writing (3.8) in the form

$$(\mathbf{A} - \lambda\,\mathbf{I})\,\mathbf{x} = \mathbf{0} \tag{3.9}$$

shows that $\mathbf{x}$ can be nonzero only for certain "special" or "characteristic" values of λ that make $(\mathbf{A} - \lambda\,\mathbf{I})$ singular, that is

$$\det\,(\mathbf{A} - \lambda\mathbf{I}) = 0 \tag{3.10}$$

We shall see in the next section that these "characteristic" values of λ are the eigenvalues and the nonzero vectors $\mathbf{x}$ that satisfy (3.8) are the eigenvectors of the matrix $\mathbf{A}$. For the time being let us accept that in the example under consideration, there are two sets of solutions for λ and $\mathbf{x}$. For $\lambda_1 = 3$, $\mathbf{x} = (1, 1)$ and for $\lambda_2 = -1$, $\mathbf{x} = (-1, 1)$. So, the corresponding solutions of (3.6) are

$$\mathbf{u} = \begin{bmatrix} u_1 \\ u_2 \end{bmatrix} = e^{3t}\begin{bmatrix} 1 \\ 1 \end{bmatrix} \text{ and } \mathbf{u} = \begin{bmatrix} u_1 \\ u_2 \end{bmatrix} = e^{-t}\begin{bmatrix} 1 \\ -1 \end{bmatrix} \tag{3.11}$$

and the general solution is

$$\mathbf{u} = c_1 e^{3t}\begin{bmatrix} 1 \\ 1 \end{bmatrix} + c_2 e^{-t}\begin{bmatrix} 1 \\ -1 \end{bmatrix} \tag{3.12}$$

where c_1 and c_2 constants.

Exercise 3.1

1. In a community of 900 people, a fraction has been infected with a virus. The rate of infection is 10% in a day. A medical team rushes with a drug which was found to cure 20% of the infected in a day. There were 500 healthy people and 400 infected people when application of the drug began. What were the distributions of healthy versus infected people in 1 and 2 days thereafter? If the infectivity of the virus and efficacy of the drug remain the same, where do you think the distribution will stabilize?
2. Solve the following systems of linear differential equations:

a. $\mathbf{u}' = \begin{bmatrix} u_1' \\ u_2' \end{bmatrix} = \begin{bmatrix} 4 & 0 \\ 0 & 7 \end{bmatrix} \begin{bmatrix} u_1 \\ u_2 \end{bmatrix}$

b. $\mathbf{u}' = \begin{bmatrix} u_1' \\ u_2' \end{bmatrix} = \begin{bmatrix} 2 & 0 \\ 1 & 3 \end{bmatrix} \begin{bmatrix} u_1 \\ u_2 \end{bmatrix}$

3.2 Eigensystem basics

Here, we are going to formalize the concept of eigenvectors and eigenvalues.

Definition 3.2.1 For a square matrix $\mathbf{A} \in \mathbb{R}^{n\times n}$, a scalar $\lambda \in \mathbb{R}$ and a nonzero vector $\mathbf{x} \in \mathbb{R}^n$ satisfying the condition, called the ***eigenvalue equation***,

$$\mathbf{A}\mathbf{x} = \lambda\mathbf{x} \tag{3.13}$$

are, respectively, called an ***eigenvalue*** and the corresponding ***eigenvector***, and any such pair $(\lambda, \mathbf{x})$ is called an ***eigenpair*** of **A**.

Example 3.2.1: If

$$\mathbf{A} = \begin{bmatrix} 4 & 3 \\ 2 & -1 \end{bmatrix} \quad \text{and} \quad \mathbf{x} = \begin{bmatrix} 3 \\ 1 \end{bmatrix}$$

it can be shown that $\mathbf{A}\mathbf{x} = 5\mathbf{x}$, so $\lambda = 5$ is an eigenvalue of **A** with corresponding eigenvector $\mathbf{x} = (3, 1)$.

Nonuniqueness of eigenvectors

We can take note of the nonuniqueness of eigenvectors. If **x** is an eigenvector of **A** corresponding to eigenvalue λ, then for any nonzero $k \in \mathbb{R}$,

$$\mathbf{A}(k\mathbf{x}) = k\mathbf{A}(\mathbf{x}) = k\lambda(\mathbf{x}) = \lambda(k\mathbf{x}) \tag{3.14}$$

which means $k\mathbf{x}$ is also an eigenvector of $\mathbf{A}$. To generalize, all vectors collinear to $\mathbf{x}$ are also eigenvectors of $\mathbf{A}$. Two vectors pointing in the same direction are said to be codirected; they are called collinear if pointing in the same or opposite direction.

Example 3.2.2: For the matrix $\mathbf{A}$ in Example 3.2.1

$$\mathbf{A}(2\mathbf{x}) = \begin{bmatrix} 4 & 3 \\ 2 & -1 \end{bmatrix} 2 \begin{bmatrix} 3 \\ 1 \end{bmatrix} = \begin{bmatrix} 4 & 3 \\ 2 & -1 \end{bmatrix} \begin{bmatrix} 6 \\ 2 \end{bmatrix} = \begin{bmatrix} 30 \\ 10 \end{bmatrix} = 5 \begin{bmatrix} 6 \\ 2 \end{bmatrix}$$

Thus (6, 2) is also an eigenvector of $\mathbf{A}$ corresponding to $\lambda = 5$. So are (9, 3), (12, 4), etc.

Besides, there can be multiple eigenvalues of a matrix $\mathbf{A} \in \mathbb{R}^{n \times n}$. So, we have the following definition.

Definition 3.2.2 For a matrix $\mathbf{A} \in \mathbb{R}^{n \times n}$, the set of all eigenvectors corresponding to an eigenvalue λ spans a subspace of $\mathbb{R}^n$ called the ***eigenspace***, denoted by E_λ, of $\mathbf{A}$ with respect to λ.

The set of distinct eigenvalues, denoted by $\sigma(\mathbf{A})$ and satisfying the eigenvalue equation, is called the ***eigenspectrum***, or simply the ***spectrum***, of $\mathbf{A}$.

In general, multiplication of a vector $\mathbf{x}$ by a square matrix $\mathbf{A}$ changes both its magnitude and direction; however, in the special case, when $\mathbf{x}$ is an eigenvector of $\mathbf{A}$, multiplication by $\mathbf{A}$ leaves the direction of $\mathbf{x}$ unchanged. The operation $\mathbf{A}\mathbf{x} = \lambda\mathbf{x}$ just "stretches" or "compresses" $\mathbf{x}$ by a factor λ; if λ is negative, the direction is reversed.

In the case of a 2×2 matrix $\mathbf{A}$, collinearity of the eigenvectors can also be described by introducing the concept of ***A-invariance***. In $\mathbb{R}^2$, a line L through the origin is called ***A-invariant*** if for each $\mathbf{x}$ in L, $\mathbf{A}\mathbf{x}$ is also in L, that is, $\mathbf{A}$ is a linear transformation $\mathbb{R}^2 \to \mathbb{R}^2$ carrying L into itself.

Computing eigenvectors

The eigenvalue equation can be rewritten as

$$\mathbf{A}\mathbf{x} = \lambda\mathbf{I}\mathbf{x}$$

or

$$(\lambda\mathbf{I} - \mathbf{A})\mathbf{x} = \mathbf{0} \quad (3.15)$$

So, the problem reduces to finding the nonzero vectors in the solution space of this linear system. This solution space is known as the ***eigenspace*** of $\mathbf{A}$ corresponding to λ. This is stated in the following theorem.

Theorem 3.2.1 For an $n \times n$ matrix $\mathbf{A}$, the λ-eigenvectors $\mathbf{x}$ are the nonzero solutions to the homogeneous system

$$(\lambda\mathbf{I} - \mathbf{A})\mathbf{x} = \mathbf{0} \quad (3.16)$$

where $(\lambda\mathbf{I} - \mathbf{A})$ is the coefficient matrix.

Nonzero solution to (3.16) requires the coefficient matrix $(\lambda\mathbf{I}-\mathbf{A})$ to be non-invertible which, in turn, requires

$$\det(\lambda\mathbf{I}-\mathbf{A}) = 0$$

as stated in the following theorem.

Theorem 3.2.2 λ is an eigenvalue of an $n \times n$ matrix $\mathbf{A}$ if and only if it satisfies the equation

$$\det(\lambda\mathbf{I}-\mathbf{A}) = 0 \tag{3.17}$$

This is called the ***characteristic equation*** of $\mathbf{A}$.

Contextually, we have the following definition.

Definition 3.2.3 The ***characteristic polynomial*** $p_A(\lambda)$ of an $n \times n$ matrix $\mathbf{A}$ is defined by

$$p_A(\lambda) = \det(\lambda\mathbf{I}-\mathbf{A}) \tag{3.18}$$

When expanded the characteristic polynomial takes the form

$$p_A(\lambda) = \lambda^n + a_1\lambda^{n-1} + \cdots + a_n \tag{3.19}$$

which is a polynomial of degree n. A polynomial of degree n has at most n distinct roots. Hence, the characteristic equation of an $n \times n$ matrix has at most n distinct solutions. Consequently, the maximum number of distinct eigenvalues of such a matrix is n.

Theorem 3.2.3 The eigenvalues $\lambda \in \mathbb{R}$ of a matrix $\mathbf{A} \in \mathbb{R}^{n \times n}$ are the roots of the characteristic polynomial $p_A(\lambda)$ of $\mathbf{A}$.

Example 3.2.3: Let us determine the characteristic polynomial, eigenvalues and eigenvectors of the 2×2 matrix

$$\mathbf{A} = \begin{bmatrix} 4 & 3 \\ 2 & -1 \end{bmatrix}$$

The characteristic polynomial is

$$\begin{aligned} p_A(\lambda) &= \det(\lambda\mathbf{I}-\mathbf{A}) \\ &= \det\begin{bmatrix} \lambda-4 & -3 \\ -2 & \lambda+1 \end{bmatrix} \\ &= (\lambda-4)(\lambda+1)-6 \\ &= \lambda^2-3\lambda-10 \\ &= (\lambda-5)(\lambda+2) \end{aligned} \tag{3.20}$$

which is a polynomial of degree 2. (3.20) shows that the roots of $p_A(\lambda)$ are $\lambda_1 = 5$ and $\lambda_2 = -2$; hence, these are the eigenvalues of $\mathbf{A}$.

The eigenvectors can be determined by solving the homogeneous system

$$(\lambda \mathbf{I} - \mathbf{A})\mathbf{x} = \begin{bmatrix} \lambda - 4 & -3 \\ -2 & \lambda + 1 \end{bmatrix} \begin{bmatrix} x_1 \\ x_2 \end{bmatrix} = \mathbf{0} \tag{3.21}$$

For $\lambda = \lambda_1 = 5$, the general solution to

$$\begin{bmatrix} 1 & -3 \\ -2 & 6 \end{bmatrix} \begin{bmatrix} x_1 \\ x_2 \end{bmatrix} = \mathbf{0}$$

is

$$\mathbf{x} = s\begin{bmatrix} 3 \\ 1 \end{bmatrix}$$

and for $\lambda = \lambda_2 = -2$, the general solution to

$$\begin{bmatrix} -6 & -3 \\ -2 & -1 \end{bmatrix} \begin{bmatrix} x_1 \\ x_2 \end{bmatrix} = \mathbf{0}$$

is

$$\mathbf{x} = t\begin{bmatrix} -1 \\ 2 \end{bmatrix}$$

where s and t are arbitrary real numbers, neither equal to zero.

Hence, $\begin{bmatrix} 3 \\ 1 \end{bmatrix}$ and $\begin{bmatrix} -1 \\ 2 \end{bmatrix}$ are the basic eigenvectors of the matrix $\mathbf{A}$ corresponding to the eigenvalues 5 and -2, respectively.

We can also write

$$E_5 = \text{span}\begin{bmatrix} 3 \\ 1 \end{bmatrix}$$

and

$$E_{-2} = \text{span}\begin{bmatrix} -1 \\ 2 \end{bmatrix}$$

The two eigenspaces E_5 and E_{-1} are one-dimensional since each is spanned by a single vector.

Example 3.2.4: Eigenpairs of a 3×3 matrix.

Let us find out the characteristic polynomial, eigenvalues, and basic eigenvectors for

$$\mathbf{A} = \begin{bmatrix} 3 & 0 & 0 \\ 1 & 2 & 1 \\ 1 & -3 & -2 \end{bmatrix}$$

The characteristic polynomial is

$$\begin{aligned} p_A(\lambda) &= \det(\lambda \mathbf{I} - \mathbf{A}) \\ &= \det \begin{bmatrix} \lambda - 3 & 0 & 0 \\ -1 & \lambda - 2 & -1 \\ -1 & 3 & \lambda + 2 \end{bmatrix} \\ &= (\lambda - 3)[(\lambda - 2)(\lambda + 2) + 3] \\ &= (\lambda - 3)(\lambda - 1)(\lambda + 1) \end{aligned}$$

Hence, the eigenvalues are $\lambda_1 = 3$, $\lambda_2 = 1$, $\lambda_3 = -1$.

The eigenvectors corresponding to $\lambda_1 = 3$ will be given by the nonzero solutions of

$$(\lambda_1 \mathrm{I} - \mathrm{A})\mathbf{x} = \mathbf{0} \tag{3.22}$$

$$\text{Now, } (\lambda_1 \mathbf{I} - \mathbf{A}) \begin{bmatrix} 0 & 0 & 0 \\ -1 & 1 & -1 \\ -1 & 3 & 5 \end{bmatrix}$$

So, the augmented matrix is

$$\begin{bmatrix} 0 & 0 & 0 & 0 \\ -1 & 1 & -1 & 0 \\ -1 & 3 & 5 & 0 \end{bmatrix}$$

which by row operations becomes

$$\begin{bmatrix} 1 & 0 & 4 & 0 \\ 0 & 1 & 3 & 0 \\ 0 & 0 & 0 & 0 \end{bmatrix}$$

Hence, the general solution to (3.22) is

$$\mathbf{x} = t \begin{bmatrix} -4 \\ -3 \\ 1 \end{bmatrix}$$

where t is arbitrary, and the basic eigenvector corresponding to $\lambda_1 = 3$ is

$$\mathbf{x}_1 = \begin{bmatrix} -4 \\ -3 \\ 1 \end{bmatrix}$$

Similarly, we can find basic eigenvectors

$$\mathbf{x}_2 = \begin{bmatrix} 0 \\ -1 \\ 1 \end{bmatrix} \text{ and } \mathbf{x}_3 = \begin{bmatrix} 0 \\ -1 \\ 3 \end{bmatrix}$$

corresponding to eigenvalues $\lambda_2 = 1$ and $\lambda_3 = -1$ respectively.

Eigenvalues of some special matrices

Let us consider an ***upper triangular*** matrix

$$\mathbf{A} = \begin{bmatrix} a_{11} & a_{12} \cdots\cdots\cdots\cdots & a_{1n} \\ 0 & a_{22} \cdots\cdots\cdots\cdots & a_{2n} \\ & \cdots\cdots\cdots\cdots\cdots\cdots & \\ & \cdots\cdots\cdots\cdots\cdots\cdots & \\ 0 & 0 \cdots\cdots\cdots\cdots & a_{nn} \end{bmatrix}$$

we have

$$\det(\lambda \mathbf{I} - \mathbf{A}) = \det \begin{bmatrix} \lambda - a_{11} & a_{12} & \cdots\cdots\cdots & a_{1n} \\ 0 & \lambda - a_{22} & \cdots\cdots\cdots & a_{2n} \\ & \cdots\cdots\cdots\cdots\cdots\cdots & & \\ & \cdots\cdots\cdots\cdots\cdots\cdots & & \\ 0 & 0 & \cdots\cdots\cdots & \lambda - a_{nn} \end{bmatrix}$$

so that the characteristic equation is

$$(\lambda - a_{11})(\lambda - a_{22}) \cdots (\lambda - a_{nn}) = 0 \tag{3.23}$$

and the eigenvalues are

$$\lambda_1 = a_{11}, \quad \lambda_2 = a_{22}, \quad \cdots \cdots, \quad \lambda_n = a_{nn} \tag{3.24}$$

which are precisely the diagonal elements of $\mathbf{A}$.

The results are similar for ***lower triangular*** and ***diagonal*** matrices. So, we have the following theorem.

Theorem 3.2.4 If A is an $n \times n$ upper triangular, lower triangular, or diagonal matrix, the eigenvalues of $\mathbf{A}$ are the entries on the main diagonal of $\mathbf{A}$.

Example 3.2.5: Eigenvalues of $\mathbf{A}^T$.

Since for a square matrix $\mathbf{A}$,

$$\det \mathbf{A}^T = \det \mathbf{A}$$

and,

$$\lambda \mathbf{I} - \mathbf{A}^T = (\lambda \mathbf{I} - \mathrm{A})^T$$

$$p_{A^T}(\lambda) = \det(\lambda \mathbf{I} - \mathbf{A}^T) = \det\left[(\lambda \mathbf{I} - \mathbf{A})^T\right] = \det(\lambda \mathbf{I} - \mathbf{A}) = p_A(\lambda)$$

Hence, $p_{A^T}(\lambda)$ and $p_A(\lambda)$ have the same roots, and $\mathbf{A}^T$ and $\mathbf{A}$ have the same eigenvalues.

Linear independence of eigenvectors

Theorem 3.2.5 If $\lambda_1, \lambda_2, \ldots, \lambda_k$ are distinct eigenvalues of a matrix $\mathbf{A} \in \mathbb{R}^{n \times n}$, and $\mathbf{v}_1, \mathbf{v}_2, \ldots, \mathbf{v}_k$ are the corresponding eigenvectors, then $\{\mathbf{v}_1, \mathbf{v}_2, \ldots, \mathbf{v}_k\}$ is a linearly independent set.

Proof. Suppose k ($k \leq n$) eigenvalues $\{\lambda_1, \lambda_2, \ldots, \lambda_k)$ of $\mathbf{A}$ are distinct, and the corresponding eigenvectors $\{\mathbf{v}_1, \mathbf{v}_2, \ldots, \mathbf{v}_k\}$ are defined by

$$\mathbf{A}\,\mathbf{v}_j = \lambda_j\,\mathbf{v}_j, \quad \mathbf{v}_j \neq 0, \quad \text{for } j = 1, 2, \ldots, k \tag{3.25}$$

It has to be shown that in the vanishing linear combination

$$\alpha_1\,\mathbf{v}_1 + \alpha_2\,\mathbf{v}_2 + \cdots + \alpha_k\,\mathbf{v}_k = \mathbf{0} \tag{3.26}$$

each $\alpha_j = 0$.

Let us consider two such eigenvectors $\mathbf{v}_1$ and $\mathbf{v}_2$. Suppose a linear combination of the two eigenvectors vanishes, that is,

$$\alpha_1\,\mathbf{v}_1 + \alpha_2\,\mathbf{v}_2 = \mathbf{0} \tag{3.27}$$

Multiplying the left side by $\mathbf{A}$ we have

$$\alpha_1\,\mathbf{A}\,\mathbf{v}_1 + \alpha_2\,\mathbf{A}\,\mathbf{v}_2 = \mathbf{0}$$

or,

$$\alpha_1\,\lambda_1\,\mathbf{v}_1 + \alpha_2\,\lambda_2\,\mathbf{v}_2 = \mathbf{0} \tag{3.28}$$

Multiplying (3.27) by λ_1 and subtracting the result from (3.28), we get

$$\alpha_2\,(\lambda_2 - \lambda_1)\,\mathbf{v}_2 = \mathbf{0} \tag{3.29}$$

Since $\lambda_1 \neq \lambda_2$ and $\mathbf{v}_2 \neq \mathbf{0}$, (3.26) implies that $\alpha_2 = 0$. Putting this back in (3.27), we have $\alpha_1 = 0$. So, $\mathbf{v}_1$ and $\mathbf{v}_2$ are linearly independent.

Similarly, we can take the eigenvectors corresponding to any two eigenvalues and show that

$$\alpha_3\,(\lambda_3 - \lambda_2)\,\mathbf{v}_3 = \mathbf{0}, \quad \alpha_4\,(\lambda_4 - \lambda_3)\,\mathbf{v}_4 = \mathbf{0}, \; \ldots, \; \alpha_k\,(\lambda_k - \lambda_{k-1})\,\mathbf{v}_k = \mathbf{0}$$

and since λ_js are distinct, $\alpha_j = 0$ for $j = 1, 2, \ldots, k$. Hence, $\{\mathbf{v}_1, \mathbf{v}_2, \ldots, \mathbf{v}_k\}$ is a linearly independent set.

According to the theorem, eigenvectors of a matrix with n distinct eigenvalues form a basis of $\mathbb{R}^n$.

Eigendecomposition

Let us extend the problem of distribution of molecules in different states as discussed earlier to a generalized case where the distribution vector $\mathbf{v} \in \mathbb{R}^n$. In such case, the transition rates are represented by a fixed matrix $\mathbf{A} \in \mathbb{R}^{n \times n}$. A sequence of column vectors $\mathbf{v}_0, \mathbf{v}_1, \mathbf{v}_2, \ldots,$ constitutes a ***linear discrete dynamical system***. The distribution $\mathbf{v}_{k+1}$ at time $t = (k + 1)$ seconds is related to the distribution $\mathbf{v}_k$ at time $t = k$ sec by

$$\mathbf{v}_{k+1} = \mathbf{A}\,\mathbf{v}_k \quad \text{for each } k = 0, 1, 2, \ldots \tag{3.30}$$

Hence, if $\mathbf{v}_0$ is known, we have

$$\mathbf{v}_1 = \mathbf{A}\,\mathbf{v}_0$$

$$\mathbf{v}_2 = \mathbf{A}\,\mathbf{v}_1$$

and, in general,

$$\mathbf{v}_k = \mathbf{A}^k \mathbf{v}_0 \quad \text{for all } k \geq 0 \tag{3.31}$$

This is called ***matrix recurrence*** for the vectors $\mathbf{v}_k$.

Direct computation of the powers of a square matrix **A** is a formidable task unless **A** is diagonal. Fast computation of powers as well as determinants and inverses are facilitated by diagonal matrices. This underscores the importance of matrix ***diagonalization***. For a diagonal matrix **D**, a matrix power $\mathbf{D}^k$ is obtained by raising each diagonal element to the power k.

Evidently, for easy and efficient computation, it is helpful to decompose a matrix **A** into factors (matrices) including the diagonalized matrix **D**. In general, matrix decomposition is factorization of a matrix into simpler and interpretable matrices. The decomposition based on the eigenvalues and eigenvectors of matrices is called ***eigendecomposition***. As we shall see later in the book, eigen-decomposition of matrices is important in statistics, particularly in principal component analysis (PCA).

As a computational practice, the set of eigenvectors of **A** is put together in a matrix **U**, each column of **U** being an eigenvector of **A**. The eigenvalues of **A** are stored in a diagonal matrix **Λ**, the diagonal elements of **Λ** being the eigenvalues of **A**. The eigenvalue in (3.13) is then rewritten as

$$\mathbf{A}\mathbf{U} = \mathbf{U}\Lambda \tag{3.32}$$

which can be also written in the form

$$\mathbf{A} = \mathbf{U}\Lambda\mathbf{U}^{-1} \tag{3.33}$$

Eigendecomposition is essentially guided by the following theorem.

Theorem 3.2.6 A matrix $\mathbf{A} \in \mathbb{R}^{n \times n}$ can be factored into

$$\mathbf{A} = \mathbf{PDP}^{-1}$$

where $\mathbf{P} \in \mathbb{R}^{n \times n}$ and $\mathbf{D} \in \mathbb{R}^{n \times n}$ is a diagonal matrix, whose diagonal entries are the eigenvalues of **A**, if and only if the eigenvectors of **A** form a basis for $\mathbb{R}^n$.

Geometric intuition for eigendecomposition

Let us consider **A** to be the transformation matrix of a linear mapping with respect to the standard basis. Theorem 3.2.6 implies that the image of a vector (say, **v**) under multiplication by a diagonalizable matrix **A** can be obtained in three steps—(1) $\mathbf{P}^{-1}$ executes a basis change from the standard basis into the eigenbasis, (2) the diagonal matrix **D** then scales the components of the vector along these axes by the corresponding eigenvalues, and (3) finally, **P** transforms the scaled vector back into the standard basis. The following example will illustrate the eigendecomposition of a matrix.

Example 3.2.6: Let the matrix to be eigendecomposed be

$$\mathbf{A} = \begin{bmatrix} 2 & 1 \\ 1 & 2 \end{bmatrix}$$

and $\mathbf{x} = (2, 1)$.

On calculation, the eigenvalues of **A** are found to be

$$\lambda_1 = 3, \ \lambda_2 = 1$$

and the corresponding eigenvectors

$$\mathbf{v}_1 = \begin{bmatrix} 1 \\ 1 \end{bmatrix} . \mathbf{v}_2 = \begin{bmatrix} -1 \\ 1 \end{bmatrix}$$

so that

$$\mathbf{P} = \begin{bmatrix} 1 & -1 \\ 1 & 1 \end{bmatrix}, \quad \mathbf{P}^{-1} = \begin{bmatrix} \frac{1}{2} & \frac{1}{2} \\ -\frac{1}{2} & \frac{1}{2} \end{bmatrix}, \quad \text{and} \quad \mathbf{D} = \begin{bmatrix} 3 & 0 \\ 0 & 1 \end{bmatrix}$$

With respect to the standard basis the given vector is

$$\mathbf{x} = \begin{bmatrix} 2 \\ 1 \end{bmatrix}$$

which in the eigenbasis is given by

$$\mathbf{P}^{-1}\mathbf{x} = \begin{bmatrix} \frac{1}{2} & \frac{1}{2} \\ -\frac{1}{2} & \frac{1}{2} \end{bmatrix} \begin{bmatrix} 2 \\ 1 \end{bmatrix} = \begin{bmatrix} \frac{3}{2} \\ -\frac{1}{2} \end{bmatrix}$$

Scaled by **D** we get

$$\mathbf{DP}^{-1}\mathbf{x} = \begin{bmatrix} 3 & 0 \\ 0 & 1 \end{bmatrix} \begin{bmatrix} \frac{3}{2} \\ -\frac{1}{2} \end{bmatrix} = \begin{bmatrix} \frac{9}{2} \\ -\frac{1}{2} \end{bmatrix}$$

Finally, transforming back to the standard basis

$$\mathbf{PDP}^{-1}\mathbf{x} = \begin{bmatrix} 1 & -1 \\ 1 & 1 \end{bmatrix} \begin{bmatrix} \frac{9}{2} \\ -\frac{1}{2} \end{bmatrix} = \begin{bmatrix} 5 \\ 4 \end{bmatrix}$$

Each of the above three steps is shown in Fig. 3.2.

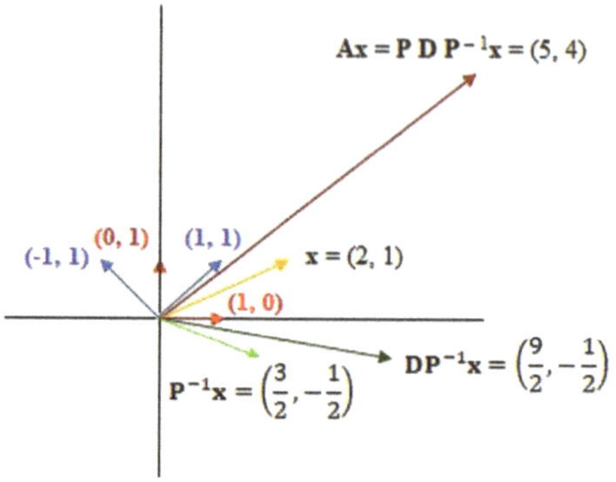

FIGURE 3.2

Eigendecomposition. Red arrows: standard basis; blue arrows: eigenbasis.

The direct product of **A** and **x** is

$$\mathbf{A}\,\mathbf{x} = \begin{bmatrix} 2 & 1 \\ 1 & 2 \end{bmatrix}\begin{bmatrix} 2 \\ 1 \end{bmatrix} = \begin{bmatrix} 5 \\ 4 \end{bmatrix}$$

Diagonalization

To diagonalize **A**, we have to find a matrix **P** such that

$$\mathbf{P}^{-1}\mathbf{A}\mathbf{P} = \mathbf{D} \tag{3.34}$$

where **D** is a diagonal matrix. From (3.34) we can derive

$$\mathbf{A} = \mathbf{P}\mathbf{D}\mathbf{P}^{-1} \tag{3.35}$$

so that

$$\mathbf{A}^2 = (\mathbf{P}\mathbf{D}\mathbf{P}^{-1})(\mathbf{P}\mathbf{D}\mathbf{P}^{-1}) = \mathbf{P}\mathbf{D}^2\mathbf{P}^{-1}$$

$$\mathbf{A}^3 = (\mathbf{P}\mathbf{D}\mathbf{P}^{-1})(\mathbf{P}\mathbf{D}^2\mathbf{P}^{-1}) = \mathbf{P}\mathbf{D}^3\mathbf{P}^{-1}$$

Continuing further, we reach the following theorem:

Theorem 3.2.7 If $\mathbf{A} = \mathbf{P}\,\mathbf{D}\,\mathbf{P}^{-1}$, then $\mathbf{A}^k = \mathbf{P}\,\mathbf{D}^k\,\mathbf{P}^{-1}$ for each $k = 1, 2, \ldots\ldots$

Matrix diagonalization using (3.34) is based on certain properties of square matrices that we are going to discuss next.

Let us consider square matrices **A** and $\mathbf{P} \in \mathbb{R}^{n \times n}$, **P** being invertible. The transformation

$$\mathbf{A} \rightarrow \mathbf{P}^{-1}\mathbf{A}\mathbf{P}$$

maps the matrix $\mathbf{A}$ into the matrix $\mathbf{P}^{-1}\mathbf{A}\,\mathbf{P}$. If an $n \times n$ matrix $\mathbf{B} = \mathbf{P}^{-1}\mathbf{A}\,\mathbf{P}$, we say that $\mathbf{A}$ and $\mathbf{B}$ are similar and write $\mathbf{A} \sim \mathbf{B}$. Such a transformation is called a **similarity transformation**. Many of the properties of a square matrix, such as determinant, invertibility, and eigenvalues, are invariant under similarity transformation. So, we have the following definition of similar matrices.

Definition 3.2.4 Two $n \times n$ matrices $\mathbf{A}$ and $\mathbf{B}$ are said to be similar if there is an invertible matrix $\mathbf{P}$ such that $\mathbf{B} = \mathbf{P}^{-1}\mathbf{A}\mathbf{P}$.

It is to be noted that the similarity relation can also be expressed as $\mathbf{A} = \mathbf{Q}^{-1}\mathbf{B}\,\mathbf{Q}$ by taking $\mathbf{Q} = \mathbf{P}^{-1}$.

A specific case of similarity transformation that we are interested in at this moment is

$$\mathbf{D} = \mathbf{P}^{-1}\mathbf{A}\,\mathbf{P} \tag{3.36}$$

that is, $\mathbf{A}$ is similar to a diagonal matrix $\mathbf{D}$, which contains the eigenvalues of $\mathbf{A}$ on the diagonal. So, we have the following definition of diagonalizability of a square matrix.

Definition 3.2.5 A matrix $\mathbf{A} \in \mathbb{R}^{n \times n}$ is said to be diagonalizable if it is similar to a diagonal matrix (say, $\mathbf{D} \in \mathbb{R}^{n \times n}$), that is, if there exists an invertible matrix $\mathbf{P} \in \mathbb{R}^{n \times n}$ such that $\mathbf{D} = \mathbf{P}^{-1}\mathbf{A}\mathbf{P}$. The matrix $\mathbf{P}$ is said to diagonalize $\mathbf{A}$.

Invertibility of matrix P

Let us consider the homogeneous system

$$\mathbf{P}\,\mathbf{x} = \mathbf{0}$$

If $\mathbf{P}^{-1}$ exists, then we have

$$\mathbf{x} = \mathbf{I}_n\mathbf{x} = \mathbf{P}^{-1}\mathbf{P}\mathbf{x} = \mathbf{P}^{-1}\mathbf{0} = \mathbf{0}$$

leading to the following theorem.

Theorem 3.2.8 If $\mathbf{P}$ is invertible, the homogeneous system $\mathbf{P}\mathbf{x} = \mathbf{0}$ has only trivial solution $\mathbf{x} = \mathbf{0}$.

It can be proved that the reverse is also true, that is, if a homogeneous system $\mathbf{P}\mathbf{x} = \mathbf{0}$ has only trivial solution $\mathbf{x} = \mathbf{0}$, the matrix $\mathbf{P}$ is invertible.

Now, let $\{\mathbf{c}_1, \mathbf{c}_2, \ldots, \mathbf{c}_n\}$ denote the columns of matrix $\mathbf{P}$. If we write

$$\mathbf{x} = \begin{bmatrix} x_1 \\ x_2 \\ \cdot \\ \cdot \\ \cdot \\ x_n \end{bmatrix}$$

then,

$$\mathbf{Px} = x_1\,\mathbf{c}_1 + x_2\,\mathbf{c}_2 + \cdots + x_n\,\mathbf{c}_n \tag{3.37}$$

from which it follows that $\{\mathbf{c}_1, \mathbf{c}_2, \ldots, \mathbf{c}_n\}$ is linearly independent if and only if $\mathbf{P\,x} = \mathbf{0}$ implies $\mathbf{x} = \mathbf{0}$. Hence, from (3.37) together with Theorem 3.2.8, it can be concluded that if the columns of **P** are linearly independent, **P** is invertible.

Diagonalizability of a matrix

Where do the linearly independent columns of **P** come from? The question, as we shall see, is connected with the diagonalizability of **A**.

We already know that **A**, **P**, and $\mathbf{D} \in \mathbb{R}^{n\times n}$. Let $\{\lambda_1, \lambda_2, \ldots, \lambda_n\}$ be a set of scalars that constitute the diagonal entries of **D** and $\{\mathbf{p}_1, \mathbf{p}_2, \ldots, \mathbf{p}_n\}$ be a set of vectors in $\mathbb{R}^n$ such that $\mathbf{P} = [\mathbf{p}_1\ \mathbf{p}_2 \ldots \mathbf{p}_n]$. From (3.31) we have

$$\mathbf{A\,P} = \mathbf{P\,D} \tag{3.38}$$

On expansion,

$$\mathbf{A\,P} = \mathbf{A}\begin{bmatrix}\mathbf{p}_1 & \mathbf{p}_2 & \cdots & \mathbf{p}_n\end{bmatrix} = \begin{bmatrix}\mathbf{A\,p}_1 & \mathbf{A\,p}_2 & \cdots & \mathbf{A\,p}_n\end{bmatrix}$$

and

$$\begin{aligned}\mathbf{P\,D} &= \begin{bmatrix}\mathbf{p}_1 & \mathbf{p}_2 & \cdots & \mathbf{p}_n\end{bmatrix}\begin{bmatrix}\lambda_1 & & & 0\\ & \cdot & & \\ & & \cdot & \\ & & & \cdot \\ 0 & & & \lambda_n\end{bmatrix}\\ &= \begin{bmatrix}\lambda_1\,\mathbf{p}_1 & \lambda_2\,\mathbf{p}_2 & \cdots & \lambda_n\,\mathbf{p}_n\end{bmatrix}\end{aligned}$$

Thus (3.38) implies

$$\begin{aligned}\mathbf{A\,p}_1 &= \lambda_1\,\mathbf{p}_1\\ \mathbf{A\,p}_2 &= \lambda_2\,\mathbf{p}_2\\ &\cdot\\ &\cdot\\ &\cdot\\ \mathbf{A\,p}_n &= \lambda_n\,\mathbf{p}_n\end{aligned}$$

Therefore the columns of the diagonalizing matrix **P** must be the eigenvectors of **A**. Furthermore, the definition of diagonalization requires the columns of **P** to be linearly independent. As we have seen earlier (Theorem 3.2.5), if eigenvalues $\{\lambda_1, \lambda_2, \ldots, \lambda_n\}$ of **A** are distinct, then the corresponding eigenvectors $\{\mathbf{p}_1, \mathbf{p}_2, \ldots, \mathbf{p}_n\}$ are linearly independent. Hence, the columns of $\mathbf{P}$ are constituted of linearly independent eigenvectors of $\mathbf{A}$. It is also implied that the linearly independent eigenvectors of $\mathbf{A}$ constitute a basis for $\mathbb{R}^n$. so, the following theorem determines the diagonalizability of $\mathbf{A}$.

Theorem 3.2.9 An $n \times n$ matrix is diagonalizable

1. if it has n distinct eigenvalues
2. if and only if $\mathbb{R}^n$ has a basis consisting of the eigenvectors of $\mathbf{A}$.

An $n \times n$ matrix with n distinct eigenvalues is ***diagonalizable***.

Example 3.2.7: Let us diagonalize the matrix

$$\mathbf{A} = \begin{bmatrix} 3 & 0 & 0 \\ 1 & 2 & 1 \\ 1 & -3 & -2 \end{bmatrix}$$

We have seen earlier (Example 3.2.4) that the eigenvalues of the matrix $\mathbf{A}$ are $\lambda_1 = 3$, $\lambda_2 = 1$, $\lambda_3 = -1$ and the corresponding eigenvectors are

$$\mathbf{x}_1 = \begin{bmatrix} -4 \\ -3 \\ 1 \end{bmatrix}, \quad \mathbf{x}_2 = \begin{bmatrix} 0 \\ -1 \\ 1 \end{bmatrix}, \quad \text{and} \quad \mathbf{x}_3 = \begin{bmatrix} 0 \\ -1 \\ 3 \end{bmatrix} \quad \text{respectively}$$

Since the eigenvalues of $\mathbf{A}$ are distinct, $\{\mathbf{x}_1, \mathbf{x}_2, \mathbf{x}_3\}$ is a linearly independent set. Hence,

$$\mathbf{P} = [\mathbf{x}_1 \;\; \mathbf{x}_2 \;\; \mathbf{x}_3] = \begin{bmatrix} -4 & 0 & 0 \\ -3 & -1 & -1 \\ 1 & 1 & 3 \end{bmatrix}$$

is invertible and it can be shown that

$$\mathbf{P}^{-1} = \frac{1}{4}\begin{bmatrix} -1 & 0 & 0 \\ 4 & -6 & -2 \\ -1 & 2 & 2 \end{bmatrix}$$

so that,

$$\mathbf{P}^{-1}\mathbf{A}\mathbf{P} = \frac{1}{4}\begin{bmatrix} -1 & 0 & 0 \\ 4 & -6 & -2 \\ -1 & 2 & 2 \end{bmatrix}\begin{bmatrix} 3 & 0 & 0 \\ 1 & 2 & 1 \\ 1 & -3 & -2 \end{bmatrix}\begin{bmatrix} -4 & 0 & 0 \\ -3 & -1 & -1 \\ 1 & 1 & 3 \end{bmatrix}$$

$$= \begin{bmatrix} 3 & 0 & 0 \\ 0 & 1 & 0 \\ 0 & 0 & -1 \end{bmatrix} = \begin{bmatrix} \lambda_1 & 0 & 0 \\ 0 & \lambda_2 & 0 \\ 0 & 0 & \lambda_3 \end{bmatrix} = \mathbf{D}$$

The eigenvalues of $\mathbf{A}$ are the diagonal entries of $\mathbf{D}$.

It is emphasized here that a matrix $\mathbf{A} \in \mathbb{R}^{n \times n}$ is diagonalizable if it has a total of n linearly independent eigenvectors; if the total is less than n, then it is not. The following example will illustrate this point.

Example 3.2.8: Let us find out if the following matrix is diagonalizable,

$$\mathbf{A} = \begin{bmatrix} 2 & 0 & 0 \\ 1 & 3 & 0 \\ -3 & 5 & 3 \end{bmatrix}$$

The characteristic polynomial of $\mathbf{A}$ is

$$\det(\lambda\mathbf{I} - \mathbf{A}) = \begin{vmatrix} \lambda - 2 & 0 & 0 \\ -1 & \lambda - 3 & 0 \\ 3 & -5 & \lambda - 3 \end{vmatrix}$$

$$= (\lambda - 2)(\lambda - 3)^2$$

and the characteristic equation is

$$(\lambda - 2)(\lambda - 3)^2 = 0$$

Clearly, the distinct eigenvalues of $\mathbf{A}$ are $\lambda_1 = 3$, and $\lambda_2 = 2$ and it can be shown that the corresponding bases for the eigenspaces are

$$\mathbf{p}_1 = \begin{bmatrix} 0 \\ 0 \\ 1 \end{bmatrix} \quad \text{and} \quad \mathbf{p}_2 = \begin{bmatrix} 1 \\ -1 \\ 8 \end{bmatrix} \quad \text{respectively}$$

Since $\mathbf{A} \in \mathbb{R}^{n \times n}$ $(n = 3)$ and there are only two $(n < 3)$ basis vectors, $\mathbf{A}$ is not diagonalizable.

Let us look at an example, which is somewhat different from the previous two examples.

Example 3.2.9: We have a matrix

$$\mathbf{A} = \begin{bmatrix} 0 & 2 & 2 \\ 2 & 0 & 2 \\ 2 & 2 & 0 \end{bmatrix}$$

and we are going to examine its diagonalizability.

We can find the characteristic equation to be

$$(\lambda - 4)(\lambda + 2)^2 = 0$$

So, there are two distinct eigenvalues $\lambda_1 = 4$ and $\lambda_2 = -2$, though λ_2 is repeated and it is said that λ_2 has a ***multiplicity*** two. For $\lambda_1 = 4$, it can be verified that the system of equation

$$(\lambda_1 \mathbf{I} - \mathbf{A})\mathbf{x} = \mathbf{0}$$

has a general solution

$$\mathbf{x} = t\begin{bmatrix} 1 \\ 1 \\ 1 \end{bmatrix}$$

so that the basic eigenvector corresponding to λ_1 is

$$\mathbf{x}_1 = \begin{bmatrix} 1 \\ 1 \\ 1 \end{bmatrix}$$

For $\lambda_2 = -2$, we can solve the system $(\lambda_2 \mathbf{I} - \mathbf{A})\,\mathbf{x} = \mathbf{0}$ and find the general solution to be

$$\mathbf{x} = s\begin{bmatrix} -1 \\ 1 \\ 0 \end{bmatrix} + t\begin{bmatrix} -1 \\ 0 \\ 1 \end{bmatrix}$$

which produces two basic eigenvectors corresponding to λ_2

$$\mathbf{x}_2 = \begin{bmatrix} -1 \\ 1 \\ 0 \end{bmatrix} \quad \text{and} \quad \mathbf{y}_2 = \begin{bmatrix} -1 \\ 0 \\ 1 \end{bmatrix}$$

The matrix

$$\mathbf{P} = \begin{bmatrix} \mathbf{x}_1 & \mathbf{x}_2 & \mathbf{y}_2 \end{bmatrix} = \begin{bmatrix} 1 & -1 & -1 \\ 1 & 1 & 0 \\ 1 & 0 & 1 \end{bmatrix}$$

is invertible and

$$\mathbf{P}^{-1}\mathbf{A}\mathbf{P} = \text{diag}\,(4, -2, -2)$$

Orthogonal diagonalization

Orthogonal bases are very useful to work with. In Section 2.8, we have seen that a vector can be very conveniently expanded as a linear combination of orthogonal basis vectors since the coefficients can be systematically determined. This requires **P** to be an orthogonal matrix so that we have the following definition.

Definition 3.2.6 If matrices **A** and **B** $\in \mathbb{R}^{n\times n}$, **B** is said to be orthogonally similar to **A** if there be a matrix **P** $\in \mathbb{R}^{n\times n}$ such that $\mathbf{B} = \mathbf{P}^T\mathbf{A}\,\mathbf{P}$.

If **A** is similar to a diagonal matrix $\mathbf{D} \in \mathbb{R}^{n\times n}$, that is,

$$\mathbf{D} = \mathbf{P}^T\mathbf{A}\mathbf{P} \tag{3.39}$$

it is said that **A** is ***orthogonally diagonalizable*** and **P** ***orthogonally diagonalizes*** A.

Since **P** is a square orthogonal matrix, $\mathbf{P}\mathbf{P}^T = \mathbf{P}^T\mathbf{P} = \mathbf{I}$. So, multiplying left side of (3.39) by **P** and the right side by $\mathbf{P}^T$, we have

$$\mathbf{A} = \mathbf{P}\mathbf{D}\mathbf{P}^T \tag{3.40}$$

Transposing both sides of (3.40)

$$\mathbf{A}^T = (\mathbf{P}\mathbf{D}\mathbf{P}^T)^T = \left(\mathbf{P}^T\right)^T\mathbf{D}^T\mathbf{P}^T$$

Since for a diagonal matrix $\mathbf{D}^T = \mathbf{D}$,

$$\mathbf{A}^T = \mathbf{P}\mathbf{D}\mathbf{P}^T = \mathbf{A} \tag{3.41}$$

(3.41) shows that to be orthogonally diagonalizable, **A** must be symmetric.

Next, let us examine the nature of eigenvectors of a symmetric matrix $\mathbf{A} \in \mathbb{R}^{n\times n}$. For such a matrix and any two column vectors ($n \times 1$ matrices) **u** and **v**, it can be shown that

$$\mathbf{A}\,\mathbf{u}\cdot\mathbf{v} = \mathbf{u}\cdot\mathbf{A}^T\,\mathbf{v}$$

Let $\mathbf{u}_i$ and $\mathbf{u}_j$ be eigenvectors corresponding to distinct eigenvalues λ_i and λ_j of the matrix **A**. Then, we can write

$$\mathbf{A}\,\mathbf{u}_i\cdot\mathbf{u}_j = \mathbf{u}_i\cdot\mathbf{A}^T\,\mathbf{u}_j = \mathbf{u}_i\cdot\mathbf{A}\,\mathbf{u}_j$$

that is,

$$\lambda_i\,\mathbf{u}_i\cdot\mathbf{u}_j = \mathbf{u}_i\cdot\lambda_j\,\mathbf{u}_j$$

or,

$$(\lambda_i - \lambda_j)(\mathbf{u}_i\cdot\mathbf{u}_j) = 0 \tag{3.42}$$

Since, $\lambda_i - \lambda_j \neq 0$, we have from (3.42), $\mathbf{u}_i \cdot \mathbf{u}_j = 0$. This proves that the eigenvectors from different eigenspaces of a symmetric matrix are orthogonal.

Theorem 3.2.10 An $n \times n$ matrix **A** with real entries is orthogonally diagonalizable if

1. it has an orthonormal set of n eigenvectors and
2. it is symmetric.

Each of the eigenvectors of **A** that form a basis for $\mathbb{R}^n$ can be normalized using the Gram-Schmidt process (described in Chapter 2). The matrix **P**, whose columns are the normalized vectors thus constructed will be an orthonormal matrix. **P** will orthogonally diagonalize **A**, and the eigenvalues of the resulting diagonal matrix **D** will be in the same order as their corresponding unit eigenvectors in the diagonalizing matrix **P**.

Example 3.2.10: Let us orthogonally diagonalize the matrix

$$\mathbf{A} = \begin{bmatrix} 2 & 1 \\ 1 & 2 \end{bmatrix}$$

The first step is to find an orthogonal matrix **P** that diagonalize **A**.
On calculation, the eigenvalues of **A** are found to be

$$\lambda_1 = 3, \quad \lambda_2 = 1$$

and the corresponding eigenvectors

$$\mathbf{u}_1 = \begin{bmatrix} 1 \\ 1 \end{bmatrix}, \quad \mathbf{u}_2 = \begin{bmatrix} -1 \\ 1 \end{bmatrix}$$

from which we can obtain the orthonormal eigenvectors

$$\mathbf{v}_1 = \begin{bmatrix} \frac{1}{\sqrt{2}} \\ \frac{1}{\sqrt{2}} \end{bmatrix}, \quad \mathbf{v}_2 = \begin{bmatrix} -\frac{1}{\sqrt{2}} \\ \frac{1}{\sqrt{2}} \end{bmatrix}$$

So, the orthogonally diagonalizing matrix

$$\mathbf{P} = \begin{bmatrix} \frac{1}{\sqrt{2}} & -\frac{1}{\sqrt{2}} \\ \frac{1}{\sqrt{2}} & \frac{1}{\sqrt{2}} \end{bmatrix}$$

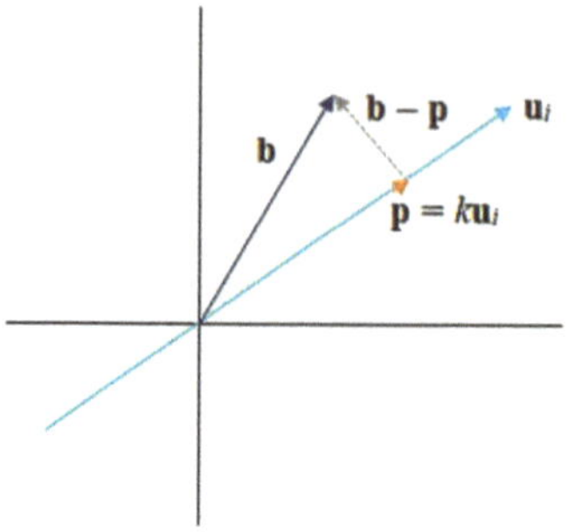

FIGURE 3.3

Orthogonal projection of **b** on **u**$_i$.

P orthogonally diagonalizes **A** as follows:

$$\mathbf{D} = \mathbf{P}^T\mathbf{A}\,\mathbf{P} = \begin{bmatrix} \frac{1}{\sqrt{2}} & \frac{1}{\sqrt{2}} \\ -\frac{1}{\sqrt{2}} & \frac{1}{\sqrt{2}} \end{bmatrix} \begin{bmatrix} 2 & 1 \\ 1 & 2 \end{bmatrix} \begin{bmatrix} \frac{1}{\sqrt{2}} & -\frac{1}{\sqrt{2}} \\ \frac{1}{\sqrt{2}} & \frac{1}{\sqrt{2}} \end{bmatrix} = \begin{bmatrix} 3 & 0 \\ 0 & 1 \end{bmatrix}$$

Projection matrix and spectral decomposition

Let us consider a vector $\mathbf{b} \in \mathbb{R}^n$ and a unit column vector $\mathbf{u}_i$ that spans a one-dimensional subspace W of $\mathbb{R}^n$. As shown in Fig. 3.3, the orthogonal projection of **b** on $\mathbf{u}_i$ can be expressed as (Section 2.8)

$$\mathbf{p} = k\,\mathbf{u}_i \tag{3.43}$$

where k is a scalar. It can also be seen from Fig. 3.3 that **p** is perpendicular to $(\mathbf{b} - \mathbf{p})$ so that we can write

$$\mathbf{u}_i^T(\mathbf{b} - \mathbf{p}) = 0$$

or,

$$\mathbf{u}_i{}^T(\mathbf{b} - k\,\mathbf{u}_i) = 0$$

and the value of the scalar multiplier is given by

$$k = \frac{\mathbf{u}_i^T\mathbf{b}}{\mathbf{u}_i^T\mathbf{u}_i} \tag{3.44}$$

Then, from (3.43)

$$\mathbf{p} = \mathbf{u}_i \frac{\mathbf{u}_i^T \mathbf{b}}{\mathbf{u}_i^T \mathbf{u}_i} = \frac{\mathbf{u}_i \mathbf{u}_i^T}{\mathbf{u}_i^T \mathbf{u}_i} \mathbf{b} \tag{3.45}$$

Since $\mathbf{u}_i$ is a unit vector, the denominator in (3.45), $\mathbf{u}_i^T \mathbf{u}_i = 1$. The numerator $\mathbf{u}_i \mathbf{u}_i^T$ is a square matrix. Therefore (3.40) can be rewritten as

$$\mathbf{p} = \mathbf{P}\mathbf{b} \tag{3.46}$$

where $\mathbf{P} = \mathbf{u}_i \mathbf{u}_i^T$ is the projection matrix.

Now, reconsidering the matrix $\mathbf{A} \in \mathbb{R}^{n \times n}$ whose eigenvalues are $\lambda_1, \lambda_2, \ldots, \lambda_n$ corresponding, respectively, to unit eigenvectors $\mathbf{u}_1, \mathbf{u}_2, \ldots, \mathbf{u}_n$ and its diagonalizing matrix

$$\mathbf{P} = [\mathbf{u}_1 \;\; \mathbf{u}_2 \;\; \cdots\cdots \;\; \mathbf{u}_n]$$

(3.40) can be rewritten as

$$\begin{aligned}
\mathbf{A} &= \mathbf{P}\mathbf{D}\mathbf{P}^T \\
&= [\mathbf{u}_1 \quad \mathbf{u}_2 \quad \cdots \quad \mathbf{u}_n] \begin{bmatrix} \lambda_1 & 0 & \cdots & 0 \\ 0 & \lambda_2 & \cdots & 0 \\ \cdot & \cdot & \cdot & \cdot \\ \cdot & \cdot & \cdot & \cdot \\ \cdot & \cdot & \cdot & \cdot \\ 0 & 0 & \cdots & \lambda_n \end{bmatrix} \begin{bmatrix} \mathbf{u}_1^T \\ \mathbf{u}_2^T \\ \cdot \\ \cdot \\ \cdot \\ \mathbf{u}_n^T \end{bmatrix} \\
&= [\lambda_1 \mathbf{u}_1 \quad \lambda_2 \mathbf{u}_2 \quad \cdots \quad \lambda_n \mathbf{u}_n] \begin{bmatrix} \mathbf{u}_1^T \\ \mathbf{u}_2^T \\ \cdot \\ \cdot \\ \cdot \\ \mathbf{u}_n^T \end{bmatrix}
\end{aligned}$$

In a compact form

$$\mathbf{A} = \lambda_1 \mathbf{u}_1 \mathbf{u}_1^T + \lambda_2 \mathbf{u}_2 \mathbf{u}_2^T + \ldots\ldots + \lambda_n \mathbf{u}_n \mathbf{u}_n^T \tag{3.47}$$

(3.47) is known as the ***spectral decomposition of A***. Since $\mathbf{u}_i$ ($i = 1, 2, \ldots, n$) is an $n \times 1$ column, $\mathbf{u}_i \mathbf{u}_i^T$ is an $n \times n$ matrix. As shown in (3.46), $\mathbf{u}_i \mathbf{u}_i^T$ is the standard matrix for the orthogonal projection of $\mathbb{R}^n$ on the subspace spanned by $\mathbf{u}_i$. (3.47) implies that the image of a vector (say, **b**) under multiplication by a symmetric matrix **A** can be obtained in three steps—(1) first, orthogonally projecting **b** on the lines (or one-dimensional subspaces) spanned by the normalized

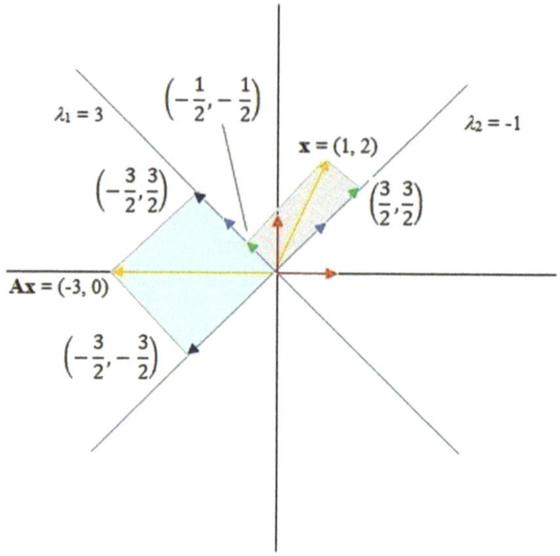

FIGURE 3.4

Spectral decomposition. Red arrows: standard basis; blue arrows: eigenbasis.

eigenvectors of **A**, (2) scaling these projections by the eigenvalues, and (3) finally adding the scaled projections. Let us illustrate this in the following example (illustrated in Fig. 3.4).

Example 3.2.11: Let us consider a 2×2 matrix.

$$\mathbf{A} = \begin{bmatrix} 1 & -2 \\ -2 & 1 \end{bmatrix}$$

and a vector

$$\mathbf{x} = \begin{bmatrix} 1 \\ 2 \end{bmatrix}$$

Direct multiplication of **x** by **A** gives

$$\mathbf{A}\mathbf{x} = \begin{bmatrix} 1 & -2 \\ -2 & 1 \end{bmatrix}\begin{bmatrix} 1 \\ 2 \end{bmatrix} = \begin{bmatrix} -3 \\ 0 \end{bmatrix}$$

Now, let us try to spectrally decompose **A**. The eigenvalues of A are found to be

$$\lambda_1 = 3,\ \lambda_2 = -1$$

and the corresponding unit (normalized) eigenvectors are

$$\mathbf{u}_1 = \begin{bmatrix} -\frac{1}{\sqrt{2}} \\ \frac{1}{\sqrt{2}} \end{bmatrix}, \quad \mathbf{u}_2 = \begin{bmatrix} \frac{1}{\sqrt{2}} \\ \frac{1}{\sqrt{2}} \end{bmatrix}$$

Standard matrices for orthogonal projections on subspaces spanned by $\mathbf{u}_1$ and $\mathbf{u}_2$ are, respectively,

$$\mathbf{u}_1\,\mathbf{u}_1^T == \begin{bmatrix} -\frac{1}{\sqrt{2}} \\ \frac{1}{\sqrt{2}} \end{bmatrix} \begin{bmatrix} -\frac{1}{\sqrt{2}} & \frac{1}{\sqrt{2}} \end{bmatrix} = \begin{bmatrix} \frac{1}{2} & -\frac{1}{2} \\ -\frac{1}{2} & \frac{1}{2} \end{bmatrix}$$

and

$$\mathbf{u}_2\,\mathbf{u}_2^T = \begin{bmatrix} \frac{1}{\sqrt{2}} \\ \frac{1}{\sqrt{2}} \end{bmatrix} \begin{bmatrix} \frac{1}{\sqrt{2}} & \frac{1}{\sqrt{2}} \end{bmatrix} = \begin{bmatrix} \frac{1}{2} & \frac{1}{2} \\ \frac{1}{2} & \frac{1}{2} \end{bmatrix}$$

The orthogonal projections of $\mathbf{x}$ on $\mathbf{u}_1$ and $\mathbf{u}_2$ scaled by corresponding eigenvectors are, respectively,

$$\lambda_1\,\mathbf{u}_1\,\mathbf{u}_1^T\mathbf{x} = 3 \begin{bmatrix} \frac{1}{2} & -\frac{1}{2} \\ -\frac{1}{2} & \frac{1}{2} \end{bmatrix} \begin{bmatrix} 1 \\ 2 \end{bmatrix} = 3 \begin{bmatrix} -\frac{1}{2} \\ \frac{1}{2} \end{bmatrix} = \begin{bmatrix} -\frac{3}{2} \\ \frac{3}{2} \end{bmatrix}$$

and

$$\lambda_2\,\mathbf{u}_2\,\mathbf{u}_2^T\mathbf{x} = -1 \begin{bmatrix} \frac{1}{2} & \frac{1}{2} \\ \frac{1}{2} & \frac{1}{2} \end{bmatrix} \begin{bmatrix} 1 \\ 2 \end{bmatrix} = -1 \begin{bmatrix} \frac{3}{2} \\ \frac{3}{2} \end{bmatrix} = \begin{bmatrix} -\frac{3}{2} \\ -\frac{3}{2} \end{bmatrix}$$

Adding them together we have

$$\mathbf{A}\mathbf{x} = \begin{bmatrix} -3 \\ 0 \end{bmatrix}$$

which is the same as that obtained by direct multiplication of $\mathbf{x}$ by $\mathbf{A}$.

Exercise 3.2

1. In each case, find the characteristic polynomial, eigenvalues, and eigenvectors of the matrix

   a. $\mathbf{A} = \begin{bmatrix} 2 & 4 \\ 1 & -1 \end{bmatrix}$

   b. $\mathbf{A} = \begin{bmatrix} 4 & 2 \\ 2 & 1 \end{bmatrix}$

   c. $\mathbf{A} = \begin{bmatrix} 1 & 0 & 1 \\ 1 & 2 & -1 \\ 1 & 3 & -2 \end{bmatrix}$

   d. $\mathbf{A} = \begin{bmatrix} -2 & 0 & 1 \\ -2 & 1 & 0 \\ 4 & 0 & 1 \end{bmatrix}$

   e. $\mathbf{A} = \begin{bmatrix} 1 & 1 & 2 \\ 1 & 2 & 1 \\ 2 & 1 & 1 \end{bmatrix}$

2. Find a 2×2 matrix, which has
   a. eigenvalues 3 and 1, and corresponding eigenvectors (1, 1) and (−1, 1)
   b. eigenvalues $\frac{3}{2}$ and 0, and corresponding eigenvectors $(-\frac{1}{2}, 1)$ and (1, 1)
3. Find the characteristic equation and eigenvalues of the following matrices just by inspection.

   a. $\mathbf{A} = \begin{bmatrix} 2 & 0 & 0 \\ 2 & 1 & 0 \\ 1 & 2 & 4 \end{bmatrix}$

   b. $\mathbf{A} = \begin{bmatrix} 2 & 4 & 1 \\ 0 & -1 & 2 \\ 0 & 0 & 1 \end{bmatrix}$

4. In each of the following cases, find a matrix P that diagonalizes matrix A and compute $\mathbf{P}^{-1}\,\mathbf{A}\,\mathbf{P}$.

   a. $\mathbf{A} = \begin{bmatrix} 4 & 2 \\ 2 & 1 \end{bmatrix}$

   b. $\mathbf{A} = \begin{bmatrix} 1 & 2 \\ 2 & 1 \end{bmatrix}$

c. $\mathbf{A} = \begin{bmatrix} 1 & 0 & 1 \\ 1 & 2 & -1 \\ 1 & 3 & -2 \end{bmatrix}$

d. $\mathbf{A} = \begin{bmatrix} 1 & 1 & 2 \\ 1 & 2 & 1 \\ 2 & 1 & 1 \end{bmatrix}$

5. In each of the following cases, find the orthogonally diagonalizing matrix **P** and diagonalize **A**.

a. $\mathbf{A} = \begin{bmatrix} 4 & 2 \\ 2 & 4 \end{bmatrix}$

b. $\mathbf{A} = \begin{bmatrix} 1 & 2 \\ 2 & 4 \end{bmatrix}$

c. $\mathbf{A} = \begin{bmatrix} -1 & -1 & 2 \\ -1 & 2 & -1 \\ 2 & -1 & -1 \end{bmatrix}$

d. $\mathbf{A} = \begin{bmatrix} 1 & 0 & -1 \\ 0 & 1 & 2 \\ -1 & 2 & 5 \end{bmatrix}$

6. Find the spectral decomposition of matrix **A**.

a. $\mathbf{A} = \begin{bmatrix} 3 & 2 \\ 2 & 6 \end{bmatrix}$

b. $\mathbf{A} = \begin{bmatrix} -1 & -1 & 2 \\ -1 & 2 & -1 \\ 2 & -1 & -1 \end{bmatrix}$

3.3 Singular value decomposition

In a sense, singular value decomposition (SVD) generalizes eigendecomposition to rectangular matrices. It uses the eigendecomposition of a positive semi-definite matrix to decompose a rectangular matrix into three simple matrices—two orthogonal and a diagonal. SVD is based on the following theorem.

Theorem 3.3.1 A matrix $\mathbf{A} \in \mathbb{R}^{m \times n}$ of rank r can be decomposed as

$$\mathbf{A}_{m \times n} = \mathbf{U}_{m \times m} \, \mathbf{\Sigma}_{m \times n} \mathbf{V}^T_{n \times n} \tag{3.48}$$

where $\mathbf{U}_{m \times m}$ and $\mathbf{V}_{n \times n}$ are orthogonal matrices and

$$\mathbf{\Sigma}_{m \times n} = \begin{bmatrix} \sigma_1 & 0 & \cdots & 0 & 0 & \cdots & 0 \\ 0 & \sigma_2 & \cdots & 0 & 0 & \cdots & 0 \\ \vdots & \vdots & \ddots & \vdots & \vdots & \cdots & \vdots \\ 0 & 0 & \cdots & \sigma_r & 0 & \cdots & 0 \\ 0 & 0 & \cdots & 0 & 0 & \cdots & 0 \\ \vdots & \vdots & \vdots & \vdots & \vdots & \ddots & \vdots \\ 0 & 0 & \cdots & 0 & 0 & \cdots & 0 \end{bmatrix}$$

The entries on the diagonal of $\mathbf{\Sigma}$, $\sigma_i \geq 0, i = 1, 2, \ldots, r$, are called the singular values. $\mathbf{u}_i$, $i = 1, 2, \ldots, m$, are called the left singular vectors and $\mathbf{v}_j$, $j = 1, 2, \ldots, n$ called the right singular vectors. By convention $\sigma_1 \geq \sigma_2 \geq \cdots \geq \sigma_r$.

Definition 3.3.1 For a vector $\mathbf{x} \in V$, a symmetric matrix $\mathbf{S} \in \mathbb{R}^{n \times n}$ that satisfies

$$\mathbf{x}^T \mathbf{S} \mathbf{x} \geq 0 \tag{3.49}$$

is called symmetric, positive semidefinite.

Let us consider a matrix $\mathbf{A}$ and define $\mathbf{S} = \mathbf{A}^T\mathbf{A}$. Then,

$$\mathbf{S} = \mathbf{A}^T\mathbf{A} = \mathbf{A}^T\left(\mathbf{A}^T\right)^T = \left(\mathbf{A}^T\mathbf{A}\right)^T = \mathbf{S}^T \tag{3.50}$$

indicating that $\mathbf{S}$ is symmetric. Furthermore,

$$\mathbf{x}^T\mathbf{S}\mathbf{x} = \mathbf{x}^T\mathbf{A}^T\mathbf{A}\mathbf{x} = \left(\mathbf{x}^T\mathbf{A}^T\right)(\mathbf{A}\,\mathbf{x}) = (\mathbf{A}\,\mathbf{x})^T(\mathbf{A}\,\mathbf{x}) \geq 0 \tag{3.51}$$

Hence, $\mathbf{S}$ is a positive semidefinite matrix.

Theorem 3.3.2 Given a matrix $\mathbf{A} \in \mathbb{R}^{m \times n}$, it is always possible to obtain a symmetric, positive semidefinite matrix $\mathbf{S} \in \mathbb{R}^{n \times n}$ by defining

$$\mathbf{S} = \mathbf{A}^T\mathbf{A} \tag{3.52}$$

In the construction of SVD, the matrix product of the form $\mathbf{A}^T\mathbf{A}$ plays an important role. It provides a symmetric, positive semidefinite matrix required for the purpose. The eigenvalues of such a matrix are always positive or null and, if the eigenvalues are different, the corresponding eigenvectors are pairwise orthogonal.

Theorem 3.3.3 If $\mathbf{A}$ is an $m \times n$ matrix,

1. the matrix product $\mathbf{A}^T\mathbf{A}$ is orthogonally diagonalizable and
2. the eigenvalues of $\mathbf{A}^T\mathbf{A}$ are nonnegative.

Furthermore, the following theorem is useful in the computation of SVD. We present it here without proof.

Theorem 3.3.4 If $\mathbf{A}$ is an $m \times n$ matrix, (1) $\mathbf{A}$ and $\mathbf{A}^T\mathbf{A}$ have the same null space, the same row space and the same rank while (2) $\mathbf{A}^T$ and $\mathbf{A}^T\mathbf{A}$ have the same column space.

As it appears, two sets of orthonormal bases—$U = \{\mathbf{u}_1, \mathbf{u}_2, \ldots, \mathbf{u}_m\}$ for the codomain $\mathbb{R}^m$ and $V = \{\mathbf{v}_1, \mathbf{v}_2, \ldots, \mathbf{v}_n\}$ for the domain $\mathbb{R}^n$—are required to be derived to construct the matrices $\mathbf{U} = [\mathbf{u}_1\ \mathbf{u}_2\ \cdots\ \mathbf{u}_m]$ and $\mathbf{V} = [\mathbf{v}_1\ \mathbf{v}_2\ \cdots\ \mathbf{v}_n]$ containing, respectively, the left-singular and right-singular vectors. Since $\mathbf{A}^T\mathbf{A} \in \mathbb{R}^{n\times n}$ is orthogonally diagonalizable, we can write

$$\mathbf{A}^T\mathbf{A} = \mathbf{P}\mathbf{D}\mathbf{P}^T = \mathbf{P}\begin{bmatrix} \lambda_1 & 0 & \cdots & 0 \\ 0 & \lambda_2 & \cdots & 0 \\ \vdots & \vdots & \ddots & \vdots \\ 0 & 0 & \cdots & \lambda_n \end{bmatrix}\mathbf{P}^T \tag{3.53}$$

where $\mathbf{P}$ is an orthonormal matrix containing the orthonormal eigenbasis. $\lambda_i \geq 0$, $i = 1, 2, \ldots, n$, are the eigenvalues of $\mathbf{A}^T\mathbf{A}$. Then, from (3.48) we have

$$\mathbf{A}^T\mathbf{A} = (\mathbf{U}\Sigma\mathbf{V}^T)^T(\mathbf{U}\Sigma\mathbf{V}^T) = \mathbf{V}\Sigma^T\mathbf{U}^T\mathbf{U}\Sigma\mathbf{V}^T \tag{3.54}$$

Since $\mathbf{U}$ and $\mathbf{V}$ are orthonormal matrices, $\mathbf{U}^T\mathbf{U} = \mathbf{I}$ and

$$\mathbf{A}^T\mathbf{A} = \mathbf{V}\boldsymbol{\Sigma}^T\boldsymbol{\Sigma}\mathbf{V}^T = \mathbf{V}\begin{bmatrix} \sigma_1^2 & 0 & \cdots & 0 \\ 0 & \sigma_2^2 & \cdots & 0 \\ \vdots & \vdots & \ddots & \vdots \\ 0 & 0 & \cdots & \sigma_n^2 \end{bmatrix}\mathbf{V}^T \tag{3.55}$$

Comparing (3.53) and (3.55)

$$\mathbf{V}^T = \mathbf{P}^T \tag{3.56}$$

and

$$\sigma_i^2 = \lambda_i \qquad i = 1, 2, \cdots, n \tag{3.57}$$

So, the column vectors of $\mathbf{V} = [\ \mathbf{v}_1\ \mathbf{v}_2\ \cdots\ \mathbf{v}_n]$ are the normalized eigenvectors of $\mathbf{A}^T\mathbf{A}$ and the diagonal entries of $\mathbf{D}$ are the eigenvalues of $\mathbf{A}^T\mathbf{A}$ corresponding in succession to these column vectors. Since $\mathbf{A}$ has been assumed to have rank r, in accordance with Theorem 3.3.4, $\mathbf{A}^T\mathbf{A}$ also has rank r. Hence, the diagonal matrix D, which is similar to $\mathbf{A}^T\mathbf{A}$, has rank r and can be expressed as

$$\mathbf{D} = \begin{bmatrix} \lambda_1 & 0 & \cdots & 0 & 0 & \cdots & 0 \\ 0 & \lambda_2 & \cdots & 0 & 0 & \cdots & 0 \\ \vdots & \vdots & \ddots & \vdots & \vdots & \cdots & \vdots \\ 0 & 0 & \cdots & \lambda_r & 0 & \cdots & 0 \\ 0 & 0 & \cdots & 0 & 0 & \cdots & 0 \\ \vdots & \vdots & \vdots & \vdots & \vdots & \ddots & \vdots \\ 0 & 0 & \cdots & 0 & 0 & \cdots & 0 \end{bmatrix} \tag{3.58}$$

where $\lambda_1 \geq \lambda_2 \geq \cdots\cdots \geq \lambda_r > 0$.

Let us consider the image $\mathbf{v}_i$ under transformation by $\mathbf{A}$. The first r elements of the orthonormal basis $\{\mathbf{v}_1, \mathbf{v}_2, \ldots, \mathbf{v}_n\}$ span the row space of $\mathbf{A}$ and the remaining $n-r$ elements span the null space of $\mathbf{A}$. The orthogonality of $\mathbf{v}_i$ and $\mathbf{v}_j$, $i = j$, implies

$$(\mathbf{A}\,\mathbf{v}_i)^T(\mathbf{A}\,\mathbf{v}_j) = \mathbf{v}_i^{\,T}(\mathbf{A}^T\,\mathbf{A})\,\mathbf{v}_j = \mathbf{v}_i^{\,T}(\lambda_j\,\mathbf{v}_j) = \lambda_j\,\mathbf{v}_i^{\,T}\,\mathbf{v}_j = 0$$

Hence, the image of $\mathbf{v}_i$ under $\mathbf{A}$ is also orthogonal and $\{\mathbf{A}\,\mathbf{v}_1, \mathbf{A}\,\mathbf{v}_2, \ldots, \mathbf{A}\,\mathbf{v}_k\}$ is a basis of an r-dimensional subspace of $\mathbb{R}^m$. If the images are normalized

$$\mathbf{u}_i = \frac{\mathbf{A}\mathbf{v}_i}{||\mathbf{A}\mathbf{v}_i||}$$

Now, $||\mathbf{A}\,\mathbf{v}_i||^2 = \mathbf{A}\,\mathbf{v}_i \cdot \mathbf{A}\,\mathbf{v}_i = \mathbf{v}_i \cdot \mathbf{A}^T\mathbf{A}\mathbf{v}_i = \mathbf{v}_i \cdot \lambda_i\,\mathbf{v}_i = \lambda_i\,(\mathbf{v}_i \cdot \mathbf{v}_i) = \lambda_i\,||\mathbf{v}_i||^2 = \lambda_i$ so that,

$$\mathbf{u}_i = \frac{1}{\sqrt{\lambda_i}}\mathbf{A}\,\mathbf{v}_i = \frac{1}{\sigma_i}\mathbf{A}\,\mathbf{v}_i \quad i = 1, 2, \ldots, r \tag{3.59}$$

which can be rewritten as

$$\mathbf{A}\,\mathbf{v}_i = \sigma_i\,\mathbf{u}_i \qquad i = 1, 2, \ldots, r \tag{3.60}$$

Evidently, $\mathbf{A}\,\mathbf{v}_i = 0$ for $i > r$

Let $\mathbf{u}_i$, $i = 1, 2, \ldots, n$, constitute the columns of an orthogonal matrix as

$$\mathbf{U} = [\mathbf{u}_1 \;\; \mathbf{u}_2 \;\cdots\cdots\; \mathbf{u}_r \;\; \mathbf{u}_{r+1} \;\cdots\cdots\; \mathbf{u}_n]$$

and consider the diagonal matrix

$$\mathbf{\Sigma} = \begin{bmatrix} \sigma_1 & 0 & \cdots & 0 & 0 & \cdots & 0 \\ 0 & \sigma_2 & \cdots & 0 & 0 & \cdots & 0 \\ \vdots & \vdots & \ddots & \vdots & \vdots & \cdots & \vdots \\ 0 & 0 & \cdots & \sigma_r & 0 & \cdots & 0 \\ 0 & 0 & \cdots & 0 & 0 & \cdots & 0 \\ \vdots & \vdots & \vdots & \vdots & \vdots & \ddots & \vdots \\ 0 & 0 & \cdots & 0 & 0 & \cdots & 0 \end{bmatrix}$$

Then,

$$\begin{aligned}\mathbf{U}\,\boldsymbol{\Sigma} &= [\sigma_1\,\mathbf{u}_1 \quad \sigma_2\,\mathbf{u}_2 \quad \cdots\cdots \quad \sigma_r\,\mathbf{u}_r \quad 0 \quad \cdots\cdots \quad 0] \\ &= [\mathbf{A}\,\mathbf{v}_1 \quad \mathbf{A}\,\mathbf{v}_2 \quad \cdots\cdots \quad \mathbf{A}\,\mathbf{v}_r \quad \mathbf{A}\,\mathbf{v}_{r+1} \quad \cdots\cdots \quad \mathbf{A}\,\mathbf{v}_n] \\ &= \mathbf{A}\,\mathbf{V}\end{aligned}$$

Since $\mathbf{V}$ is orthogonal, multiplying right-multiplication with $\mathbf{V}^T$ gives the SVD of A as

$$\mathbf{A} = \mathbf{U}\,\boldsymbol{\Sigma}\,\mathbf{V}^T \tag{3.61}$$

It is to be noted that the left-singular vectors $\mathbf{U}$ can be obtained directly following a procedure similar to that for the right-singular vectors $\mathbf{V}$. In this case, we compute the SVD of the symmetric matrix $\mathbf{A}\,\mathbf{A}^T$ (instead of $\mathbf{A}^T\,\mathbf{A}$). From (3.48) we have

$$\mathbf{A}\,\mathbf{A}^T = (\mathbf{U}\,\boldsymbol{\Sigma}\,\mathbf{V}^T)(\mathbf{U}\,\boldsymbol{\Sigma}\,\mathbf{V}^T)^T = \mathbf{U}\,\boldsymbol{\Sigma}\,\mathbf{V}^T\,\mathbf{V}\,\boldsymbol{\Sigma}^T\,\mathbf{U}^T = \mathbf{U}\,\boldsymbol{\Sigma}\,\boldsymbol{\Sigma}^T\,\mathbf{U}^T = \mathbf{U}\begin{bmatrix}\sigma_1^2 & 0 & \cdots & 0\\ 0 & \sigma_2^2 & \cdots & 0\\ \vdots & \vdots & \ddots & \vdots\\ 0 & 0 & \cdots & \sigma_m^2\end{bmatrix}\mathbf{U}^T$$

The orthonormal eigenvectors of $\mathbf{A}\,\mathbf{A}^T$ are the left-singular vectors U. These column vectors form an orthonormal bases set in the codomain. $\mathbf{A}^T\mathbf{A}$ and $\mathbf{A}\,\mathbf{A}^T$ have the same nonzero eigenvalues.

Example 3.3.1: Let us consider a 2×3 matrix

$$\mathbf{A} = \begin{bmatrix}2 & 1 & 0\\ -1 & 0 & 1\end{bmatrix}$$

and determine, step by step, its SVD.

$$\mathbf{A}^T = \begin{bmatrix}2 & -1\\ 1 & 0\\ 0 & 1\end{bmatrix}$$

and

$$\mathbf{A}^T\mathbf{A} = \begin{bmatrix}5 & 2 & -1\\ 2 & 1 & 0\\ -1 & 0 & 1\end{bmatrix}$$

Each of the matrices $\mathbf{A}$, $\mathbf{A}^T$, and $\mathbf{A}^T\mathbf{A}$ has rank 2.

The eigenvalues are determined from the characteristic polynomial of $\mathbf{A}^T\mathbf{A}$ and found to be

$$\lambda_1 = 6, \quad \lambda_2 = 1, \quad \lambda_3 = 0$$

The singular values of $\mathbf{A}$ in decreasing order are

$$\sigma_1 = \sqrt{6}, \quad \sigma_2 = 1, \quad \sigma_3 = 0$$

The corresponding normalized eigenvectors are obtained as

$$\mathbf{v}_1 = \begin{bmatrix} -\frac{5}{\sqrt{30}} \\ -\frac{2}{\sqrt{30}} \\ \frac{1}{\sqrt{30}} \end{bmatrix}, \quad \mathbf{v}_2 = \begin{bmatrix} 0 \\ \frac{1}{\sqrt{5}} \\ \frac{2}{\sqrt{5}} \end{bmatrix}, \quad \mathbf{v}_3 = \begin{bmatrix} \frac{1}{\sqrt{6}} \\ -\frac{2}{\sqrt{6}} \\ \frac{1}{\sqrt{6}} \end{bmatrix}$$

from which we have the right-singular vectors

$$\mathbf{V} = \begin{bmatrix} -\frac{5}{\sqrt{30}} & 0 & \frac{1}{\sqrt{6}} \\ -\frac{2}{\sqrt{30}} & \frac{1}{\sqrt{5}} & -\frac{2}{\sqrt{6}} \\ \frac{1}{\sqrt{30}} & \frac{2}{\sqrt{5}} & \frac{1}{\sqrt{6}} \end{bmatrix}$$

Using (3.59) we can calculate the left-singular vectors as

$$\mathbf{U} = [\mathbf{u}_1 \quad \mathbf{u}_2] = \begin{bmatrix} -\frac{2}{\sqrt{5}} & \frac{1}{\sqrt{5}} \\ \frac{1}{\sqrt{5}} & \frac{2}{\sqrt{5}} \end{bmatrix}$$

and the singular value matrix

$$\boldsymbol{\Sigma} = \begin{bmatrix} \sqrt{6} & 0 & 0 \\ 0 & 1 & 0 \end{bmatrix}$$

The parts can be assembled together to get the expression for SVD of **A**

$$\mathbf{A} = \mathbf{U}\boldsymbol{\Sigma}\mathbf{V}^T = \begin{bmatrix} -\frac{2}{\sqrt{5}} & \frac{1}{\sqrt{5}} \\ \frac{1}{\sqrt{5}} & \frac{2}{\sqrt{5}} \end{bmatrix} \begin{bmatrix} \sqrt{6} & 0 & 0 \\ 0 & 1 & 0 \end{bmatrix} \begin{bmatrix} -\frac{5}{\sqrt{30}} & -\frac{2}{\sqrt{30}} & \frac{1}{\sqrt{30}} \\ 0 & \frac{1}{\sqrt{5}} & \frac{2}{\sqrt{5}} \\ \frac{1}{\sqrt{6}} & -\frac{2}{\sqrt{6}} & \frac{1}{\sqrt{6}} \end{bmatrix}$$

Eigendecomposition and singular value decomposition compared

SVD exists for any matrix $\in \mathbb{R}^{m \times n}$ whereas eigendecomposition is defined only for square matrices (though not all). In general, there are several differences between the two. A key difference is that unlike eigendecomposition, in SVD domain and codomain can be vector spaces of different dimensions. However, for a symmetric matrix $\mathbf{A} \in \mathbb{R}^{n \times n}$, the eigendecomposition is the same as its SVD. The following examples will highlight this point.

Example 3.3.2: Let us consider a nonsymmetric square matrix

$$\mathbf{A} = \begin{bmatrix} -1 & 2 \\ 1 & -2 \end{bmatrix}$$

We can calculate the eigenvalues and corresponding eigenvectors of $\mathbf{A}$, find an invertible matrix $\mathbf{P}$, and write its eigendecomposition as

$$\mathbf{A} = \mathbf{P}\mathbf{D}\mathbf{P}^{-1} = \begin{bmatrix} -1 & 2 \\ 1 & 1 \end{bmatrix} \begin{bmatrix} -3 & 0 \\ 0 & 0 \end{bmatrix} \begin{bmatrix} -\frac{1}{3} & \frac{2}{3} \\ \frac{1}{3} & \frac{1}{3} \end{bmatrix}$$

Following the procedure we have already discussed, the SVD of $\mathbf{A}$ is found to be

$$\mathbf{A} = \mathbf{U}\boldsymbol{\Sigma}\mathbf{V}^T = \begin{bmatrix} \frac{1}{\sqrt{2}} & \frac{1}{\sqrt{2}} \\ -\frac{1}{\sqrt{2}} & \frac{1}{\sqrt{2}} \end{bmatrix} \begin{bmatrix} \sqrt{10} & 0 \\ 0 & 0 \end{bmatrix} \begin{bmatrix} -\frac{1}{\sqrt{5}} & \frac{2}{\sqrt{5}} \\ \frac{2}{\sqrt{5}} & \frac{1}{\sqrt{5}} \end{bmatrix}$$

Example 3.3.3: Let us now consider a symmetric matrix

$$\mathbf{A} = \begin{bmatrix} 2 & 1 \\ 1 & 2 \end{bmatrix}$$

$\mathbf{A}$ has been decomposed in Example 3.2.6. If we further normalize the eigenvectors of $\mathbf{A}$, its eigendecomposition is given as

$$\mathbf{A} = \mathbf{P}\mathbf{D}\mathbf{P}^{-1} = \begin{bmatrix} \frac{1}{\sqrt{2}} & -\frac{1}{\sqrt{2}} \\ \frac{1}{\sqrt{2}} & \frac{1}{\sqrt{2}} \end{bmatrix} \begin{bmatrix} 3 & 0 \\ 0 & 1 \end{bmatrix} \begin{bmatrix} \frac{1}{\sqrt{2}} & \frac{1}{\sqrt{2}} \\ -\frac{1}{\sqrt{2}} & \frac{1}{\sqrt{2}} \end{bmatrix}$$

The SVD of $\mathbf{A}$ is also found to be

$$\mathbf{A} = \mathbf{U}\boldsymbol{\Sigma}\mathbf{V}^T = \begin{bmatrix} \frac{1}{\sqrt{2}} & -\frac{1}{\sqrt{2}} \\ \frac{1}{\sqrt{2}} & \frac{1}{\sqrt{2}} \end{bmatrix} \begin{bmatrix} 3 & 0 \\ 0 & 1 \end{bmatrix} \begin{bmatrix} \frac{1}{\sqrt{2}} & \frac{1}{\sqrt{2}} \\ -\frac{1}{\sqrt{2}} & \frac{1}{\sqrt{2}} \end{bmatrix}$$

Exercises 3.3

1. Determine the SVD of each of the following matrices:

a. $\begin{bmatrix} 1 & 1 & -1 \\ -1 & 2 & 1 \end{bmatrix}$

b. $\begin{bmatrix} 1 & 2 & -1 \\ 2 & -1 & 2 \end{bmatrix}$

c. $\begin{bmatrix} 1 & 2 \\ 2 & 3 \\ 3 & -1 \end{bmatrix}$

d. $\begin{bmatrix} 0 & -1 \\ 1 & 0 \\ -1 & 0 \end{bmatrix}$

2. For each of the following matrices, find if the SVD is same as the eigendecomposition:

a. $\begin{bmatrix} 2 & 0 & 1 \\ 0 & 2 & 0 \\ 1 & 0 & 2 \end{bmatrix}$

b. $\begin{bmatrix} 1 & 0 & 1 \\ 0 & 2 & 0 \\ 2 & 0 & 2 \end{bmatrix}$

Further reading

Anton, H., Rorres, C., 2014. Elementary Linear Algebra. Wiley.

Deisenroth, M.P., Faisal, A.A., Ong, C.S., 2020. Mathematics for Machine Learning. Cambridge University Press.

Nicholson, W.K., 2021. Linear Algebra with Applications, Open Edition. Creative Common License (CC BY-NC-SA).

CHAPTER

Vector calculus

4

Two paradigms of molecular structural biology are the potential energy surface (PES) and collective atomic motions. In general, biomolecular structures correspond to the positions of the minima on a PES and functional mode analysis (FMA) identifies the collective motions of atoms in a protein related to a specific function. Protein conformations are represented by vectors and the PES is described as a function of the vectors. The inherent complexity of the system requires approximation of the potential energy by Taylor series expansion about a "starting geometry." When computation of second- and higher-order derivates is too much taxing, the analysis proceeds by linear approximation (linearization). For FMA, the equation of motion is written in terms of the second-order derivatives.

4.1 Derivatives of univariate functions

We have already familiarized ourselves with the concept of a function in the earlier chapters. To rephrase, a function assigns a quantity (input) in the domain, $\mathbf{x} \in \mathbb{R}^D$, to an output (or function value) in the range or codomain, $f(\mathbf{x}) \in \mathbb{R}^C$. This can be expressed as

$$f\colon \mathbb{R}^D \rightarrow \mathbb{R}^C \tag{4.1}$$

where D and C denote domain and codomain respectively. If the input is a single independent variable, $x \in \mathbb{R}$, then the function $f(x) \in \mathbb{R}$, which defines a one-to-one mapping between the input and output, is a ***univariate function***. So, (4.1) is expressed as

$$f\colon \mathbb{R} \rightarrow \mathbb{R} \tag{4.2}$$

Example 4.1.1: An example of a simple univariate function is

$$f(x) = 2x^3 - 3x^2 - 12x \tag{4.3}$$

Graphically, such a function can be represented by a line (Fig. 4.1). We have also seen (Chapter 1) that the derivative of $f(x)$, denoted by $f'(x)$, can be

Mathematical Approaches to Molecular Structural Biology. DOI: https://doi.org/10.1016/B978-0-323-90397-4.00005-6
© 2023 Elsevier Inc. All rights reserved.

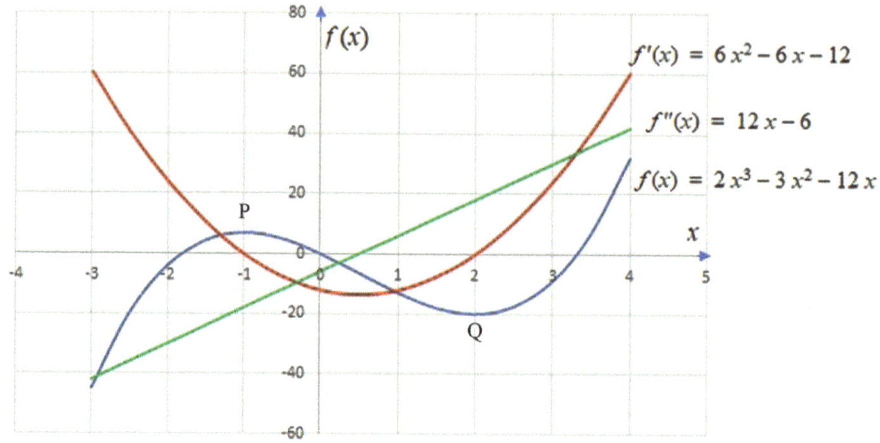

FIGURE 4.1

Graphical representation of a univariate function.

visualized graphically as the gradient or slope of $f(x)$ at x. The slope is also given by tan ϕ, where ϕ is the angle made by the tangent to $f(x)$ at x.

Example 4.1.2: The derivative of the function $f(x)$ in Example 4.1.1 is given by

$$f'(x) = 6x^2 - 6x - 12 \tag{4.4}$$

Here, $f'(x)$ is also a function of x and graphically shown in Fig. 4.1. It can be seen in the figure that there is a local maximum of $f(x)$ at P and a local minimum at Q. At both points, called the ***stationary points*** (also called ***critical points***), the tangent to $f(x)$ is horizontal and $f'(x)$ vanishes. The roots of the equation $f'(x)=0$ are -1 and 2 which are also the points of intersection of $f'(x)$ with the x-axis. Clearly, as the figure shows, $f(x)$ is maximum at $x=-1$ and minimum at $x = 2$.

Further, at $x=-1$, $f(x)$ "concaves down," that is, the change in the slope of the tangent to the curve is negative, whereas at $x=2$, $f(x)$ "concaves up." We have the derivative of $f'(x)$ [or the second derivative of $f(x)$] as

$$f''(x) = 12x - 6 \tag{4.5}$$

and find that $f''(-1) < 0$ and $f''(2) > 0$. Hence, it can be concluded that

a. for a maximum, $f'(x) = 0$ and $f''(x) < 0$ and
b. for a minimum, $f'(x) = 0$ and $f''(x) > 0$.

This is the "second derivative test" for a univariate function.

In univariate calculus, there can be a "chain" of relations between variables. For example, $f(x)$ is a differentiable function of x and $g\ (f(x))$ is a differentiable function of $f(x)$. In such case, the derivative of $g\ (f(x))$ with respect to x can be calculated by applying the chain rule as

$$\frac{dg(f(x))}{dx} = \frac{dg(f(x))}{df(x)}\frac{df(x)}{dx} \tag{4.6}$$

or, more compactly as

$$\frac{dg}{dx} = \frac{dg}{df}\frac{df}{dx} \tag{4.7}$$

We shall see later how the chain rule is extended to multivariate functions.

4.2 Derivatives of multivariate functions

A ***multivariate function***, as the name suggests, depends on more than one variable. For example, the volume of a cylinder with the radius of cross-section r and length l is given by

$$V = \pi r^2 l$$

which can also be expressed as a function of two variables

$$V(r, l) = \pi r^2 l$$

In general, a multivariate function can be expressed as a mapping of n real input variables to a real output.

$$f: \mathbb{R}^n \rightarrow \mathbb{R} \tag{4.8}$$

Here, we are going to consider some examples of functions of two variables, $f(x, y)$. It can be found that the graph in each case is no longer a curve in $\mathbb{R}^2$, but a surface in $\mathbb{R}^3$. The input variables x and y are represented by points on the xy-plane while $f(x, y)$ is represented by the height of the surface above the point (x, y) (Figs. 4.2 and 4.3).

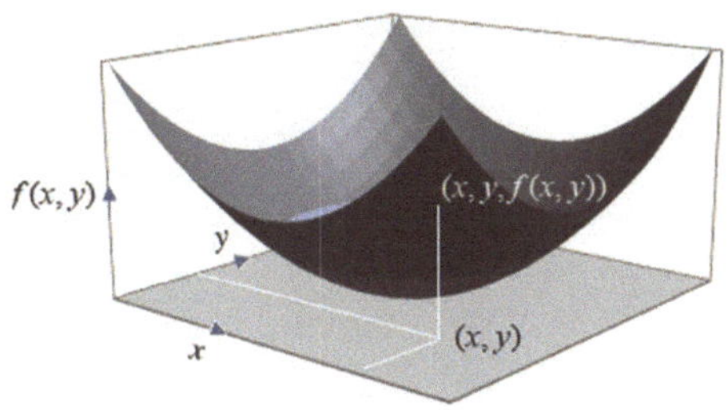

FIGURE 4.2

Graphical representation of $f(x, y) = x^2 + y^2$.

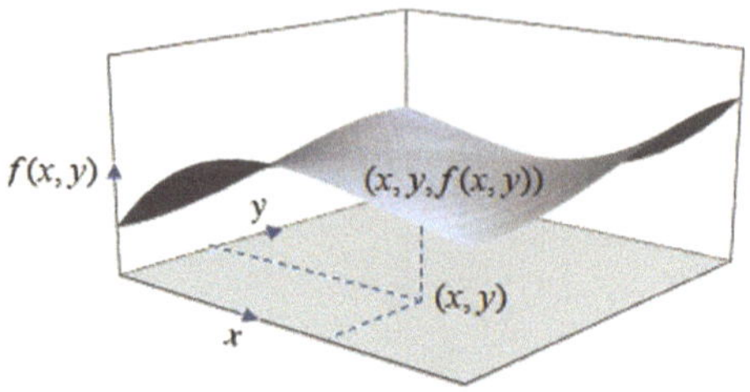

FIGURE 4.3

Graphical representation of $f(x, y) = 3x^3 + 6x^2y - 2y^3 - 150y$.

Example 4.2.1:

$$f(x, y) = x^2 + y^2 \tag{4.9}$$

Example 4.2.2:

$$f(x, y) = 3x^3 + 6x^2y - 2y^3 - 150y \tag{4.10}$$

Partial derivatives

Partial derivatives of a multivariate function have been introduced in Chapter 1. Here, let us examine the partial derivatives of some two-variable functions. The partial derivatives of $f(x, y)$ are going to be denoted as

$$f_x = \frac{\partial f}{\partial x} \text{ and } f_y = \frac{\partial f}{\partial y} \tag{4.11}$$

for the first derivatives, and

$$f_{xx} = (f_x)_x = \frac{\partial}{\partial x}\left(\frac{\partial f}{\partial x}\right) = \frac{\partial^2 f}{\partial x^2}$$

$$f_{yx} = (f_y)_x = \frac{\partial}{\partial x}\left(\frac{\partial f}{\partial y}\right) = \frac{\partial^2 f}{\partial x \partial y} \tag{4.12}$$

$$f_{xy} = (f_x)_y = \frac{\partial}{\partial y}\left(\frac{\partial f}{\partial x}\right) = \frac{\partial^2 f}{\partial y \partial x}$$

$$f_{yy} = (f_y)_y = \frac{\partial}{\partial y}\left(\frac{\partial f}{\partial y}\right) = \frac{\partial^2 f}{\partial y^2} \tag{4.13}$$

for the second derivatives.

Critical points and local extrema

The concept of critical points, as in the case of single-variable functions, can be extended to two-variable functions as well. Furthermore, the second derivatives can be used to determine the nature of the critical points.

Definition 4.2.1: In the case of a function $f(x, y)$ whose partial derivatives $\partial f/\partial x$ and $\partial f/\partial y$ exist everywhere, a point (a, b) is said to be a ***critical point*** if the tangent plane to the graph of $f(x, y)$ at (a, b) is horizontal, that is $f_x(a, b) = f_y(a, b) = 0$.

Example 4.2.3: For the function used in Example 4.2.1, that is,

$$f(x, y) = x^2 + y^2$$

we have

$$f_x = 2x$$

and

$$f_y = 2y$$

The critical point is given by setting $f_x = 0$ and $f_y = 0$ which gives $x = 0$ and $y = 0$. So, $(x, y) = (0, 0)$ is the critical point of the given $f(x, y)$.

Example 4.2.4: Let us consider the function

$$f(x, y) = \sin x \cos y$$

The critical points of this function are given by

$$\frac{\partial f}{\partial x} = 0 \text{ and } \frac{\partial f}{\partial y} = 0$$

from which we have

$$\cos x \cos y = 0 \quad \text{and} \quad \sin x \sin y = 0$$

so that ($\cos x = 0$ or $\cos y = 0$) and ($\sin x = 0$ or $\sin y = 0$)

$$\left(\mathrm{x} = \frac{\pi}{2} + m\pi \text{ or } y = \frac{\pi}{2} + n\pi\right) \text{ and } (x = r\pi \text{ or } y = s\pi)$$

where m, n, r, and s are all integers. Therefore, the given function has an infinite number of critical points

$$\left(\frac{\pi}{2} + m\pi, \frac{\pi}{2} + n\pi\right), (r\pi, s\pi)$$

This is apparent from the graph of the function in Fig. 4.4.

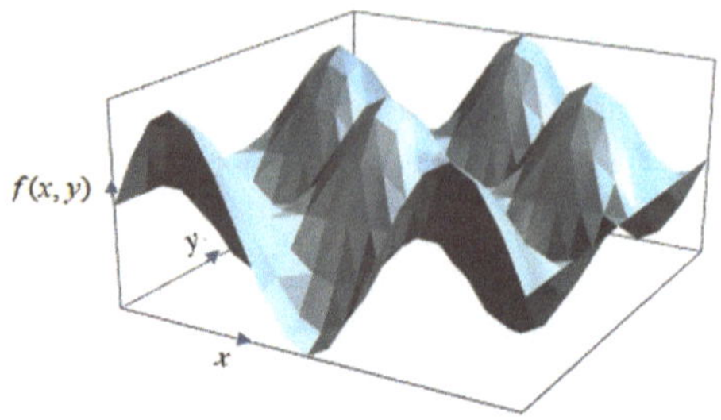

FIGURE 4.4

Graph of function $f(x, y) = \sin x \cos y$.

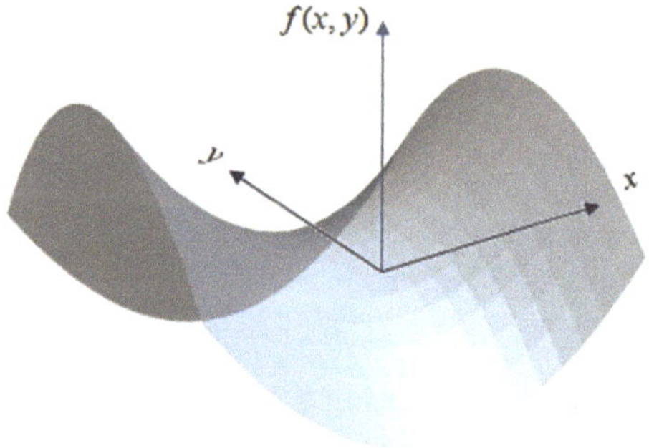

FIGURE 4.5

Graph of function $f(x, y) = x^2 - y^2$

The procedure to determine the nature of critical points of a two-variable function is analogous to, but somewhat more involved than, that of a univariate function. Let us look at the following example.

Example 4.2.5: $f(x, y) = x^2 - y^2$

The partial derivatives of this function (graphically shown in Fig. 4.5) are

$$f_x = 2x \quad f_y = -2y$$

$$f_{xx} = 2 \quad f_{yy} = -2$$

$$f_{xy} = 0 \quad f_{yx} = 0$$

At the critical points, $f_x = f_y = 0$. So, $x = 0$ and $y = 0$. indicating that there is a single critical point (0, 0). However, the second derivatives are both zero and, as the graph shows (Fig. 4.5), the function concaves up in some directions while concaving down in other directions. The point (0, 0) in this case is called a ***saddle***. To generalize, there is a "second derivative test" for two-variable functions provided by the following theorem.

Theorem 4.2.1: Let a two-variable function $f(x, y)$ be twice differentiable and (x_0, y_0) a critical point of the function, the nature of the critical point is determined as follows:

1. if $f_{xx}f_{yy} - f_{xy}^2 < 0$ at (x_0, y_0), then (x_0, y_0) is a saddle point.
2. if $f_{xx}f_{yy} - f_{xy}^2 > 0$ at (x_0, y_0), and
   a. if $f_{xx} < 0, f_{yy} < 0$ at (x_0, y_0), then (x_0, y_0) is a maximum point
   b. if $f_{xx} > 0, f_{yy} > 0$ at (x_0, y_0), then (x_0, y_0) is a minimum point.

In Example 4.2.5, it can be seen that $f_{xx}f_{yy} - f_{xy}^2 < 0$. This is consistent with the earlier conclusion.

Example 4.2.6: $f(x, y) = x^3 + 3y^2 + 6xy - 18x - 9y + 12$

The function is graphically shown in Fig. 4.6. The partial derivatives of the function are

$$f_x = 3x^2 + 6y - 18 \qquad f_y = 6y + 6x - 9$$

$$f_{xx} = 6x \qquad f_{yy} = 6$$

$$f_{xy} = 6 \qquad f_{yx} = 6$$

At critical points, $f_x = f_y = 0$, that is,

$$3x^2 + 6y - 18 = 0$$

and

$$6y + 6x - 9 = 0$$

On solving the equations, it can be found that $x = -1, 3$ and the corresponding values of $y = 5/2, -3/2$. Therefore the critical points are $(-1, 5/2)$ and $(3, -3/2)$.

(1) For the point $(-1, 5/2)$, $f_{xx}f_{yy} - f_{xy}^2 = (-6)(6) - (6)^2 = -72 < 0$, and $f_{xx} = -6$ and $f_{yy} = 6$

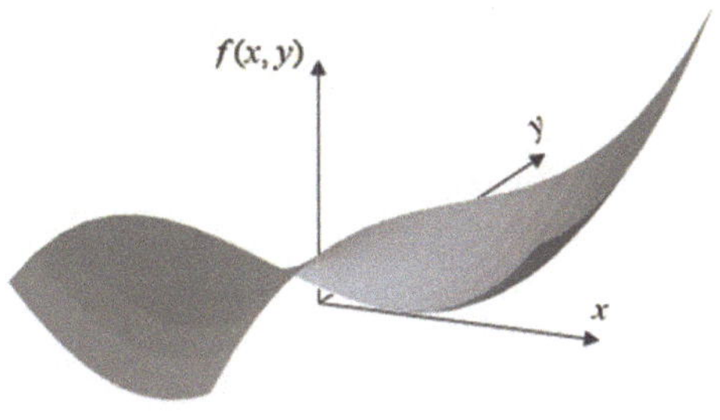

FIGURE 4.6

Graph of function $f(x, y) = x^3 + 3y^2 + 6xy - 18x - 9y + 12$.

Hence, it is a saddle point.
(2) For the point $(3, -3/2)$, $f_{xx}f_{yy} - f_{xy}^2 = (18)(6) - (6)^2 = 72 > 0$,

and $f_{xx} = 18$ and $f_{yy} = 6$

Hence, it is a local minimum.

4.3 Gradients of scalar- and vector-valued functions

The output of the multivariate function defined in (4.8) is a set of real numbers. Hence, such functions are also called ***scalar-valued functions***. If the input variables, which are also real numbers, are organized into a vector

$$\mathbf{x} = \begin{bmatrix} x_1 \\ x_2 \\ \vdots \\ x_n \end{bmatrix}$$

then, we have a scalar-valued function $f(\mathbf{x})$ of a vector $\mathbf{x}$. With

$$\mathbf{x} = \begin{bmatrix} x \\ y \end{bmatrix}$$

(4.9) can be rewritten as

$$f(\mathbf{x}) = x^2 + y^2 \tag{4.14}$$

Example 4.3.1: The norm of a vector, as defined in (2.15),

$$f(\mathbf{u}) = \|\mathbf{u}\| = \left(\mathbf{u}^T\mathbf{u}\right)^{1/2}$$

is a scalar-valued function of a vector. The distance between two vectors **u** and **v** is also a scalar-valued function of **u** and **v**.

$$d(\mathbf{u}, \mathbf{v}) = \|\mathbf{u} - \mathbf{v}\| = \sqrt{(u_1 - v_1)^2 + (u_2 - v_2)^2 + \cdots\cdots + (u_n - v_n)^2}$$

Example 4.3.2: The electrostatic potential at a point $\mathbf{r} \in \mathbb{R}^3$ due to a point charge q at the origin is given by

$$V(\mathbf{r}) = \frac{q}{4\pi\epsilon_0}\frac{1}{\|\mathbf{r}\|}$$

$V(\mathbf{r})$, which is an example of a function $f: D \to \mathbb{R}$, where D is an open subset of $\mathbb{R}^3$, defines a *scalar field*.

We have seen in Chapter 1 that for an n-variable function $f(x_1, x_2, \ldots, x_n)$, the partial derivatives are given by

$$\frac{\partial f}{\partial x_i} = \lim_{\Delta x_i \to 0} \frac{[f(x_1, x_2, \ldots, x_i + \Delta x_i, \ldots, x_n) - f(x_1, x_2, \ldots, x_i, \ldots, x_n)]}{\Delta x_i}$$

These partial derivatives can be collected in a row vector to give the gradient of f

$$\nabla f = \text{grad}\, f = \left[\frac{\partial f}{\partial x_1} \quad \frac{\partial f}{\partial x_2} \cdots \frac{\partial f}{\partial x_i} \cdots \frac{\partial f}{\partial x_n} \right] \in \mathbb{R}^{1 \times n} \tag{4.15}$$

The row vector (4.15), which is a vector-valued function of f, is also called the ***Jacobian***.

Example 4.3.3: Calculating the gradient of the function

$$f(x_1, x_2, x_3) = x_1^2 x_2 x_3 + x_1 x_2^2 x_3 + x_1 x_2 x_3^2$$

The partial derivatives of f are

$$\frac{\partial f(x_1, x_2, x_3)}{\partial x_1} = 2x_1 x_2 x_3 + x_2^2 x_3 + x_2 x_3^2$$

$$\frac{\partial f(x_1, x_2, x_3)}{\partial x_2} = x_1^2 x_3 + 2x_1 x_2 x_3 + x_1 x_3^2$$

$$\frac{\partial f(x_1, x_2, x_3)}{\partial x_3} = x_1^2 x_2 + x_1 x_2^2 + 2x_1 x_2 x_3$$

Hence, the gradient is

$$\nabla f = \text{grad}\, f = \left[\frac{\partial f(x_1, x_2, x_3)}{\partial x_1} \quad \frac{\partial f(x_1, x_2, x_3)}{\partial x_2} \quad \frac{\partial f(x_1, x_2, x_3)}{\partial x_2} \right]$$

$$= \left[2x_1 x_2 x_3 + x_2^2 x_3 + x_2 x_3^2 \quad x_1^2 x_3 + 2x_1 x_2 x_3 + x_1 x_3^2 \quad x_1^2 x_2 + x_1 x_2^2 + 2x_1 x_2 x_3 \right] \in \mathbb{R}^{1\text{x}3}$$

In (4.6) and (4.7), we have seen how the chain rule of differentiation is applied to univariate functions. In the case of a two-variable differential function $f(x_1, \text{x}_2)$, where each of the variables x_1 and x_2 is a differentiable function of t, the chain rule can be applied to compute the gradient of f with respect to t as

$$\nabla_t f = \text{grad}_t\, f = \frac{\partial f}{\partial x_1}\frac{\partial x_1}{\partial t} + \frac{\partial f}{\partial x_2}\frac{\partial x_2}{\partial t} \tag{4.16}$$

or, as a product of two vectors

$$\nabla_t f = \left[\frac{\partial f}{\partial x_1} \quad \frac{\partial f}{\partial x_2} \right] \begin{bmatrix} \dfrac{\partial x_1}{\partial t} \\ \dfrac{\partial x_2}{\partial t} \end{bmatrix} \tag{4.17}$$

On the other hand, if $f(x_1, x_2)$ is a two-variable function and each of x_1 and x_2 is a two-variable function, say, $x_1(s, t)$ and $x_2(s, t)$, the partial derivatives obtained by the application of the chain rule are

$$\frac{\partial f}{\partial s} = \frac{\partial f}{\partial x_1}\frac{\partial x_1}{\partial s} + \frac{\partial f}{\partial x_2}\frac{\partial x_2}{\partial s} \tag{4.18}$$

$$\frac{\partial f}{\partial t} = \frac{\partial f}{\partial x_1}\frac{\partial x_1}{\partial t} + \frac{\partial f}{\partial x_2}\frac{\partial x_2}{\partial t} \tag{4.19}$$

so that the gradient is obtained by matrix multiplication

$$\nabla_{s,t} = \frac{\partial f}{\partial \mathbf{x}}\frac{\partial \mathbf{x}}{\partial (s, t)} = \begin{bmatrix} \frac{\partial f}{\partial x_1} & \frac{\partial f}{\partial x_2} \end{bmatrix} \begin{bmatrix} \frac{\partial x_1}{\partial s} & \frac{\partial x_1}{\partial t} \\ \frac{\partial x_2}{\partial s} & \frac{\partial x_2}{\partial t} \end{bmatrix} \tag{4.20}$$

A number of concepts in physics, such as electromagnetic forces, which have already found applications in molecular structural biology, are aptly described using a mathematical concept called vector fields. It must be remembered that these forces are not just one single vector – both magnitude and direction may vary with the position in 3D space and even with time.

A ***vector-valued function***, or ***vector function***, is essentially a function whose domain is a set of real numbers and codomain or range is a set of vectors. A vector-valued function, $\mathbf{F}(x, y, z)$: $\mathbb{R}^3 \rightarrow \mathbb{R}^3$, which assigns to each point (x, y, z) a vector $\mathbf{F}(x, y, z)$, defines a ***vector field***. The general form of a vector field in $\mathbb{R}^3$ is given by

$$\mathbf{F}(x,y,z) = F_x(x,y,z)\mathbf{i} + F_y(x,y,z)\mathbf{j} + F_z(x,y,z)\mathbf{k}$$

where F_x, F_y and F_z are scalar-valued functions.

The concept of vector-valued functions can be generalized as $\mathbf{F}$: $\mathbb{R}^n \rightarrow \mathbb{R}^m$, where $n \geq 1$ and $m \geq 1$. So, for a vector $\mathbf{x} \in \mathbb{R}^n$, which can be represented as a column

$$\mathbf{x} = \begin{bmatrix} x_1 \\ x_2 \\ \vdots \\ x_n \end{bmatrix}$$

the corresponding vector of function values is

$$\mathbf{F}(\mathbf{x}) = \begin{bmatrix} F_1(\mathbf{x}) \\ F_2(\mathbf{x}) \\ \vdots \\ F_m(\mathbf{x}) \end{bmatrix} \in \mathbb{R}^m$$

The partial derivative of a vector-valued function $\mathbf{F}: \mathbb{R}^n \rightarrow \mathbb{R}^m$ with respect to $x_i \in \mathbb{R}$, $i = 1, 2, \ldots, n$, is given as the vector

$$\frac{\partial \mathbf{F}}{\partial x_i} = \begin{bmatrix} \frac{\partial F_1}{\partial x_i} \\ \frac{\partial F_2}{\partial x_i} \\ \vdots \\ \frac{\partial F_m}{\partial x_i} \end{bmatrix} = \begin{bmatrix} \lim_{h\to 0} \frac{F_1(x_1, x_2, \cdots, x_{i-1}, x_i + \mathrm{h}, x_{i+1}, \cdots, x_n) - F_1(\mathbf{x})}{h} \\ \lim_{h\to 0} \frac{F_2(x_1, x_2, \cdots, x_{i-1}, x_i + \mathrm{h}, x_{i+1}, \cdots, x_n) - F_2(\mathbf{x})}{h} \\ \vdots \\ \lim_{h\to 0} \frac{F_m(x_1, x_2, \cdots, x_{i-1}, x_i + \mathrm{h}, x_{i+1}, \cdots, x_n) - F_m(\mathbf{x})}{h} \end{bmatrix} \mathbb{R}^m \quad (4.21)$$

Evidently, in (4.21). every partial derivative $\partial \mathbf{F}/\partial x_i$ is a column vector. Therefore the gradient of $\mathbf{F}: \mathbb{R}^n \rightarrow \mathbb{R}^m$ with respect to $\mathbf{x}$ is obtained by arranging the partial derivatives in the following manner:

$$\nabla_{\mathbf{x}} \mathbf{F}(\mathbf{x}) = \begin{bmatrix} \frac{\partial \mathbf{F}(\mathbf{x})}{\partial x_1} & \cdots & \frac{\partial \mathbf{F}(\mathbf{x})}{\partial x_n} \end{bmatrix}$$

$$= \begin{bmatrix} \frac{\partial F_1(\mathbf{x})}{\partial x_1} & \cdots & \frac{\partial F_1(\mathbf{x})}{\partial x_n} \\ \vdots & \vdots & \vdots \\ \frac{\partial F_m(\mathbf{x})}{\partial x_1} & \cdots & \frac{\partial F_m(\mathbf{x})}{\partial x_n} \end{bmatrix} \quad (4.22)$$

and we have the formal definition of a ***Jacobian*** as follows.

Definition 4.3.1: The ***Jacobian*** is an $m \times n$ matrix which is the collection of all first-order partial derivatives of a vector-valued function $\mathbf{F}: \mathbb{R}^n \rightarrow \mathbb{R}^m$, and expressed as

$$\mathbf{J} = \nabla_{\mathbf{x}} \mathbf{F}(\mathbf{x}) = \begin{bmatrix} \frac{\partial F_1(\mathbf{x})}{\partial x_1} & \cdots & \frac{\partial F_1(\mathbf{x})}{\partial x_n} \\ \vdots & & \vdots \\ \frac{\partial F_m(\mathbf{x})}{\partial x_1} & \cdots & \frac{\partial F_m(\mathbf{x})}{\partial x_n} \end{bmatrix} \quad (4.23)$$

$$\text{where } \mathbf{x} = \begin{bmatrix} x_1 \\ x_2 \\ \vdots \\ x_n \end{bmatrix}, J(i,j) = \frac{\partial F_i(\mathbf{x})}{\partial x_j}$$

As it can be seen, (4.15) is a special case of (4.23) where $m = 1$.

Example 4.3.4: Let F: $\mathbb{R}^2 \rightarrow \mathbb{R}^3$ be given by

$$\mathbf{F}(\mathbf{x}) = \begin{bmatrix} F_1(\mathbf{x}) \\ F_2(\mathbf{x}) \\ F_3(\mathbf{x}) \end{bmatrix} = \begin{bmatrix} 4x_1 + x_2^2 \\ 2x_1^2 + 3x_2 \\ 3x_1 + e^{x_2} \end{bmatrix}$$

Then,

$$\nabla_{\mathbf{x}}\mathbf{F}(\mathbf{x}) = \begin{bmatrix} \frac{\partial F_1(\mathbf{x})}{\partial x_1} & \frac{\partial F_1(\mathbf{x})}{\partial x_2} \\ \frac{\partial F_2(\mathbf{x})}{\partial x_1} & \frac{\partial F_2(\mathbf{x})}{\partial x_2} \\ \frac{\partial F_3(\mathbf{x})}{\partial x_1} & \frac{\partial F_3(\mathbf{x})}{\partial x_2} \end{bmatrix} = \begin{bmatrix} 4 & 2x_2 \\ 4x_1 & 3 \\ 3 & e^{x_2} \end{bmatrix}$$

Example 4.3.5: Let us consider the vector-valued function **F**: $\mathbb{R}^2 \rightarrow \mathbb{R}^2$

$$\mathbf{F}(\mathbf{x}) = \begin{bmatrix} F_1(\mathbf{x}) \\ F_2(\mathbf{x}) \end{bmatrix} = \begin{bmatrix} (x_1 - 1)^2 + x_2^2 + 4 \\ x_1^2 + x_2 + 1 \end{bmatrix}$$

Here,

$$\frac{\partial F_1(\mathbf{x})}{\partial x_1} = 2(x_1 - 1), \quad \frac{\partial F_1(\mathbf{x})}{\partial x_2} = 2x_2, \quad \frac{\partial F_2(\mathbf{x})}{\partial x_1} = 2x_1, \quad \frac{\partial F_2(\mathbf{x})}{\partial x_2} = 1$$

Hence, the Jacobian is

$$\mathbf{J} = \begin{bmatrix} 2(x_1 - 1) & 2x_2 \\ 2x_1 & 1 \end{bmatrix}$$

Vector-valued function expressed as a matrix transformation

Let vector $\mathbf{x} \in \mathbb{R}^n$ and the vector-valued function $\mathbf{F}(\mathbf{x}) = \mathbf{A}\mathbf{x} \in \mathbb{R}^m$, where the matrix $\mathbf{A} \in \mathbb{R}^{m \times n}$, and compute the Jacobian. Now, since

$$F_i(\mathbf{x}) = \sum_{j=1}^{n} A_{ij}x_j$$

the partial derivatives of $\mathbf{F}(\mathbf{x})$ with respect to x_j are

$$\frac{\partial F_i(\mathbf{x})}{\partial x_j} = A_{ij}$$

The partial derivatives can be collected in the Jacobian as

$$\mathbf{J} = \begin{bmatrix} \frac{\partial F_1(\mathbf{x})}{\partial x_1} & \cdots & \frac{\partial F_1(\mathbf{x})}{\partial x_n} \\ \vdots & & \vdots \\ \frac{\partial F_m(\mathbf{x})}{\partial x_1} & \cdots & \frac{\partial F_m(\mathbf{x})}{\partial x_n} \end{bmatrix} = \begin{bmatrix} A_{11} & \cdots & A_{1n} \\ \vdots & & \vdots \\ A_{m1} & \cdots & A_{mn} \end{bmatrix} = \mathbf{A} \in \mathbb{R}^{m \times n} \tag{4.24}$$

4.4 Gradients of matrices

The derivative of a matrix with respect to a scalar variable is quite straightforward – it is a matrix of the same size containing the derivatives of the individual elements. Let a matrix $\mathbf{A} \in \mathbb{R}^{m \times n}$ contain the elements $A_{ij}(x)$, $i = 1, 2, \ldots, m$ and $j = 1, 2, \ldots, n$, then the derivative of $\mathbf{A}$ with respect of x is

$$\frac{d\mathbf{A}}{dx} = \begin{bmatrix} \frac{dA_{11}}{dx} & \cdots & \frac{dA_{1n}}{dx} \\ \vdots & & \vdots \\ \frac{dA_{m1}}{dx} & \cdots & \frac{dA_{mn}}{dx} \end{bmatrix} \tag{4.25}$$

The operation of differentiation of a matrix with respect to a vector is a logical extension of the above operation carried out with respect to a scalar. Let the matrix $\mathbf{A} \in \mathbb{R}^{m \times n}$ be a function of the vector $\mathbf{x} \in \mathbb{R}^s$. $\mathbf{A}$ can be represented as a column vector

$$\mathbf{A}(\mathbf{x}) = \begin{bmatrix} \mathbf{A}_1(\mathbf{x}) \\ \vdots \\ \mathbf{A}_m(\mathbf{x}) \end{bmatrix} \tag{4.26}$$

where each $\mathbf{A}_i$ is a row vector

$$\mathbf{A}_i(\mathbf{x}) = [A_{i1}(\mathbf{x}) \quad \cdots \quad A_{in}(\mathbf{x})] \in \mathbb{R}^s \tag{4.27}$$

so that the gradient of $\mathbf{A} \in \mathbb{R}^{m \times n}$ with respect to $\mathbf{x} \in \mathbb{R}^s$ is given by

$$\nabla_{\mathbf{x}}\mathbf{A}(\mathbf{x}) = \begin{bmatrix} \nabla_{\mathbf{x}}\mathbf{A}_1(\mathbf{x}) \\ \vdots \\ \nabla_{\mathbf{x}}\mathbf{A}_m(\mathbf{x}) \end{bmatrix} \tag{4.28}$$

where the gradient of $\mathbf{A}_i(\mathbf{x}) \in \mathbb{R}^s$ is given by the Jacobian

$$\mathbf{J}_i = \nabla_{\mathbf{x}}\mathbf{A}_i(\mathbf{x}) = \begin{bmatrix} \frac{\partial A_{i1}(\mathbf{x})}{\partial x_1} & \cdots & \frac{\partial A_{i1}(\mathbf{x})}{\partial x_s} \\ \vdots & & \vdots \\ \frac{\partial A_{in}(\mathbf{x})}{\partial x_1} & \cdots & \frac{\partial A_{in}(\mathbf{x})}{\partial x_s} \end{bmatrix} \tag{4.29}$$

Evidently, (4.28) is a three-dimensional array of partial derivatives and said to be a ***third-order tensor*** whose entries are given by $J_{ijk} = \partial A_{ij} / \partial x_k$, $i = 1, \ldots, m$; $j = 1, \ldots, n$; and $k = 1, \ldots, s$.

The operation can be further extended to compute the gradient of a matrix $\mathbf{A} \in \mathbb{R}^{m \times n}$ with respect to a matrix $\mathbf{B} \in \mathbb{R}^{s \times t}$. The resulting Jacobian (gradient) $\mathbf{J} \in \mathbb{R}^{(m \times n) \times (s \times t)}$ will be a four-dimensional array of partial derivatives, that is, a ***fourth-order tensor***, with entries $J_{ijkl} = \partial A_{ij}/\partial B_{kl}$, $i = 1, \ldots, m$; $j = 1, \ldots, n$; $k = 1, \ldots, s$; and $l = 1, \ldots, t$.

Example 4.4.1: Let us consider the matrix $\mathbf{A}(\mathbf{x}) \in \mathbb{R}^{4 \times 2}$, $\mathbf{x} \in \mathbb{R}^3$,

$$\mathbf{A}(\mathbf{x}) = \begin{bmatrix} 2x_1^2 + 3x_2 + x_3 & x_1 + x_2 - 2x_3 \\ x_1 + x_2^2 - 2x_3^2 & -x_1^2 - x_2 + x_3^2 \\ -3x_1^2 + x_2^2 + x_3^2 & 4x_1 + x_2^2 - 2x_3 \\ x_1^2 - 2x_2 - x_3 & 2x_1 + x_2 - 3x_3 \end{bmatrix}$$

and calculate some of the entries of the Jacobian $\mathbf{J} \in \mathbb{R}^{4 \times 2 \times 3}$. We can find

$$J_{213} = \frac{\partial A_{21}}{\partial x_3} = -4x_3$$

$$J_{321} = \frac{\partial A_{32}}{\partial x_1} = 4$$

$$J_{222} = \frac{\partial A_{22}}{\partial x_2} = -1$$

4.5 Higher-order derivates – Hessian

We have seen earlier in Theorem 4.2.1 how the nature of the critical point, in the case of a bivariate function $f(x, y)$, is determined by the value of $f_{xx}f_{yy} - f_{xy}^2$. The partial derivatives can be collected in a matrix

$$\mathbf{M} = \begin{bmatrix} f_{xx} & f_{xy} \\ f_{yx} & f_{yy} \end{bmatrix}$$

(it is to be noted that $f_{xy} = f_{yx}$), where $\mathbf{M}$ is called the Hessian matrix of $f(x, y)$, and the criteria as laid down in Theorem 4.2.1 can be restated as follows:

1. If det $[\mathbf{M}] < 0$ at (x_0, y_0), then (x_0, y_0) is a saddle point and
2. If det $[\mathbf{M}] > 0$ at (x_0, y_0), then (x_0, y_0) is either a maximum or a minimum.

The concept of ***Hessian*** can be extended to a scalar-valued function f: $\mathbb{R}^n \rightarrow \mathbb{R}$ of a vector $\mathbf{x} \in \mathrm{R}^n$. The ***Hessian matrix*** of $f(\mathbf{x})$ is

$$\mathbf{H} = \nabla_{\mathbf{x}}^2 f(\mathbf{x}) = \begin{bmatrix} \frac{\partial^2 f}{\partial x_1^2} & \frac{\partial^2 f}{\partial x_1 \partial x_2} & \cdots & \frac{\partial^2 f}{\partial x_1 \partial x_n} \\ \frac{\partial^2 f}{\partial x_2 \partial x_1} & \frac{\partial^2 f}{\partial x_2^2} & \cdots & \frac{\partial^2 f}{\partial x_2 \partial x_n} \\ \vdots & \vdots & \vdots & \vdots \\ \frac{\partial^2 f}{\partial x_n \partial x_1} & \frac{\partial^2 f}{\partial x_n \partial x_2} & \cdots & \frac{\partial^2 f}{\partial x_n^2} \end{bmatrix} \tag{4.30}$$

which is a symmetric and $n \times n$ matrix.

Extending the concept further to a vector field

$$\mathbf{F}(\mathbf{x}) = \begin{bmatrix} F_1(\mathbf{x}) \\ F_2(\mathbf{x}) \\ \cdot \\ \cdot \\ \cdot \\ F_m(\mathbf{x}) \end{bmatrix} \in \mathbb{R}^m$$

so that $\mathbf{F}$: $\mathbb{R}^n \rightarrow \mathbb{R}^m$, we can find that Hessian is an $(m \times n \times n)$ tensor.

Example 4.5.1: Let us calculate the Hessian of

$$f(x_1, x_2, x_3) = x_1^2 + x_2^2 + x_3^2 + 2x_1x_2x_3 + 5$$

$$\frac{\partial f}{\partial x_1} = 2x_1 + 2x_2x_3; \quad \frac{\partial^2 f}{\partial x_1^2} = 2; \quad \frac{\partial^2 f}{\partial x_2 \partial x_1} = 2x_3; \quad \frac{\partial^2 f}{\partial x_3 \partial x_1} = 2x_2$$

$$\frac{\partial f}{\partial x_2} = 2x_2 + 2x_1x_3; \quad \frac{\partial^2 f}{\partial x_1 \partial x_2} = 2x_3; \quad \frac{\partial^2 f}{\partial x_2^2} = 2; \quad \frac{\partial^2 f}{\partial x_3 \partial x_2} = 2x_1$$

$$\frac{\partial f}{\partial x_3} = 2x_3 + 2x_1x_2; \quad \frac{\partial^2 f}{\partial x_1 \partial x_3} = 2x_2; \quad \frac{\partial^2 f}{\partial x_2 \partial x_3} = 2x_1; \quad \frac{\partial^2 f}{\partial x_3^2} = 2$$

Therefore the Hessian is

$$\mathbf{H} = \begin{bmatrix} \frac{\partial^2 f}{\partial x_1^2} & \frac{\partial^2 f}{\partial x_1 \partial x_2} & \frac{\partial^2 f}{\partial x_1 \partial x_3} \\ \frac{\partial^2 f}{\partial x_2 \partial x_1} & \frac{\partial^2 f}{\partial x_2^2} & \frac{\partial^2 f}{\partial x_2 \partial x_3} \\ \frac{\partial^2 f}{\partial x_3 \partial x_1} & \frac{\partial^2 f}{\partial x_3 \partial x_2} & \frac{\partial^2 f}{\partial x_3^2} \end{bmatrix} = \begin{bmatrix} 2 & 2x_3 & 2x_2 \\ 2x_3 & 2 & 2x_1 \\ 2x_2 & 2x_1 & 2 \end{bmatrix}$$

Optimization

In Section 4.1, we have seen how derivatives of first order and second order are used to optimize (determine the critical points and their nature of) a univariate function $f(x)$. In Section 4.2, optimization of a bivariate function $f(x, y)$ involved partial derivatives of first order and second order. Now, we shall be concerned with the optimization problem for a function $f(\mathbf{x})$ where $\mathbf{x} \in \mathbb{R}^n$. In Chapter 1, we have introduced the concept of definiteness of a square matrix. Here, we enunciate the rules for optimization based on the definiteness of the Hessian.

For a scalar-valued function $f(\mathbf{x})$, $\mathbf{x} \in \mathbb{R}^n$, the gradient can be expressed as

$$\nabla f(\mathbf{x}) = \begin{bmatrix} \dfrac{\partial f(\mathbf{x})}{\partial x_1} \\ \dfrac{\partial f(\mathbf{x})}{\partial x_2} \\ \vdots \\ \dfrac{\partial f(\mathbf{x})}{\partial x_n} \end{bmatrix} \tag{4.31}$$

$\mathbf{x}_0$ is a critical point if and only if $\nabla f(\mathbf{x}) = \mathbf{0}$, that is,

$$\nabla f(\mathbf{x}_0) = \begin{bmatrix} \dfrac{\partial f(\mathbf{x}_0)}{\partial x_1} \\ \dfrac{\partial f(\mathbf{x}_0)}{\partial x_2} \\ \vdots \\ \dfrac{\partial f(\mathbf{x}_0)}{\partial x_n} \end{bmatrix} = \mathbf{0} \tag{4.32}$$

To determine the nature of the critical point, we have to consider the Hessian of $f(\mathbf{x})$

$$\mathbf{H}(\mathbf{x}) = \begin{bmatrix} \dfrac{\partial^2 f(\mathbf{x})}{\partial x_1^2} & \dfrac{\partial^2 f(\mathbf{x})}{\partial x_1 \partial x_2} & \cdots & \dfrac{\partial^2 f(\mathbf{x})}{\partial x_1 \partial x_n} \\ \dfrac{\partial^2 f(\mathbf{x})}{\partial x_2 \partial x_1} & \dfrac{\partial^2 f(\mathbf{x})}{\partial x_2^2} & \cdots & \dfrac{\partial^2 f(\mathbf{x})}{\partial x_2 \partial x_n} \\ \vdots & \vdots & \vdots & \vdots \\ \dfrac{\partial^2 f(\mathbf{x})}{\partial x_n \partial x_1} & \dfrac{\partial^2 f(\mathbf{x})}{\partial x_n \partial x_2} & \cdots & \dfrac{\partial^2 f(\mathbf{x})}{\partial x_n^2} \end{bmatrix}$$

Then, the second-order conditions for optimization are as follows:

a. If $\mathbf{H}(\mathbf{x}_0)$ is positive definite, $f(\mathbf{x}_0)$ is a local minimum;
b. If $\mathbf{H}(\mathbf{x}_0)$ is negative definite, $f(\mathbf{x}_0)$ is a local maximum;
c. If $\mathbf{H}(\mathbf{x}_0)$ is indefinite, $f(\mathbf{x}_0)$ is a saddle point.

Example 4.5.2: Let us find the critical points and the nature of the critical points of the function given in Example 4.5.1.

Setting $\nabla f(x_1, x_2, x_3) = \mathbf{0}$, we have

$$\nabla f(x_1,\ x_2,\ x_3) = \begin{bmatrix} \dfrac{\partial f}{\partial x_1} \\ \dfrac{\partial f}{\partial x_2} \\ \dfrac{\partial f}{\partial x_3} \end{bmatrix} = \begin{bmatrix} 2x_1 + 2x_2x_3 \\ 2x_2 + 2x_1x_3 \\ 2x_3 + 2x_1x_2 \end{bmatrix} = \begin{bmatrix} 0 \\ 0 \\ 0 \end{bmatrix}$$

from which we obtain

$$x_1 = -x_2x_3, \qquad x_2 = -x_1x_3, \qquad x_3 = -x_1x_2$$

We have five solutions to these equations (for the critical points)

$$(0,0,0),\ (1,1,-1),\ (1,-1,1).(-1,1,1),\ (-1,-1,-1)$$

The Hessian has already been calculated in Example 4.5.1 as

$$\mathbf{H} = \begin{bmatrix} 2 & 2x_3 & 2x_2 \\ 2x_3 & 2 & 2x_1 \\ 2x_2 & 2x_1 & 2 \end{bmatrix}$$

At (0, 0, 0)

$$\mathbf{H}(0,\ 0,\ 0) = \begin{bmatrix} 2 & 0 & 0 \\ 0 & 2 & 0 \\ 0 & 0 & 2 \end{bmatrix}$$

and the values of the leading principal minors are

$$\det[2] = 2, \quad \det\begin{bmatrix} 2 & 0 \\ 0 & 2 \end{bmatrix} = 4, \quad \det\begin{bmatrix} 2 & 0 & 0 \\ 0 & 2 & 0 \\ 0 & 0 & 2 \end{bmatrix} = 8$$

which are all positive. Therefore **H**(0, 0, 0) is positive definite in nature and $f(0,\ 0,\ 0) = 5$ is a local minimum.

At all the other points we can find the values of the leading principal minors to be 2, 0, and -32, so that the Hessian is indefinite in nature. Critical points (1, 1, -1), (1, -1, 1), (-1, 1, 1), and (-1, -1, -1) are saddle points.

4.6 Linearization and multivariate Taylor series

It has been mentioned in Chapter 2 that since nonlinear dynamical equations are relatively difficult to solve, nonlinear systems are usually approximated by linear equations (a process called linearization). This involves computation of functions which are "close" to the exact solution. For a function $f(x)$ to approximate a function $g(x)$ near x_0 it is not sufficient that $f(x_0) = g(x_0)$, although it is a necessary condition. It should be supplemented at least by the condition that $f'(x_0) = g'(x_0)$. In such case one can say that the function $f(x)$ is a first-order (or linear) approximation to the function $g(x)$ at $x = x_0$. The most basic local approximation can be worked out based on Taylor's series which we are going to discuss now.

In Chapter 1, we had introduced Taylor's series for a univariate function $f(x)$. Recalling it here

$$f(x) = f(x_0) + \frac{f'(x_0)}{1!}(x - x_0) + \frac{f''(x_0)}{2!}(x-x_0)^2 + \cdots\cdots$$
$$= \sum_{n=0}^{\infty} \frac{f^{(n)}(x_0)}{n!}(x-x_0)^n \tag{4.33}$$

If x_0 is close to x such that $(x-x_0)$ is very small, the contributions of the second order and higher terms towards the sum will be insignificant, and $f(x)$ can be approximated by the first and second terms:

$$f(x) = f(x_0) + f'(x_0)(x - x_0) \tag{4.34}$$

This approximation is linear in x; hence, it is known as linearization.

Example 4.6.1: Let us linearize $f(x) = 2x^2-3x+5$ at $x = 2$

Here, $x_0 = 2$. Therefore,

$$f(2) = 7$$
$$f'(2) = 5$$
$$f(x) = 7 + 5 * (x - 2) = 5x - 3$$

Clearly, $f(x) = 5x - 3$ is linear in x.

Now, the concept of linearization will be extended to multivariate functions. Let us first consider a bivariate function $f(x, y)$. Then, the bivariate Taylor expansion about $(x, y) = (x_0, y_0)$ is given by

$$f(x, y) = f(x_0, y_0) + \left.\frac{\partial f}{\partial x}\right|_{x_0,y_0}(x - x_0) + \left.\frac{\partial f}{\partial y}\right|_{x_0,y_0}(y - y_0) + \frac{1}{2}\left.\frac{\partial^2 f}{\partial x^2}\right|_{x_0,y_0}(x-x_0)^2$$
$$+ \left.\frac{\partial^2 f}{\partial x \partial y}\right|_{x_0,y_0}(x - x_0)(y - y_0) + \frac{1}{2}\left.\frac{\partial^2 f}{\partial y^2}\right|_{x_0,y_0}(y-y_0)^2 + \cdot \quad \cdot \quad \cdot \quad \cdot \tag{4.35}$$

Writing $\mathbf{h}$ as a column vector

$$\mathbf{h} = \begin{bmatrix}(x - x_0)\\(y - y_0)\end{bmatrix}$$

we have

$$\nabla f\big|_{x_0,y_0}\cdot \mathbf{h} = \begin{bmatrix}\frac{\partial f}{\partial x} & \frac{\partial f}{\partial y}\end{bmatrix}_{x_0,y_0}\begin{bmatrix}(x-x_0)\\(y-y_0)\end{bmatrix} = \frac{\partial f}{\partial x}\bigg|_{x_0,y_0}(x-x_0)+\frac{\partial f}{\partial y}\bigg|_{x_0,y_0}(y-y_0) \tag{4.36}$$

The second-order partial derivatives can be collected in a Hessian

$$\mathbf{H}_f = \begin{bmatrix}\frac{\partial^2 f}{\partial x^2} & \frac{\partial^2 f}{\partial x\partial y}\\ \frac{\partial^2 f}{\partial y\partial x} & \frac{\partial^2 f}{\partial y^2}\end{bmatrix}$$

so that we have

$$\mathbf{h}^T\mathbf{H}_f\big|_{x_0,y_0}\mathbf{h} = [(x-x_0)\quad (y-y_0)]\left.\begin{bmatrix}\frac{\partial^2 f}{\partial x^2} & \frac{\partial^2 f}{\partial x\partial y}\\ \frac{\partial^2 f}{\partial y\partial x} & \frac{\partial^2 f}{\partial y^2}\end{bmatrix}\right|_{x_0,y_0}\begin{bmatrix}(x-x_0)\\(y-y_0)\end{bmatrix}$$

$$= \frac{\partial^2 f}{\partial x^2}\bigg|_{x_0,y_0}(x-x_0)^2 + 2\frac{\partial^2 f}{\partial x\partial y}\bigg|_{x_0,y_0}(x-x_0)(y-y_0)+\frac{\partial^2 f}{\partial y^2}\bigg|_{x_0,y_0}(y-y_0)^2 \tag{4.37}$$

and (4.35) can be rewritten as

$$f(x, y) = f(x_0, y_0) + \nabla f\big|_{x_0,y_0}\cdot \mathbf{h} + \frac{1}{2}\mathbf{h}^T\mathbf{H}_f\big|_{x_0,y_0}\mathbf{h} + \cdots\cdot \tag{4.38}$$

Ignoring second order and higher terms in (4.38) we have the linearized bivariate function about $(x, y) = (x_0, y_0)$

$$f(x, y) = f(x_0, y_0) + \frac{\partial f}{\partial x}\bigg|_{x_0,y_0}(x-x_0)+\frac{\partial f}{\partial y}\bigg|_{x_0,y_0}(y-y_0) \tag{4.39}$$

Example 4.6.2: The bivariate function

$$f(\mathbf{x}) = f(x_1, x_2) = 2x_1^2 + x_2^2 - 3x_1 + 2x_2 - 3$$

is to be linearized at (1, 2).

We can calculate

$$f(1,2) = 4$$

$$\frac{\partial f}{\partial x_1}\bigg|_{1,\,2}(x_1-1) = 1 * (x_1-1)$$

$$\frac{\partial f}{\partial x_2}\bigg|_{1,\,2}(x_2-2) = 6 * (x_2-2)$$

Hence, we have the linearized function

$$f(\mathbf{x}) = f(x_1, x_2) = x_1 + 6x_2 - 9$$

(4.38) can now be extended to a scalar-valued function $f(\mathbf{x})$, $\mathbf{x} \in \mathbb{R}^n$ and written as

$$f(\mathbf{x}) = f(\mathbf{x}_0) + \nabla f\big|_{\mathbf{x}_0} \cdot \mathbf{h} + \frac{1}{2}\mathbf{h}^T \mathbf{H}_f\big|_{\mathbf{x}_0}\mathbf{h} + \cdots \tag{4.40}$$

where

$$\mathbf{h} = \begin{bmatrix} x_1 - x_{1_0} \\ x_2 - x_{2_0} \\ \vdots \\ x_n - x_{n_0} \end{bmatrix}$$

and the Hessian

$$\mathbf{H}_f = \begin{bmatrix} \dfrac{\partial^2 f}{\partial x_1^2} & \dfrac{\partial^2 f}{\partial x_1 \partial x_2} & \cdots & \dfrac{\partial^2 f}{\partial x_1 \partial x_n} \\ \dfrac{\partial^2 f}{\partial x_2 \partial x_1} & \dfrac{\partial^2 f}{\partial x_2^2} & \cdots & \dfrac{\partial^2 f}{\partial x_2 \partial x_n} \\ \vdots & \vdots & \vdots & \vdots \\ \dfrac{\partial^2 f}{\partial x_n \partial x_1} & \dfrac{\partial^2 f}{\partial x_n \partial x_2} & \cdots & \dfrac{\partial^2 f}{\partial x_n^2} \end{bmatrix}$$

Exercise 4

1. Find the critical points of the following functions and determine their nature:
   a. $f(x) = x^3 - x$
   b. $f(x) = 2x^3 + 3x^2 - 12x + 12$
   c. $f(x) = x^3 - x$
   d. $f(x) = x^3 - x^2 - x - 1$
   e. $f(x) = 2x + 4 \sin x$
2. Find and classify the critical points of the following functions:
   a. $f(x, y) = x^2 + y^2$
   b. $f(x, y) = x^3 + 3xy^2 - 2y^3 - 12x$
   c. $f(x, y) = x^2 + 4x - 6xy + 2y^3$
3. Calculate the gradient of the following functions:
   a. $f(x_1, x_2, x_3) = (x_1 - 1)(x_2 - 1)(x_3 - 1)$
   b. $f(x, y) = e^{xy} + \cos x$
   c. $f(x, y, z) = x\, e^{yz}$

4. Find the Jacobian of the following vector-valued functions:

   a.

$$\mathbf{F}(\mathbf{x}) = \begin{bmatrix} 2x_1 + \sqrt{x_2} \\ e^{x_1} + 3x_2 \\ x_1^2 + x_2^2 \end{bmatrix}$$

   b.

$$\mathbf{F}(\mathbf{x}) = \begin{bmatrix} \sqrt{x_1} + x_2 \\ x_1 + \sqrt{x_2} \end{bmatrix}$$

   c.

$$\mathbf{F}(\mathbf{x}) = \begin{bmatrix} 4x_1 + 2x_2 \\ x_1^2 + x_2^2 \\ \sin x_1 + \cos x_2 \end{bmatrix}$$

5. Consider the matrix $\mathbf{A}(\mathbf{x}) \in \mathbb{R}^{3\times 2}$, $\mathbf{x} \in \mathbb{R}^2$,

$$\mathbf{A}(\mathbf{x}) = \begin{bmatrix} x_1^2 + 2x_2 & x_1 + x_2^2 \\ 2x_1 + \sin x_2 & x_1^2 + \cos x_2 \\ x_1^2 x_2 & x_1 x_2^2 \end{bmatrix}$$

   Calculate the following entries of the Jacobian $\mathbf{J} \in \mathbb{R}^{3\times 2\times 2}$:

$$J_{212} \quad J_{121} \quad J_{311} \quad J_{112} \quad J_{321}$$

6. Compute the Hessian of the following functions:
   a. $f(x, y) = x^2 - y^2$
   b. $f(x, y, z) = xy^2z^3$
   c. $f(x, y) = \cos x + \sin(x + 2y)$
   d. $f(x, y) = e^x \sin y$
7. Linearize the following functions:
   a. $f(x, y) = x^3y^2 + x + y^3$ at (1, 1)
   b. $f(x, y) = \sin x + \cos y$ at (0, 0)
   c. $f(x, y) = e^{x+y}$ at (0, 0)
   d. $f(x, y) = 4x^3 - 6x^2y^3 + y^5$ at (1, 1)
   e. $f(x, y, z) = \sin x + \cos y + e^z$ at (0, 0, 0)

Further Reading

Deisenroth, M.P., Faisal, A.A., Ong, C.S., 2020. Mathematics for Machine Learning. Cambridge University Press.

Fong, F.T.-H., 2015. Multivariable and Vector Calculus: Lecture Notes for MATH 0200. Brown University.

Stewart, J., 2012. Multivariable Calculus. Brooks/Cole.

CHAPTER

Integral transform

5

X-ray crystallography of biomacromolecules has laid the foundation of molecular structural biology. As it is known, a crystal is a periodic layout of a motif in a lattice – the motif can be a single atom, a small molecule, a macromolecule, or a combination of two or more of these. Any repetitive and periodic arrangement of points or motifs is said to constitute a direct lattice (or a real lattice). In X-ray crystallography, the detailed atomic arrangement within a crystal is obtained by diffraction of X-rays. The diffraction by a lattice produces a lattice as a diffraction pattern (reciprocal lattice), which is in the reciprocal space. Nevertheless, the direct space of the crystal lattice and the reciprocal space of the diffraction pattern are related by the Fourier transform. The concept of the reciprocal lattice will be further discussed later in the chapter on X-ray crystallography (Chapter 7).

Fourier transform is the most widely used integral transform. In general, an integral transform can be conceived as follows. If we have a function $f(x)$ and another function $k(x, s)$ such that the product of $f(x)$ and $k(x, s)$ is integrable in the domain of x under consideration, then it is possible to form a function of s

$$F(s) = \int_a^b f(x)k(x,s)\,dx \tag{5.1}$$

where $k(x, s)$ is called the ***kernel***. It can be said that a function $f(x)$ has been "transformed" into a function $F(s)$ by means of an "integral transform." The necessity or usefulness of integral transform arises from the fact that sometimes in the original formulation (in the ***direct*** or ***physical space***), either a problem is relatively more difficult to solve or the desired information cannot be obtained. In such a situation, an effective strategy is to formulate and solve the problem in the transform space, following which the solution is transformed back to the direct space.

5.1 Fourier transform

We have seen in Chapter 1 that a function $f(x)$ with period L can be expanded in a complex Fourier series

$$f(x) = \sum_{n=-\infty}^{\infty} c_n e^{i2\pi nx/L} \tag{5.2}$$

Mathematical Approaches to Molecular Structural Biology. DOI: https://doi.org/10.1016/B978-0-323-90397-4.00004-4
© 2023 Elsevier Inc. All rights reserved.

where Fourier coefficients are given by

$$c_n = \frac{1}{L}\int_{-L/2}^{L/2} f(x)\mathrm{e}^{-i2\pi nx/L}dx \tag{5.3}$$

where the period has been chosen to be $[-L/2, L/2]$. We can generalize the Fourier series by including in it functions defined over an infinite interval and no apparent periodicity. A nonperiodic function can be considered as a periodic function where $L\to\infty$. So, even if $f(x)$ is not periodic, it can be defined as

$$f_L(x) = \sum_{n=-\infty}^{\infty} c_n\, e^{i2\pi nx/L} \tag{5.4}$$

with the same c_n as given in (5.3). By definition, $f_L(x)$ with period L behaves like $f(x)$ within $[-L/2, L/2]$. As $L\to\infty$, Δx becomes infinitesimally small and $f_L(x)$ converges to $f(x)$, and the Fourier sum goes over to Fourier integral. This generalization leads to the definition of ***Fourier (integral) transform***.

Definition 5.1.1 The ***Fourier transform*** of the function $f(x)$ is defined as

$$\tilde{f}(k) = \frac{1}{\sqrt{(2\pi)}}\int_{-\infty}^{\infty} f(x)\mathrm{e}^{-ikx}dx \tag{5.5}$$

provided $f(x)$ is absolutely integrable.

A function f, defined on $\mathbb{R}^n$, is absolutely integrable if its norm is

$$\|f\|_1 = \int_{\mathbb{R}^n} |f(x)|dx < \infty \tag{5.6}$$

In (5.5), e^{-ikx} is the kernel of the transform. Furthermore, it should be noted that for a transform to be useful, one should be able to "reverse" or "invert" its effect. In the case of the Fourier transform, the inversion formula does exist.

Definition 5.1.2 If $f(x)$ is also continuous, the ***inverse Fourier transform*** is defined as

$$f(x) = \frac{1}{\sqrt{(2\pi)}}\int_{-\infty}^{\infty} \tilde{f}(k)\mathrm{e}^{ikx}dk \tag{5.7}$$

Example 5.1.1: Let us consider a function

$$f(x) = \begin{cases} \sin x, & \text{for } |x| < 4\pi \\ 0, & \text{otherwise} \end{cases}$$

The Fourier transform of $f(x)$ is given by

$$\tilde{f}(k) = \frac{1}{\sqrt{(2\pi)}}\int_{-\infty}^{\infty} \sin x \, e^{-ikx} dx$$
$$= \frac{1}{\sqrt{(2\pi)}}\int_{-4\pi}^{4\pi} \sin x \, e^{-ikx} dx$$

Since $e^{i\phi} = \cos\phi + i\sin\phi$ and $e^{-i\phi} = \cos\phi - i\sin\phi$ we have

$$\tilde{f}(k) = \frac{1}{2i\sqrt{(2\pi)}}\int_{-4\pi}^{4\pi} (e^{ix} - e^{-ix})\, e^{-ikx} dx$$
$$= \frac{1}{2i\sqrt{(2\pi)}}\int_{-4\pi}^{4\pi} \left[e^{i(1-k)x} - e^{-i(1+k)x}\right] dx$$

On further calculation, we can find

$$\tilde{f}(k) = \frac{1}{2i\sqrt{(2\pi)}}\left[\frac{i}{1-k}\left(e^{-ikx} - e^{ikx}\right) + \frac{i}{1+k}\left(e^{-ikx} - e^{ikx}\right)\right]$$
$$= \frac{i\sin 4\pi k}{\sqrt{(2\pi)}}\left[\frac{1}{1-k} + \frac{1}{1+k}\right]$$
$$= \sqrt{\frac{2}{\pi}}\frac{i\sin 4\pi k}{1-k^2}$$

Example 5.1.2: An exponential decay function

$$f(t) = \begin{cases} 0 & \text{for } t < 0 \\ Ae^{-\lambda t} & \text{for } t \geq 0,\ \lambda > 0 \end{cases}$$

The Fourier transform of $f(t)$ is given by

$$\tilde{f}(\omega) = \frac{1}{\sqrt{(2\pi)}}\int_{-\infty}^{\infty} f(t) e^{-i\omega t} dt$$

From the given conditions, the integral can be divided into two parts, so that

$$\tilde{f}(\omega) = \frac{1}{\sqrt{(2\pi)}}\int_{-\infty}^{0} (0)\, e^{-i\omega t} dt + \frac{1}{\sqrt{(2\pi)}}\int_{0}^{\infty} Ae^{-\lambda t} e^{-i\omega t} dt$$
$$= \frac{A}{\sqrt{(2\pi)}}\int_{0}^{\infty} e^{-\lambda t} e^{-i\omega t} dt = \frac{A}{\sqrt{(2\pi)}}\int_{0}^{\infty} e^{-(\lambda + i\omega)t} dt$$
$$= \frac{A}{\sqrt{(2\pi)}}\left[-\frac{e^{-(\lambda+i\omega)t}}{\lambda + i\omega}\right]_{0}^{\infty}$$
$$= \frac{A}{\sqrt{2\pi}(\lambda + i\omega)}$$

Example 5.1.3: Fourier transform of a Gaussian function.

Gaussian functions, of the form e^{-at^2}, with a >0, are of considerable importance in mathematics and statistics. The Fourier transform of

$$f(t) = e^{-at^2}, \ a > 0$$

is given by

$$\tilde{f}(\omega) = \frac{1}{\sqrt{(2\pi)}} \int_{-\infty}^{\infty} e^{-at^2} e^{i\omega t} dt$$

The integral can be evaluated to show that

$$\tilde{f}(\omega) = \frac{1}{\sqrt{2a}} \exp\left(-\frac{\omega^2}{4a}\right) \quad (5.8)$$

5.2 Dirac delta function

The Dirac delta (δ) function is a very useful concept in mathematical analyses of physical phenomena particularly when localized entities are involved. For example, the delta function can represent the charge density associated with a point charge. As we shall see in the chapter on X-ray crystallography (Chapter 7) that scattering centers can be represented by the delta function. In one-dimensional space,

$$\delta(x) = 0, \ x \neq 0 \quad (5.9)$$

Apparently, $\delta(x)$ is not a function in the ordinary mathematical sense; it is zero everywhere except at a single point (taken as the origin), where it can be visualized as a very sharp narrow pulse. Nevertheless, it produces an integrated effect with a definite magnitude. Hence, the basic defining equation is

$$\int_{-\infty}^{\infty} \delta(x) f(x) dx = f(0) \quad (5.10)$$

It follows that for a continuous function $f(x)$

$$\int_a^b \delta(x) f(x) dx = \begin{cases} f(0) & \text{if } a < 0 < b \\ 0 & \text{otherwise} \end{cases} \quad (5.11)$$

If the same narrow pulse is at $x = c$, instead of $x = 0$, and $a \leq c \leq b$, then we have

$$\int_a^b \delta(x) f(x - c) dx = f(c) \quad (5.12)$$

From (5.12), we can further derive the following properties of the Dirac δ-function:

$$\delta(-x) = \delta(x) \tag{5.13}$$

$$\delta(ax) = \frac{1}{|a|}\delta(x) \tag{5.14}$$

$$x\delta(x) = 0 \tag{5.15}$$

Derivative of the δ-function

The derivative of the δ-function can be obtained by the application of integration by parts (Chapter 1)

$$\int_{-\infty}^{\infty} f(x)\delta'(x)dx = [f(x)\delta(x)]_{-\infty}^{\infty} - \int_{-\infty}^{\infty} f'(x)\delta(x)\,dx$$

The first term on the right-hand side disappears and we have

$$\int_{-\infty}^{\infty} f(x)\delta'(x)dx = -f'(0) \tag{5.16}$$

If the procedure is repeated n times, the general result for the nth-order derivative can be obtained

$$\int_{-\infty}^{\infty} f(x)\delta^{(n)}(x)dx = (-1)^n f^n(0) \tag{5.17}$$

As a special case, if $f(x) = x$, we have from (5.16)

$$\int_{-\infty}^{\infty} x\delta'(x)dx = -1 = -\int_{-\infty}^{\infty} \delta(x)dx$$

so that

$$x\delta'(x) = -\delta(x) \tag{5.18}$$

If $f(x) = x^n$, $f^{(n)}(x) = n!$, so that (5.18) can be generalized based on (5.17) as

$$\int_{-\infty}^{\infty} x^n\delta'(x)dx = (-1)^n n! = (-1)^n\, n!\int_{-\infty}^{\infty} \delta(x)dx$$

to give

$$x^n\delta^{(n)}(x) = (-1)^n n!\,\delta(x) \tag{5.19}$$

Based on (5.11) and (5.12), we have the following integrations.

Example 5.2.1:

$$\int_{-5}^{5} 10e^{x^2}\cos(x)\delta(x)dx = 10$$

Example 5.2.2:

$$\int_{0}^{10} 10e^{x^3}\cos(x)\delta(x-2)dx = 10e^8$$

Example 5.2.3:

$$\int_{-1}^{1} 10e^{x^3}\cos(x)\delta(x-2)dx = 0$$

The δ-function in 3D

The definition of the δ-function can be extended to higher-dimensional spaces. In particular, in three-dimensional vector space, (5.9) and (5.10) can be written as

$$\delta(\mathbf{r}) = 0, \quad \mathbf{r} \neq 0 \tag{5.20}$$

and

$$\int_{-\infty}^{\infty} \delta^3(\mathbf{r})f(\mathbf{r})d\mathbf{r} = f(\mathbf{0}) \tag{5.21}$$

where $\delta^3\mathbf{r}$ is the product of three one-dimensional δ-functions

$$\delta^3\mathbf{r} = \delta(x)\delta(y)\delta(z)$$

so that (5.21) can be written as

$$\int_{-\infty}^{\infty} \delta^3(\mathbf{r})f(\mathbf{r})d\mathbf{r} = \int_{-\infty}^{\infty}\int_{-\infty}^{\infty}\int_{-\infty}^{\infty} f(x,\ y,\ z)\delta(x)\delta(y)\delta(z)dx\,dy\,dz = f(0,\ 0,\ 0) \tag{5.22}$$

If the "peak" of the δ-function occurs at $\mathbf{r}_0 = (x_0,\ y_0,\ z_0)$, (5.22) becomes

$$\begin{aligned}
&\int_{-\infty}^{\infty} \delta^3(\mathbf{r}-\mathbf{r_0})f(\mathbf{r})d\mathbf{r} \\
&= \int_{-\infty}^{\infty}\int_{-\infty}^{\infty}\int_{-\infty}^{\infty} f(x,\ y,\ z)\delta(x-x_0)\delta(y-y_0)\delta(z-z_0)dx\,dy\,dz \\
&= f(x_0,\ y_0,\ z_0)
\end{aligned} \tag{5.23}$$

In such case, (5.14) becomes

$$\delta(a\mathbf{r}) = \frac{1}{|a|^3}\delta(\mathbf{r}) \tag{5.24}$$

where a is a scalar constant.

Fourier series and the δ-function

In Chapter 1, we have seen that for a function $f(x)$ of period L, the Fourier series can be written as

$$f(x) = \sum_{n=-\infty}^{\infty} c_n e^{i2\pi nx/L} \tag{5.25}$$

where the Fourier coefficients are given by

$$c_n = \frac{1}{L}\int_{-L/2}^{L/2} f(x)e^{-i2\pi nx/L}dx \tag{5.26}$$

To expand the δ-function in a Fourier series, we have

$$c_n = \frac{1}{L}\int_{-L/2}^{L/2} \delta(x)e^{-i2\pi nx/L}dx = \frac{1}{L} \tag{5.27}$$

so that

$$\delta(x) = \frac{1}{L}\sum_{n=-\infty}^{\infty} e^{i2\pi nx/L} \tag{5.28}$$

Fourier transform and the δ-function

Using (5.5), we can obtain the Fourier transform of the δ-function as

$$\tilde{\delta}(x) = \frac{1}{\sqrt{(2\pi)}}\int_{-\infty}^{\infty} \delta(x)\mathrm{e}^{-ikx}dx = \frac{1}{\sqrt{(2\pi)}} \tag{5.29}$$

Similarly, the Fourier transform of $\delta(x-x_0)$ is given by

$$\tilde{\delta}(x-x_0) = \frac{1}{\sqrt{(2\pi)}}\int_{-\infty}^{\infty} \delta(x-x_0)\mathrm{e}^{-ikx}dx = \frac{1}{\sqrt{(2\pi)}}\mathrm{e}^{-ikx_0} \tag{5.30}$$

Eq. (5.30) provides an alternative definition the delta function by the application of inverse Fourier transform.

$$\begin{aligned}\delta\,(x-x_0) &= FT^{-1}\left\{\frac{1}{\sqrt{(2\pi)}}\mathrm{e}^{-ikx_0}\right\}\\ &= \frac{1}{\sqrt{(2\pi)}}\int_{-\infty}^{\infty}\left(\frac{1}{\sqrt{(2\pi)}}\mathrm{e}^{-ikx_0}\right)\mathrm{e}^{ikx}dk\\ &= \frac{1}{2\pi}\int_{-\infty}^{\infty}\mathrm{e}^{ik(x-x_0)}dk\end{aligned} \tag{5.31}$$

Dirac comb

Sometimes, we may be interested in not a δ-function at a single point but a set of δ-functions that define a one-dimensional lattice of spacing a. Its Fourier transform will be another set of δ-functions of spacing proportional to $1/a$ defining the corresponding one-dimensional reciprocal lattice. Such a set of equally spaced

δ-functions is called a ***Dirac comb***. An ***infinite comb*** contains an infinite number of δ-functions. A ***finite comb*** with $-N$ δ-functions to the left of the origin and $+N$ δ-functions to the right of the origin, besides one at the origin itself, can be written as

$$\mathrm{comb}_N(x) = \sum_{r=-N}^{N} \delta(x - ra) \tag{5.32}$$

where a is the spacing between the δ-functions. Evidently, the comb contains a total of $N_{tot} = 2N + 1$ δ-functions. The Fourier transform of the comb function is given by

$$\begin{aligned} FT\{\mathrm{comb}_N(x)\} &= \frac{1}{\sqrt{(2\pi)}} \int_{-\infty}^{\infty} \sum_{r=-N}^{N} \delta(x - ra)\, \mathrm{e}^{-iqx} dx \\ &= \frac{1}{\sqrt{(2\pi)}} \sum_{r=-N}^{N} \mathrm{e}^{-iqra} \end{aligned} \tag{5.33}$$

On simplification (not shown here), we can find

$$FT\{\mathrm{comb}_N(x)\} = \frac{1}{\sqrt{(2\pi)}} \frac{\sin\left(\frac{N_{tot}qa}{2}\right)}{\sin\left(\frac{qa}{2}\right)} \tag{5.34}$$

5.3 Convolution and deconvolution

Convolution operates on two signals or two mages; mathematically, it operates on two functions – one can be considered as the "input" while the other, called the kernel, as the "filter" or "modulator." In other words, convolution takes two images as input and produces a third as output.

In the language of calculus, convolution is an integral, say, of two functions f and g. The aim of the operation is to move g over f, the resulting overlap being the convolution of f and g. So, the integral is a "blending" of the two functions.

The convolution of two functions $f(x)$ and $g(x)$, denoted by $f * g$, over a finite range $[0, t]$, is given by

$$(f * g)(t) = \int_{x=0}^{t} f(x)g(t - x)dx \tag{5.35}$$

Often, convolution is taken over an interval $[-\infty, \infty]$. In such case (5.35) becomes

$$(f * g)(t) = \int_{-\infty}^{\infty} f(x)g(t - x)dx \tag{5.36}$$

Example 5.3.1: Let us find the convolution of the function $f(x) = e^{3x}$ with the function $g(x) = e^{5x}$ over the range $[0, t]$.

The convolution is given by

$$\begin{aligned}
(f * g)(t) &= \int_{x=0}^{t} f(x) g(t-x) dx \\
&= \int_{x=0}^{t} e^{3x} e^{5(t-x)} dx \\
&= \int_{x=0}^{t} e^{3x} e^{5t} e^{-5x} dx \\
&= e^{5t} \int_{x=0}^{t} e^{-2x} dx \\
&= e^{5t} \left[\frac{-1}{2} e^{-2x} \right]_{x=0}^{t} \\
&= \frac{1}{2} \left[e^{5t} - e^{3t} \right]
\end{aligned}$$

Example 5.3.2: Let $f(x) = \frac{1}{2\sqrt{x}}$ and $g(x) = \sqrt{2}x$. We want to compute the convolution of f with g over the range [0, 2].

The convolution is given by

$$\begin{aligned}
(f * g)(t) &= \int_{x=0}^{t} f(x) g(t-x) dx \\
&= \int_{x=0}^{t} \frac{1}{2\sqrt{x}} \sqrt{2} (t-x)^2 dx
\end{aligned}$$

On simplification, we have

$$(f * g)(t) = \frac{8\sqrt{2}}{15} t^{5/2}$$

Since $t = 2$,

$$(f * g)(t) = \frac{64}{15}$$

The Fourier transform of the convolution as defined in (5.35) is given by

$$\begin{aligned}
\tilde{h}(k) &= \frac{1}{\sqrt{(2\pi)}} \int_{-\infty}^{\infty} dt\, e^{-ikt} \left\{ \int_{-\infty}^{\infty} [f(x) g(t-x) dx] \right\} \\
&= \frac{1}{\sqrt{(2\pi)}} \int_{-\infty}^{\infty} dx\, f(x) \left\{ \int_{-\infty}^{\infty} g(t-x) e^{-ikt} dt \right\}
\end{aligned}$$

Taking $u = t - x$, we have

$$\tilde{h}(k) = \frac{1}{\sqrt{(2\pi)}} \int_{-\infty}^{\infty} dx\, f(x) \left\{ \int_{-\infty}^{\infty} g(u) e^{-ik(u+x)} du \right\}$$

Rearranging,

$$\begin{aligned}\tilde{h}(k) &= \frac{1}{\sqrt{(2\pi)}}\int_{-\infty}^{\infty} f(x)\,\mathrm{e}^{-ikx}\,dx\int_{-\infty}^{\infty} g(u)\mathrm{e}^{-iku}\,du \\ &= \frac{1}{\sqrt{(2\pi)}}\cdot\sqrt{2\pi}\tilde{f}(k)\cdot\sqrt{2\pi}\tilde{g}(k) = \sqrt{2\pi}\tilde{f}(k)\tilde{g}(k)\end{aligned} \tag{5.37}$$

Eq. (5.37) shows that the Fourier transform of a convolution $f * g$ is equal to the product of the Fourier transforms of f and g separately, multiplied by $\sqrt{2\pi}$ – this is the ***convolution theorem***.

Now, from the discussion on convolution so far, it can be inferred that if f is the incoming signal and g is the modulator in the form of a filter or noise, then what we receive is h. Usually, some statistical assumptions are made on the nature of the modulator to find the true signal following a process known as ***deconvolution***, which is the inverse of convolution.

From the convolution theorem (5.37) we obtain

$$\tilde{f}(k) = \frac{1}{\sqrt{(2\pi)}}\frac{\tilde{h}(k)}{\tilde{g}(k)} \tag{5.38}$$

The inverse Fourier transform of (5.38) gives

$$f(x) = \frac{1}{\sqrt{(2\pi)}}FT^{-1}\left[\frac{\tilde{h}(k)}{\tilde{g}(k)}\right] \tag{5.39}$$

In words, the true signal can be extracted from the modulated signal using (5.39) if the modulating function is known.

5.4 Discrete Fourier transform

Both $f(x)$ and $\tilde{f}(k)$ in (5.5) and (5.7) are functions of continuous variables. Nevertheless, physical problems are often concerned with values sampled at a discrete set (sequence) of points. In such cases, the Fourier transform needs to be considered as an operation with N number of inputs and N number of outputs and converted into a "computable" form. So, the continuous variables must be sampled and the integration changed into finite summation. As a result, (5.5) takes the form

$$X(k) = \sum_{n=0}^{N-1} x(n)\mathrm{e}^{-2\pi ikn/N} \quad k = 0,\ 1,\ 2, \ldots,\ N-1 \tag{5.40}$$

Eq. (5.40) is called the ***discrete Fourier transform*** (***DFT***), $X(k)$ being the DFT of $x(n)$.

Example 5.4.1: Computation of the 4-point DFT of the sequence $x(n) = \{1, 0, 1, 0\}$.

Here, the length of the sequence $N=4$ and the sum (5.40) is to be carried out for each value of $k=0, 1, 2, 3$. We have

$$\begin{aligned}\text{For } k=0,\quad X(0) &= x(0)\,e^{-i0} + x(1)\,e^{-i0} + x(2)\,e^{-i0} + x(3)\,e^{-i0}\\ &= 1 + 0 + 1 + 0 = 2\end{aligned}$$

$$\begin{aligned}\text{For } k=1,\quad X(1) &= x(0)\,e^{-i0} + x(1)\,e^{-i\pi/2} + x(2)\,e^{-i\pi} + x(3)\,e^{-i3\pi/2}\\ &= 1 + 0 + (-1) + 0 = 0\end{aligned}$$

Similarly, putting the appropriate values of k and n, one can calculate

$$\text{For } k=2,\quad X(2)=2, \text{and}$$

$$\text{For } k=3,\quad X(3)=0$$

So, the DFT values are $X(k)=\{2, 0, 2, 0\}$.

It may be emphasized that the DFT is a "legitimate" transform. This implies that it is possible to invert the transformation defined by (5.40). The ***inverse discrete Fourier transform*** (***IDFT***) is defined as

$$x(n) = \frac{1}{N}\sum_{k=0}^{N-1} X(k)e^{2\pi ikn/N} \qquad n=0, 1, 2, \ldots, N-1 \tag{5.41}$$

IDFT converts $X(k)$ for $k=0, 1, 2, \ldots, N-1$ back to $x(n)$ for $n=0, 1, 2, \ldots, N-1$.

Example 5.4.2: Let us compute the IDFT of the 4-point DFT sequence $X(k)=\{2, 0, 2, 0\}$ as obtained in Example 5.4.1 using (5.41).

Here, (5.41) is to be carried out for each value of $n=0, 1, 2, 3$.

$$\begin{aligned}\text{For } n=0,\quad x(0) &= \frac{1}{4}\{X(0)e^{i0} + X(1)e^{i0} + X(2)e^{i0} + X(3)e^{i0}\}\\ &= \frac{1}{4}\{2 + 0 + 2 + 0\} = 1\end{aligned}$$

$$\begin{aligned}\text{For } n=1,\quad x(1) &= \frac{1}{4}\{X(0)e^{i0} + X(1)e^{i\pi/2} + X(2)e^{i\pi} + X(3)e^{i3\pi/2}\}\\ &= \frac{1}{4}\{2 + 0 + 2(-1) + 0\} = 0\end{aligned}$$

Similarly, we can calculate

$$\text{For } n=2,\quad x(2)=1, \text{and}$$

$$\text{For } n=3,\quad x(3)=0$$

Thus we have recovered the length-4 sequence $x(n)=\{1, 0, 1, 0\}$ from the 4-point DFT values $X(k)=\{2, 0, 2, 0\}$.

5.5 Laplace transform

Besides the Fourier transform, another which has been considered to be reasonably important, is the Laplace transform. One disadvantage with the latter has been that unlike the Fourier transform, the inversion of the Laplace transform is relatively difficult to perform. Nevertheless, it has been very useful in solving differential equations. The Laplace transformation can convert differential equations into more easily solvable algebraic equations.

The Laplace transformation of a function $f(t)$ can be denoted by $\bar{f}(s)$ or $\mathcal{L}[f(t)]$. It is defined as follows.

Definition 5.5.1 The ***Laplace transform*** of a function $f(t)$ is defined as

$$\bar{f}(s) = \int_0^\infty f(t)e^{-st}\,dt \tag{5.42}$$

provided the integral exists.

Example 5.5.1: $f(t) = 1,\ s > 0$

Using (5.42) we have

$$\bar{f}(1) = \int_0^\infty e^{-st}\,dt = \left[\frac{-1}{s}e^{-st}\right]_0^\infty = \frac{1}{s}, \quad s > 0$$

Example 5.5.2: $f(t) = e^{\lambda t},\ s > 0$

Again using (5.42)

$$\bar{f}(e^{\lambda t}) = \int_0^\infty e^{\lambda t}e^{-st}\,dt = \int_0^\infty e^{(\lambda - s)t}\,dt = \left[\frac{e^{(\lambda - s)t}}{\lambda - s}\right]_0^\infty = \frac{1}{s - \lambda}, \text{ if } \quad s > \lambda$$

Example 5.5.3: ***Dirac delta function***

Using (5.12) and (5.42), we have the Laplace transform of the δ-function

$$\bar{\delta}(t - t_0) = \int_0^\infty e^{-st}\delta(t - t_0)dt = e^{-st_0}, \quad \text{for } t_0 > 0 \tag{5.43}$$

As it has been already mentioned, the Laplace transform is useful in solving differential equations. Hence, we can look at the Laplace transforms of derivatives.

The Laplace transform of the first derivative of $f(t)$ is

$$\bar{f}\left(\frac{df}{dt}\right) = \int_0^\infty \frac{df}{dt}e^{-st}\,dt$$

Taking $du/dt = df/dt$ and $v = e^{-st}$, we have

$$\bar{f}\left(\frac{df}{dt}\right) = \left[f(t)e^{-st}\right]_0^{\infty} + s\int_0^{\infty} f(t)e^{-st}dt = -f(0) + s\bar{f}(s), \text{ for } s > 0 \tag{5.44}$$

The Laplace transformation of the second derivative of $f(t)$ is

$$\bar{f}\left(\frac{d^2f}{dt^2}\right) = \int_0^{\infty} \frac{d^2f}{dt^2} e^{-st}\, dt = \left[\frac{df}{dt} e^{-st}\right]_0^{\infty} + s\int_0^{\infty} \frac{df}{dt} e^{-st}\, dt$$

Using (5.44) we have

$$\begin{aligned} \bar{f}\left(\frac{d^2f}{dt^2}\right) &= -\frac{df}{dt}(0) + s\left[s\bar{f}(s) - f(0)\right] \\ &= s^2\bar{f}(s) - sf(0) - \frac{df}{dt}(0), \text{ for } s > 0 \end{aligned} \tag{5.45}$$

Furthermore, the Laplace transformation of the integral is simply obtained as

$$\begin{aligned} \bar{f}\int_0^t f(x)\,dx &= \int_0^{\infty} dt\, e^{-st} \int_0^t f(x)\,dx \\ &= \left[-\frac{1}{s}e^{-st}\int_0^t f(x)\,dx\right]_0^{\infty} + \int_0^{\infty} \frac{1}{s}e^{-st} f(t)\,dt \end{aligned}$$

The first term on the RHS of the above equation vanishes at both limits, and we have

$$\bar{f}\int_0^t f(x)\,dx = \frac{1}{s}\bar{f}(s) \tag{5.46}$$

Exercise 5

1. Find the Fourier transform of $f(x) = \begin{cases} 1 & \text{for } a \leq x \leq \leq b \\ 0 & \text{otherwise} \end{cases}$
2. Find the Fourier transform of $f(x) = \begin{cases} 1 - x & \text{for } |x| < 1 \\ 0 & \text{for } |x| > 1 \end{cases}$
3. Show that the Fourier transform $\tilde{f}(k)$ of $e^{-x^2/2}$ is given just by changing the variable x to k.
4. Prove that $FT[f'(t)] = i\omega\tilde{f}(\omega)$ provided $f(t) \to 0$ at $t = \pm\infty$.
5. Evaluate the following:
   a. $\int_0^{\infty} x^2\delta(x-3)dx$
   b. $\int_{-5}^{5} (10-x)^5\delta(x)dx$

6. Show from the basic definition of the δ-function
   a. $\delta(ax) = \frac{1}{|a|}\delta(x)$
   b. $(x-a)\,\delta(x-a) = 0$
7. Let the function $f(x) = \frac{1}{\sqrt{x}}$ and the function $g\,(x) = x^2$. Compute the convolution of f with g over the range [0, 2].
8. Let the function $f(x) = \mathrm{e}^{3x}$ and the function $g\,(x) = \mathrm{e}^{5x}$. Show that the convolution $(f * g)(t)$ over the range $[0, t]$ is commutative.
9. Calculate the 4-point DFT of the periodic sequence $x\,(n) = \{0, 1, 0, -1\}$.
10. Calculate the 4-point DFT of the sequence $x\,(n) = \{1, 1, 0, 0\}$ and then calculate the IDFT of the DFT thus obtained to verify if the original length-4 sequence can be recovered.
11. Find the Laplace transform of the following functions:
    a. $f(t) = \mathrm{e}^{3t} - 5$
    b. $f(t) = \cos\,(4t)$
    c. $f(t) = \mathrm{e}^{2t}\,\sin\,(2\mathrm{t})$
12. If $f(t) = \cos\,(2t)$, show by calculating the Laplace transforms of the function and its derivative separately that $\bar{f}\left(\frac{df}{dt}\right) = -f(0) + s\bar{f}(s)$, for $s > 0$.

Further reading

Arfken, G.B., Weber, H.J., Harris, F.E., 2013. Mathematical Methods for Physicists: A Comprehensive Guide, seventh ed. Elsevier Academic Press.

Riley, K.F., Hobson, M.P., Bence, S.J., 2006. Mathematical Methods for Physics and Engineering, third ed. Cambridge University Press, Cambridge, UK.

CHAPTER

Probability and statistics

6

Needless to reemphasize, in the investigations on the structure and dynamics of biomolecules, two features need to be indispensably taken into considerations—the large number of entities involved (e.g., the number of atoms in a molecule) and randomness of the behavior (variation and movement). This chapter introduces some basic concepts of probability and (multivariate) statistics to help understand the structural biology of biomolecules discussed in the later chapters.

6.1 Probability—definitions and properties

Probability is an appropriate answer to the challenges of uncertainty. It provides robust mathematical rules to understand the uncertainty and randomness in the data generated by observations and experiments on different physical phenomena and arrives at meaningful conclusions. In an atom, there exists certain uncertainty as regards the position of a quantum particle such as an electron. Hence, the location of an electron is known, not with absolute precision, but in terms of "probability density."

6.1.1 Probability function

In Chapter 2, we have discussed the basic notion of set theory which has been found to be very useful in the description of the probability model. A formal definition of probability must be preceded by that of a sample space.

Definition 6.1.1 The ***sample space***, denoted by S, is the set of all possible outcomes or responses of an experiment or a situation. The size of the sample space can be *finite*, *countably infinite* or *uncountably infinite*.

A probability model also requires a *collection of events* to which probabilities can be assigned.

Definition 6.1.2 An ***event*** is a subset of S. It can be defined explicitly by giving its outcomes $A = \{u_1, u_2, \cdots, u_n\}$ or a description $A = \{u\colon u \text{ has the property } P\}$.

Mathematical Approaches to Molecular Structural Biology. DOI: https://doi.org/10.1016/B978-0-323-90397-4.00003-2

© 2023 Elsevier Inc. All rights reserved.

Most importantly, a probability model requires a *probability measure*. Each event is assigned a probability which can, therefore, be considered a *function*.

Definition 6.1.3 A ***probability function*** P on a finite sample space S assigns to each event A in S a number $P(A)$ in [0, 1] such that

1. $P(S) = 1$,
2. $P(\varnothing) = 0$, that is, if A is an empty subset $\varnothing$, $P(A) = 0$
3. $P(A \cup B) = P(A) + P(B)$, if A and B are disjoint.

The number $P(A)$ is called the probability of the event A.

Property (1) implies that the outcome of the experiment is always an element of the sample space. Further, the additivity of the probability function (3) can be extended over more than two subsets. If $A_1, A_2, \ldots$ is a finite or countable sequence of disjoint events, then P is (countably) additive, that is

$$P(A_1 \cup A_2 \cup \ldots) = P(A_1) + P(A_2) + \cdots \tag{6.1}$$

Example 6.1.1: Every nucleotide in a DNA molecule has one of the four bases linked to it—adenine (A), cytosine (C), guanine (G), and thymine (T). An experiment to determine the base linked to the first nucleotide of a polynucleotide chain produces the sample space $S = \{A, C, G, T\}$. Clearly, the probability of finding any of the four bases at the first position is

$$P(\mathrm{A}) = P(\mathrm{C}) = P(\mathrm{G}) = P(\mathrm{T}) = 1/4$$

If we look at the first two nucleotides, then

$$S = \{\mathrm{AA, AC, AG, AT, CA, CC, CG, CT, GA, GC, GG, GT, TA, TC, TG, TT}\}$$

so that

$$P(\mathrm{AA}) = P(\mathrm{AC}) = P(\mathrm{AG}) = P(\mathrm{AT}) = \cdots = 1/16$$

A complement

If A is any event in the sample space, then A complement, denoted by A^c, is the event that A does not occur. A and A^c are always disjoint and, further,

$$A \cup A^c = S$$

Then by the additivity property, we have

$$P(A) + P(A^c) = P(S) = 1$$

Therefore

$$P(A^c) = 1 - P(A) \tag{6.2}$$

Uniform probability measure

If a finite sample space S consists of $|S|$ elements and all outcomes are equally likely (that is, all single-element events have the same probability), then each outcome can be assigned a probability $1/|S|$. It follows that the probability of any event A is given by

$$P(A) = \frac{\text{Number of elements in } A}{|S|} = \frac{|A|}{|S|} \tag{6.3}$$

6.1.2 Conditional probability

Let us consider an example from protein structures.

Example 6.1.2: Catalytic triads of three amino acids are found in the active site of several enzymes. One such triad consists of acidic amino acid residues aspartate and glutamate, denoted, respectively, by D and E. Experimental observations on the arrangements of Ds and Es in the triads from different enzymes lead to the outcome written as

$$S = \{D-D-D, D-D-E, D-E-D, D-E-E, E-D-D, E-D-E,$$
$$E-E-D, E-E-E\}$$

Here, the number of elements in the sample space S is $|S| = 8$ and each outcome can be assumed to have the same probability of 1/8. Hence, $P(\{D\text{–}E\text{-}D\}) = 1/8$, $P(\{E\text{–}D\text{–}E, D\text{–}D\text{–}E\}) = 1/4$, etc.

If we look for D in the first position of the triad, the probability of finding it is

$$P(\{D-D-D, D-D-E, D-E-D, D-E-E\}) = 4/8 = 1/2$$

However, if there is prior information that Ds are present in two of the three positions, then the outcome would be one of $D\text{–}D\text{–}E$, $D\text{–}E\text{–}D$, and $E\text{–}D\text{–}D$. Consequently, the probability of finding a D in the first position would be 2/3. This is a “conditional probability.” It can be said that “conditional” on the information (or given) that Ds are in two of the three positions, the ***conditional probability*** of finding a D in the first position is 2/3. Denoting “conditionality” by a vertical bar |, this can be written a mathematical form as

$$P(\text{first position } D \mid \text{two positions } D) = 2/3.$$

Generalizing the argument for two events A and B with $P(B) > 0$, we have the following definition of conditional probability.

Definition 6.1.4 The *conditional probability* of event A given event B is

$$P(A|B) = \frac{P(A \cap B)}{P(B)} \tag{6.4}$$

assuming $P(B) > 0$.

It can be checked that in Example 6.1.2 using (6.4) that

$$P(A|B) = \frac{P(A \cap B)}{P(B)} = \frac{2/8}{3/8} = 2/3$$

Independence of events

(6.4) is useful when $P(A|B)$ needs to be calculated from $P(A \cap B)$ and $P(B)$. Often, we may need to calculate $P(A \cap B)$ if $P(A|B)$ and $P(B)$ are known. For this a simple rearrangement of (6.4), known as the ***multiplication rule***, is used.

$$P(A \cap B) = P(B)\,P(A\,|\,B) \tag{6.5}$$

If event B has no impact on event A, $P(A|B) = P(A)$. In that case (6.4) becomes

$$P(A \cap B) = P(A)\,P(B) \tag{6.6}$$

giving the property of independence.

Similar to (6.4), the definition of $P(B|A)$ leads to the multiplication rule

$$P(A \cap B) = P(A)\,P(B\,|\,A) \tag{6.7}$$

Definition 6.1.5 Two events A and B are independent if

$$P(A \cap B) = P(A)\,P(B)$$

A partition of a sample space S is its decomposition into a collection of mutually exclusive events $A_1, A_2,\ldots,$ each with a positive probability. $A_1, A_2,\ldots, A_n$ are disjoint and further

$$S = A_1 \cup A_2 \cup \cdots \cup A_n = \bigcup_{i=1}^{n} A_i \tag{6.8}$$

The probability of an event B in the same sample space can be decomposed into the sum of probabilities of the sets $A_i \cap B$, $i = 1, 2,\ldots, n$.

Theorem 6.1.1 *Law of total probability.* If $A_1, A_2,\ldots, A_n$ be events that form a partition of the sample space S, and B be any event in S, then

$$\begin{aligned} P(B) &= P(A_1 \cap B) + P(A_2 \cap B) + \cdots + P(A_n \cap B) \\ &= \sum_{i=1}^{n} P(A_i \cap B) \\ &= \sum_{i=1}^{n} P(A_i)\,P(B\,|\,A_i) \end{aligned} \tag{6.9}$$

Bayes' theorem

Combining (6.5) and (6.7), we have $P(B)\ P(A|B) = P(A)\ P(B|A)$ and hence the following theorem which is extremely useful when we need to work out $P(A|B)$ from $P(B|A)$.

Theorem 6.1.2 ***Bayes' theorem***. For two events A and B in a sample space S, each with a positive probability,

$$P(A|B) = \frac{P(A)}{P(B)} P(B|A) \tag{6.10}$$

6.2 Random variables and distribution

The previous section has explained how to construct a probability model with a sample space S and a probability measure P. Now we have to define *random variables* for the probability model. Let us begin with an example.

Example 6.2.1: An experiment is looking at several proteins with similar structure and function. One of the three basic amino acids, histidine (H), lysine (K), and arginine (R), is found at the tip of an active site loop. Therefore the sample space is $S = \{H, K, R\}$. We may define a random variable X such that $X(H) = 2$ if it is an H, $X(K) = 4$ if it is a K and $X(R) = 6$ if it is an R. So, we have the following definition.

Definition 6.2.1 A random variable is a function from the sample space S to the set $\mathbb{R}$ of real numbers.

We denote the probability of the event $\{X = x\}$ by $P(X = x)$.

6.2.1 Discrete random variable

For many random variables X, we have $P(X = x) > 0$ for certain x values. In Example 6.2.1,

$$P(X = 2) = P(X = 4) = P(X = 6) = 1/3$$

If we consider a fourth amino acid glycine (G), and define $X(G) = 8$, then

$$P(X = 8) = 0$$

However,

$$\sum_{x \in \mathbb{R}} P(X = x) = 1 \tag{6.11}$$

Definition 6.2.2 Discrete random variable

In a sample space S, a *discrete random variable* is a function $X: S \to \mathbb{R}$ that takes on a finite sequence $x_1, x_2, \ldots, x_n$ or infinite sequence $x_1, x_2, \ldots$ of distinct real numbers.

Once a discrete random variable X is defined, it is sufficient to describe the outcome of an experiment by simply listing the possible values of X and their corresponding probabilities. This information is presented as the probability mass function (pmf) of X.

Definition 6.2.3 Probability mass function

For a discrete random variable X, the *probability mass function* is the function $p: \mathbb{R} \to [0, 1]$ defined by

$$p(x) = P(X = x) \quad \text{for } -\infty < x < \infty$$

If X takes on values $x_1, x_2, \ldots$, then

$$p(x_i) > 0, \quad \sum_i p(x_i) = 1, \quad p(x) = 0 \text{ for all other } x$$

Two of the discrete probability functions most widely used in bioinformatics and computational biology are introduced here next.

The Bernoulli and binomial distributions

A Bernoulli random variable has only two possible values, referred to as "success" and "failure," labeled numerically as 1 and 0, respectively. The probability of success is denoted by p and that of failure by $1-p$.

Definition 6.2.4 A discrete random variable X has a ***Bernoulli distribution***, denoted by *Ber* (p), with parameter p ($0 \le p \le 1$), if its *probability mass function* is given by

$$p_X(1) = P(X = 1) = p \quad \text{and} \quad p_X(0) = P(X = 0) = 1 - p.$$

The binomial random variable is the number of "successes" in a fixed number n of independent experiments (called *Bernoulli trials*) and the *binomial distribution* is valid if all four of the following conditions are satisfied:

1. each (Bernoulli) trial must result in one of two possible outcomes - "success" and "failure";
2. the trials must be independent;
3. the probability of success, denoted by p (or failure denoted by $1 - p$) must be the same on all trials; and
4. the number of trials must be fixed in advance, and not determined by the outcomes of the trials as they proceed.

n and p are called the index and parameter of the distribution. We have the following definition of a binomial distribution.

Definition 6.2.5 A discrete random variable X has a ***binomial distribution***, denoted by *Bin* (n, p), with index n and parameter p respectively, where $n = 1, 2, \ldots$ and $0 \le p \le 1$, if its *probability mass function* is given by

$$p_X(k) = P(X = k) = \binom{n}{k} p^k (1-p)^{n-k} \quad \text{for} \quad k = 0, 1, \ldots, n \tag{6.12}$$

It is to be noted that

$$\binom{n}{k} = \frac{n!}{k!(n-k)!}$$

Example 6.2.2: There are four kinds of bases present in the DNA—adenine (A), cytosine (C), guanine (G), and thymine. Let the GC-content of the genome of a particular organism be 41%, that is, the probability (p) of finding either a C or a G at a particular site in the genome sequence is 0.41. We want to find the probability that a sequence 22 bases long randomly chosen from the genome will have exactly 10 G/C bases.

Here, $n = 22$, $p = 0.41$ and $k = 10$, so that from (6.12) we have

$$P(X = 10) = \frac{22!}{10!12!} \cdot 0.41^{10} \cdot 0.59^{12} \approx 0.154$$

A plot of *Bin* (22, 0.41) is shown in Fig. 6.1.

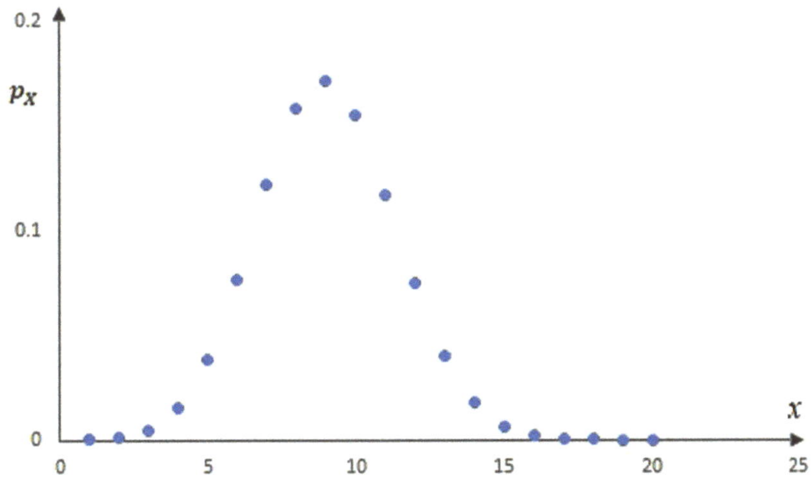

FIGURE 6.1

Plot of *Bin* (22, 0.41).

The Poisson distribution

A Poisson process is characterized by the following features:

1. The outcomes are discrete.
2. The probability of success in a certain subinterval depends only on the length of the subinterval, and not on the location of the subinterval within the overall interval.
3. The number of successes, k, in any subinterval is independent of the number of successes in any other interval.
4. The probability of two or more successes over a sufficiently small subinterval is almost zero.

A Poisson process generates a ***Poisson distribution*** which arises mainly in two contexts—(1) when events occur "randomly" in time or space and (2) as a limiting form of the binomial distribution when the number of trials n is large and the probability of success p on each trial is small.

The Poisson distribution is associated with a parameter λ which gives the average number of times an event occurs in a subinterval.

Definition 6.2.6 A discrete random variable X has a ***Poisson distribution*** (with parameter $\lambda > 0$), denoted by $X \sim \text{Poisson}(\lambda)$, if

$$p_X(k) = P(X = k) = \frac{e^{-\lambda}\lambda^k}{k!} \quad \text{for } k = 0, 1, 2, \ldots \tag{6.13}$$

Now, by definition $\lambda = np$. Substituting in the binomial distribution (Definition 6.2.5), we have

$$p_X(k) = P(X = k) = \binom{n}{k}\left(\frac{\lambda}{n}\right)^k \left(1 - \frac{\lambda}{n}\right)^{n-k}$$

It can be shown that

$$\lim_{n \to \infty} \binom{n}{k}\left(\frac{\lambda}{n}\right)^k \left(1 - \frac{\lambda}{n}\right)^{n-k} = \frac{e^{-\lambda}\lambda^k}{k!}$$

which gives the Poisson distribution. It can also be seen that the sample size n is not a part of the expression.

Example 6.2.3: Let us assume that a particular mutant gene is present at a frequency of 1 in 100,000 human population. If 50,000 persons are randomly screened, what is the probability of finding exactly one person carrying this mutant gene?

Here, the expected number of persons carrying the mutant is

$$\lambda = np = (50,000)\left(\frac{1}{100,000}\right) = 0.5$$

Using (6.13), the probability of exactly one person, in a randomly selected population of 50,000, carrying the mutant gene is found to be

$$p_X(1) = P(X = 1) = \frac{e^{-0.5}0.5^1}{1!} = 0.303$$

6.2.2 Continuous random variable

Some experiments have continuous random variables associated with them. These variables cannot be specified by giving a *probability mass function* since $P(X = x) = 0$ for all $x \in \mathbb{R}$. However, we can define *probability density function* (or *probability density*) to describe the properties of continuous random variables.

Definition 6.2.7 Probability density function

A function f: $\mathbb{R} \to \mathbb{R}$ is called a *probability density function* (or simply *probability density*) if $f(x) \geq 0$ for all $x \in \mathbb{R}$, and $\int_{-\infty}^{\infty} f(x)dx = 1$.

Definition 6.2.8 Continuous random variable

A random variable X is continuous if there is a probability density function f such that

$$P(a \leq X \leq b) = \int_a^b f(x)\, dx \tag{6.14}$$

whenever $a \leq b$.

Eq. (6.14) has been represented graphically in Fig. 6.2. As shown, the probability that X lies in an interval $[a, b]$ is given by the area under the probability density

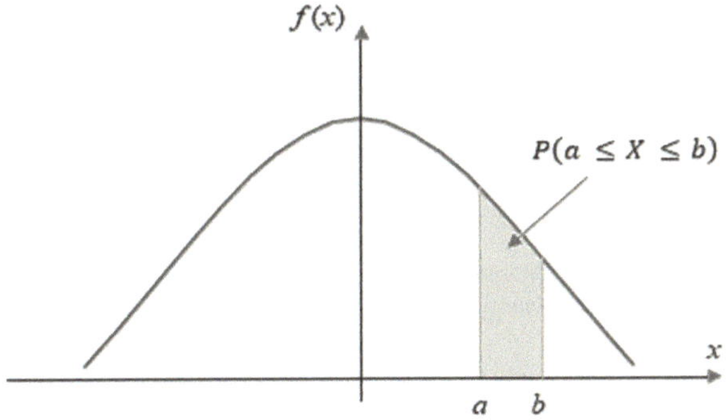

FIGURE 6.2

Area under a probability density function.

function f of x over $[a, b]$. For a small positive number ε, (6.14) can be written as

$$P(a-\varepsilon \leq X \leq a+\varepsilon) = \int_{a-\varepsilon}^{a+\varepsilon} f(x)dx$$

Evidently, for any a, as $\varepsilon \to 0$,

$$P(X = a) = 0$$

It is to be noted that although discrete random variables do not have a probability density function f and continuous random variables do not have a *probability mass function p*, both have a distribution function as defined next.

Definition 6.2.9 Given a random variable X, the distribution function is the function F: $\mathbb{R} \to [0, 1]$ is defined by

$$F(x) = P(X \leq x) \quad \text{for} \quad -\infty < x < \infty.$$

Cumulative distribution function

A special function that can be associated with any discrete random variable as well as continuous random variable is the *cumulative distribution function (cdf)*. For any $x \in \mathbb{R}$, the cdf, denoted by $F_X(x)$, is the probability that the random variable is less than or equal to x. That is,

$$F_X(x) = P(X \leq x) \tag{6.15}$$

In the case of a discrete random variable, the cdf can be written as

$$F_X(x) = \sum_{y \leq x} p_X(y) \tag{6.16}$$

For a continuous random variable with probability density function f,

$$F_X(x) = \int_{-\infty}^{x} f_X(y)\,dy \tag{6.17}$$

A few of the important continuous distributions are introduced here.

The uniform distribution

The uniform distribution arises when the probability density function associated with an experiment is constant in an interval $[a, b]$

Definition 6.2.10 A continuous random variable X has a *uniform distribution* on the interval $[a, b]$ and denoted by $U(a, b)$ if its probability density function is given by

$$f_X(x) = \frac{1}{b-a} \quad \text{for } a \leq x \leq b$$

and

$$f_X(x) = 0 \quad \text{for } x \text{ not in } [a, b]$$

The exponential distribution

In computational genetics, the exponential distribution function is used to model the time gap between successive changes such as DNA mutations, or the distance between successive events such as that between successive repeat sequences along the genome. It is defined as follows.

Definition 6.2.11 A continuous random variable X has an *exponential distribution* with parameter λ and denoted $Exp\ (\lambda)$ if its probability density function is given by

$$f_X(x) = \lambda e^{-\lambda x} \quad \text{for } x \geq 0$$

and

$$f_X(x) = 0 \quad \text{for } x < 0$$

From (6.17), we can find that the cdf of the exponential distribution function is

$$F_X(x) = \int_0^x \lambda e^{-\lambda x} dx = 1 - e^{-\lambda x} \tag{6.18}$$

Example 6.2.4: Assuming the average mammalian genome mutation rate to be 2×10^{-9} per base pair per year let us find out the probability that the interval between successive mutation at a particular base pair is 70 years or less.

This is a case of exponential distribution where $\lambda = 2 \times 10^{-9}$ per base pair per year and $x = 70$ years. The probability that a mutation will occur in 70 years or less is given by the cdf as

$$F_X(x) = 1 - e^{-(2\times10^{-9})\cdot(70)} \sim 1.4 \times 10^{-7}$$

The normal distribution

In normal, or Gaussian, distribution the frequency of variables or probability density $f_X(x)$ falls off proportionally to the difference between the score and the mean (μ) and to the frequency itself. Expressed mathematically

$$\frac{df_X(x)}{dx} = -k(x-\mu)f_X(x) \tag{6.19}$$

Since $f_X(x)$ is a density function,

$$\int_{-\infty}^{\infty} f_X(x)dx = 1 \tag{6.20}$$

Based on this and introducing the term variance σ such that

$$\sigma^2 = \int_{-\infty}^{\infty} (x-\mu)^2 f_X(x)dx \tag{6.21}$$

one can derive from (6.19)

$$f_X(x) = \frac{1}{\sigma\sqrt{2\pi}} e^{-\frac{1}{2}\left(\frac{x-\mu}{\sigma}\right)^2} \tag{6.22}$$

Definition 6.2.12 A continuous random variable X has a *normal distribution* with parameters μ and $\sigma^2 > 0$, and denoted by $N(\mu, \sigma^2)$ if its probability density function is given by

$$f_X(x) = \frac{1}{\sigma\sqrt{2\pi}} e^{-\frac{1}{2}\left(\frac{x-\mu}{\sigma}\right)^2} \quad \text{for } -\infty < x < \infty$$

A special case of the normal distribution is with $\mu = 0$ and $\sigma = 1$. The probability density function of this $N(0, 1)$ distribution is expressed as

$$\phi(x) = \frac{1}{\sqrt{2\pi}} e^{-\frac{1}{2}x^2} \quad \text{for } -\infty < x < \infty \tag{6.23}$$

As it can be seen from Fig. 6.3, the distribution is centered at zero. It is called the ***standard normal distribution*** and provides a standard way to present the Gaussian distribution. Here, the distance from the origin is measured by the variable z and, therefore, called a ***z-score***. We shall see in the following section that

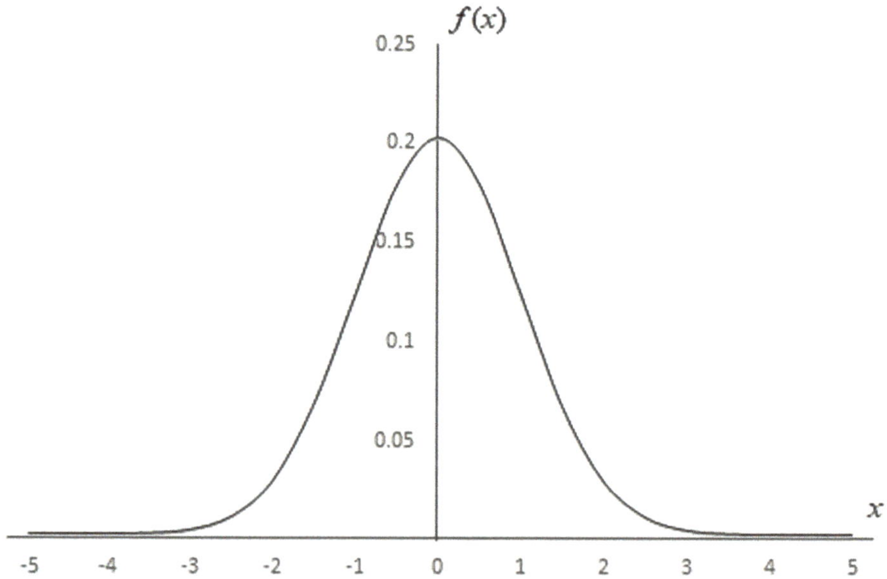

FIGURE 6.3

A standard normal distribution.

the z-score can be related to any "nonstandard" normal distribution through the "transformation"

$$z = \frac{x - \mu}{\sigma}$$

6.2.3 Transformation of random variables

In statistics, sometimes creation of new random variables from old ones becomes necessary for solving practical problems. Suppose, we have a random variable X, discrete or continuous, that takes values in a set S, and a function g: $\mathbb{R} \rightarrow \mathbb{R}$ that maps S into another set T. Then, $Y = g(X)$ is also a random variable and g can be considered a transformation. The distribution (probability density function), mean, variance, etc. of Y will differ from that of X. However, if the distribution of X is known it is possible to find the distribution of Y.

Example 6.2.5:

1. $Y = X^2$
2. $Y = 4 - X^2$
3. $Y = \sqrt{X}$

It can be seen from (6.17) that the potential density function (pdf) is the derivative of the cdf. Therefore one way to derive the pdf of the new variable is to find its cdf and take the derivative to obtain the pdf. This is illustrated in the following example.

Example 6.2.6:

Let us consider a random variable X, the pdf of which is given as follows:

$$f_X(x) = \begin{cases} 4 - x & \text{for } 0 < x < 4 \\ 0 & \text{otherwise} \end{cases}$$

and a transformation $Y = X^2$

Let $F_X(x)$ be the cdf of X. Then from (6.15), the cdf of Y is given be

$$F_Y(y) = P(Y \leq y) = P\left(X^2 \leq x\right) = P\left(X \leq \sqrt{y}\right) = F_X\left(\sqrt{y}\right)$$

That is, the cdf $F_Y(y)$ is the cdf $F_X(x)$ evaluated at $\sqrt{y}$. Hence, the pdf $f_Y(y)$ is

$$f_Y(y) = \frac{d}{dy}F_Y(y) = \frac{d}{dy}F_X\left(\sqrt{y}\right) = f_X\left(\sqrt{y}\right)\frac{d}{dy}\sqrt{y} = \left(4 - \sqrt{y}\right)\frac{1}{2\sqrt{y}} = \frac{1}{2}\left(\frac{4}{\sqrt{y}} - 1\right)$$

Linear transformations of random variables

It is to be noted that none of the transformations enlisted in Example 6.2.5 is a linear transformation. However, if X is a random variable and a and b are constants ($b > 0$), then, $g(X) = a + bX$ is a ***linear transformation*** of X. Clearly, the function scales X by b and shifts it by a.

Further, If X be a random variable with an $N(\mu, \sigma^2)$ distribution given by (6.22), then for any a and any $b \neq 0$, the random variable $a + bX$ has an $N(a + b\mu, b^2\sigma^2)$ distribution. Therefore, if we make a linear transformation of X by taking $a = -\mu/\sigma$ and $b = 1/\sigma$, the new random variable

$$Z = \left(-\frac{\mu}{\sigma}\right) + \frac{1}{\sigma}X = \frac{X - \mu}{\sigma} \tag{6.24}$$

has an $N(0, 1)$ standard normal distribution with the probability density function

$$f_Z(z) = \frac{1}{\sqrt{2\pi}} e^{-\frac{1}{2}z^2} \quad \text{for } -\infty < z < \infty \tag{6.25}$$

6.2.4 Expectation and variance

Two concepts fundamental to the probability theory and random variables are that of *expected value* and *variance*. The expected value, also called the *expectation* or *mean*, indicates the "center" of the distribution of a random variable. The variance, on the other hand, is a measure of the "spread" of its distribution.

Expectation of a discrete random variable

Definition 6.2.13 If X be a discrete random variable taking the values $x_1, x_2, \ldots$ and has a *probability mass function* $p(x)$, then the expectation (or expected value or mean) of X, denoted by $E(X)$ or μ_X, is defined by

$$E(X) = \sum_i x_i P(X = x_i) = \sum_i x_i p(x_i)$$

Example 6.2.7: If $P(X = -2) = 0.3$, $P(X = 3) = 0.5$, and $P(X = 5) = 0.2$, then

$$E(X) = (-2)(0.3) + (3)(0.5) + (5)(0.2) = -0.6 + 1.5 + 1.0 = 1.9$$

Example 6.2.8: Let us find the expectation of a Poisson (λ) distribution given in Definition 6.2.4. We have

$$E(X) = \sum_{k=0}^{\infty} k e^{-\lambda} \frac{\lambda^k}{k!} = \sum_{k=1}^{\infty} e^{-\lambda} \frac{\lambda^k}{(k-1)!} = \lambda e^{-\lambda} \sum_{k=1}^{\infty} \frac{\lambda^{k-1}}{(k-1)!}$$

Substituting $l = k-1$,

$$E(X) = \lambda e^{-\lambda} \sum_{l=0}^{\infty} \frac{\lambda^l}{l!} = \lambda e^{-\lambda} e^{\lambda} = \lambda$$

Expectation of a continuous random variable

Definition 6.2.14

If X be a continuous random variable with probability density function f, then the expectation of X is given by

$$E(X) = \int_{-\infty}^{\infty} x f(x) dx$$

Example 6.2.9: Let us find the expectation of an exponential distribution, $Exp(\lambda)$, given in Definition 6.2.11. We have

$$E(X) = \int_{0}^{\infty} x\lambda e^{-\lambda x} dx$$

Let $u = x$ and $dv = \lambda\, e^{-\lambda x}\, dx$, so that $du = dx$ and $v = -e^{-\lambda x}$. Then, integrating by parts

$$E(X) = -xe^{-\lambda x}\Big|_0^{\infty} + \int_{0}^{\infty} e^{-\lambda x} dx = -\frac{e^{-\lambda x}}{\lambda}\Big|_0^{\infty} = \frac{1}{\lambda}$$

Example 6.2.10: Let us find the expectation of a standard normal $N(0, 1)$ distribution given in (6.25). We have

$$E(Z) = \int_{-\infty}^{\infty} z\frac{1}{\sqrt{2\pi}} e^{-\frac{1}{2}z^2} dz$$

$$= \int_{-\infty}^{0} z\frac{1}{\sqrt{2\pi}} e^{-\frac{1}{2}z^2} dz + \int_{0}^{\infty} z\frac{1}{\sqrt{2\pi}} e^{-\frac{1}{2}z^2} dz \tag{6.26}$$

If we substitute u = −z, we find

$$\int_{-\infty}^{0} z\frac{1}{\sqrt{2\pi}} e^{-\frac{1}{2}z^2} dz = \int_{0}^{\infty} (-u)\frac{1}{\sqrt{2\pi}} e^{-\frac{1}{2}u^2} du$$

so that the two integrals in (6.26) cancel each other. Therefore $E(Z) = 0$.

Example 6.2.11: Expectation of $N(\mu, \sigma^2)$ distribution.

$$E(X) = \int_{-\infty}^{\infty} x\frac{1}{\sigma\sqrt{2\pi}} e^{-\frac{1}{2}\left(\frac{x-\mu}{\sigma}\right)^2} dx$$

Since normal density integrates to 1, we can write

$$E(X) = \mu + \int_{-\infty}^{\infty} (x-\mu)\frac{1}{\sigma\sqrt{2\pi}} e^{-\frac{1}{2}\left(\frac{x-\mu}{\sigma}\right)^2} dx$$

Substituting $z = (x - \mu)/\sigma$.

$$E(X) = \mu + \sigma \int_{-\infty}^{\infty} z \frac{1}{\sqrt{2\pi}} e^{-\frac{1}{2}z^2} dz = \mu + 0 = \mu$$

As mentioned earlier, the variance tells us about how far a random variable X deviates from the expected value or mean $E(X)$. Hence, we have the following definition.

Definition 6.2.15 The variance of a random variable X is the quantity

$$\sigma_x^2 = \text{Var}(X) = E\left[(X - \mu_X)^2\right]$$

where $\mu_X = E(X)$ is the mean of X.

In practical situations the standard deviation defined by $\sqrt{\text{Var}(X)}$, which has the same dimension as $E(X)$, is found to be more useful.

Definition 6.2.16 The ***standard deviation*** of a random variable X is the quantity

$$\sigma_X = \text{Sd}(X) = \sqrt{\text{Var}(X)} = \sqrt{E\left((X - \mu_X)^2\right)}$$

Some important properties of Var are given next without proof. These are found to be very useful in computing variance.

Theorem 6.2.1 If X be any random variable with expected value $\mu_X = E(X)$ and variance Var (X), then the following are true:

1. $\text{Var}(X) \geq 0$;
2. $\text{Var}(aX + b) = a^2 \text{Var}(X)$, where $a, b \in \mathbb{R}$
3. $\text{Var}(X) = E(X^2) - (\mu_X)^2 = E(X^2) - E(X)^2$
4. $\text{Var}(X) \leq E(X^2)$

Example 6.2.12: Variance of exponential distribution *Exp* (λ).

For a random variable X with an exponential distribution, the probability density function $f(x) = \lambda e^{-\lambda x}$ and expectation $E(X) = 1/\lambda$. Further,

$$E(X^2) = \int_0^{\infty} x^2 \lambda e^{-\lambda x} dx$$

Let $u = x^2$ and $dv = \lambda e^{-\lambda x} dx$, so that $du = 2x\,dx$ and $v = -e^{-\lambda x}$. Then integrating by parts,

$$E(X^2) = -x^2 e^{-\lambda x}\Big|_0^{\infty} + \int_0^{\infty} 2x e^{-\lambda x} dx = \frac{2}{\lambda}\int_0^{\infty} x\lambda e^{-\lambda x} dx = \frac{2}{\lambda} E(X) = \frac{2}{\lambda^2}$$

Hence, using Theorem 6.2.1 (no. 3),

$$\mathrm{Var}(X) = E\left(X^2\right) - E(X)^2 = 1/\lambda^2$$

Example 6.2.13: Variance of normal $N(\mu, \sigma^2)$ distribution.

If a random variable X has a normal distribution, then by Definition 6.2.15

$$\begin{aligned}\mathrm{Var}(X) &= E\left[(X - E[X])^2\right] \\ &= \int_{-\infty}^{\infty} (x-\mu)^2 \frac{1}{\sigma\sqrt{2\pi}} e^{-\frac{1}{2}\left(\frac{(x-\mu)}{\sigma}\right)^2} dx\end{aligned}$$

Substituting $z = (x - \mu)/\sigma$ and integrating by parts, it can be shown that

$$\mathrm{Var}(X) = \sigma^2 \int_{-\infty}^{\infty} z^2 \frac{1}{\sqrt{2\pi}} e^{-\frac{1}{2}z^2} dz = \sigma^2$$

6.3 Multivariate distribution

Studies in physical, chemical, biological, and even social sciences are usually based on measurements of several variables simultaneously. For example, a molecular dynamics trajectory is essentially multivariate data, where a large number of variables representing atomic positions are found to be interrelated, or interdependent.

The discussion in Section 6.2 has been around the case of a single random variable (the *univariate* case). In this section, we shall enlarge the mathematical structure to the *multivariate* case. To begin with, the *bivariate* case, in which pairs of random variables are involved, is going to be considered.

6.3.1 Bivariate distribution

In Section 6.2, a random variable was defined as a function from the sample space S to the set $\mathbb{R}$ of real numbers. Here, X and Y are a pair of random variables on the same sample space S, defined in relation to the same experiment and it is interesting to discuss how they may interact or affect one another.

Alternatively, we can consider X_1 and X_2 as an ordered pair of random variables and define $\mathbf{X} = [X_1, X_2]$ as a ***random vector*** in two-dimension (2-space, $\mathbb{R}^2$).

Accordingly, the joint distribution of two discrete random variables X and Y can be given by extending Definition 6.2.3, that is, by defining the probabilities of all possible values of the pair (X, Y).

Definition 6.3.1 The ***joint probability mass function*** (in the present case, *bivariate probability function*) of two random variables X and Y is the function p: $\mathbb{R}^2 \rightarrow [0, 1]$ defined by

$$p\,(x, y) = P(X = x, Y = y) \quad \text{for} - \infty < x, y < \infty$$

From the joint *probability mass function* (pmf), we have the following definition for the *joint distribution function*.

Definition 6.3.2 The *joint distribution function* F: $\mathbb{R}^2 \rightarrow [0, 1]$ is defined as

$$F\,(x, y) = P(X \leq x, Y \leq y) \quad \text{for} - \infty < x, y < \infty$$

With $p\ (x, y)$ representing the mass assigned to the point (x, y), we have

$$\sum_{\text{all } x} \sum_{\text{all } y} p\,(x, y) = 1 \tag{6.27}$$

Example 6.3.1: Let us consider two polypeptides consisting of amino acid residues lysine (K) and arginine (R) randomly distributed. An experiment is carried out to randomly pick three dipeptide sequences from each of the polypeptides. Let the random variables

X = Number of Ks for the first polypeptide
Y = Number of Rs for the second polypeptide

Since two different (independent) polypeptides have been taken, for any pair of possible values (x, y) of X and Y

$$p\,(x, y) = P\left(\{X = x\} \cap \{Y = y\}\right) = P(X = x) \cdot P(Y = y) = p(x) \cdot p(y) \tag{6.28}$$

The bivariate probability distribution of X and Y is summarized in Table 6.1.

Table 6.1 Bivariate probability function of X and Y.

		y			
		0	**1**	**2**	**3**
	0	1/64	3/64	3/64	1/64
	1	3/64	9/64	9/64	3/64
x	2	3/64	9/64	9/64	3/64
	3	1/64	3/64	3/64	1/64
		p (*x*, *y*)			

In the case of continuous random variables X and Y, we define an event as X observed in $[x, x+dx]$ and an event as Y observed in $[y, y+dy]$. So, the probability of the joint event, that is, X observed in $[x, x+dx]$ as well as Y observed in $[y, y+dy]$ is written as

$$P(x \le X \le x+dx, y \le Y \le y+dy) = f(x,y)\,dx\,dy$$

The function f must satisfy two properties:

1. $f(x, y) \ge 0$ for all x and y; and
2. $\int_{\infty}^{\infty} \int_{-\infty}^{\infty} f(x,y)\,dx\,dy = 1$

f is called the ***joint probability density function*** of X and Y. Then, we can define the joint continuous distribution of X and Y.

Definition 6.3.3 If X takes values in $[a, b]$ and Y takes values in $[c, d]$ so that the pair (X, Y) takes values in $[a, b] \times [c, d]$ giving the joint probability density function $f(x, y)$: $\mathbb{R}^2 \to \mathbb{R}$ at (x, y), then

$$P(a \le X \le b, c \le Y \le d) = \int_a^b \int_c^d f(x,y)dx\,dy$$

Example 6.3.2: Let us consider the bivariate density function

$$f(x,y) = 4xy \quad \text{for } 0<x<1 \text{ and } 0<y<1$$
$$= 0 \quad \text{otherwise}$$

Clearly, $f(x, y) \ge 0$ for all x and y. Further, to confirm that $f(x, y)$ is a valid joint pdf, we calculate the total probability

$$\int_0^1 \int_0^1 4xy\,dx\,dy = \int_0^1 \left[2x^2 y\right]_0^1 dy = \int_0^1 2y\,dy = 1$$

So, the conditions for a joint pdf are satisfied.

Now, let us calculate $P(0.25 \le X \le 0.75, 0.25 \le Y \le 0.75)$ as follows:

$$P(0.25 \le X \le 0.75, 0.25 \le Y \le 0.75) = \int_{0.25}^{0.75} \int_{0.25}^{0.75} 4xy\,dx\,dy$$

$$= \int_{0.25}^{0.75} \left[2x^2 y\right]_{0.25}^{0.75} dy = \int_{0.25}^{0.75} y\,dy = \left[y^2\right]_{0.25}^{0.75} = \frac{1}{2}$$

As in the case of single random variable, cdf is defined for jointly distributed random variables as well.

Definition 6.3.4 For jointly distributed random variables X and Y, the ***joint cumulative distribution function*** $F_{X,Y}$: $\mathbb{R}^2 \to [0, 1]$ is defined as

$$F_{X,Y} = P(X \le x, Y \le y)$$

For discrete random variables X and Y with joint pmf $p(x, y)$, the joint cdf is given by

$$F_{X,Y} = \sum_{x_i \leq x} \sum_{y_i \leq y} p(x_i, y_i) \tag{6.29}$$

If X and Y are continuous random variables with joint probability density function $f(x, y)$, then the joint cdf is given by

$$F_{X,Y} = \int_{-\infty}^{x} \int_{-\infty}^{y} f(x_i, y_i) dx_i \, dy_i \tag{6.30}$$

The joint pdf can be recovered from the joint cdf as

$$f(x, y) = \frac{\partial^2}{\partial x \, \partial y} F(x, y) \tag{6.31}$$

Example 6.3.3: The joint cdf of the random variables in Example 6.3.2 is

$$F_{X,Y} = \int_{-\infty}^{x} \int_{-\infty}^{y} f(x_i, y_i) dx_i \, dy_i = \int_0^y \int_0^x f(x_i, y_i) dx_i \, dy_i = x^2 y^2$$

Marginal distribution

When X and Y are jointly distributed random variables, and joint pmf/pdf or cdf is accessible, it may be necessary in some cases to consider each variable's probability distribution individually, that is, the *marginal distribution*. As shown next, it is possible to compute marginal pmf/pdf from joint pmf/pdf and marginal cdf from joint cdf.

Theorem 6.3.1 If X and Y are two jointly distributed discrete random variables, with joint pmf $p(x, y)$, then the marginal pmfs of X and Y are given, respectively, by

$$p_X(x) = \sum_y p(x, y)$$

$$p_Y(y) = \sum_x p(x, y)$$

Theorem 6.3.2 When X and Y are two jointly distributed continuous random variables, with joint probability density function $f(x, y)$, then the marginal probability densities of X and Y are given, respectively, by

$$f_X(x) = \int_{-\infty}^{\infty} f(x, y) dy$$

$$f_Y(y) = \int_{-\infty}^{\infty} f(x, y) dx$$

Example 6.3.4: Let us consider the bivariate density function

$$\begin{aligned} f(x,y) &= 2xe^{-y} \quad \text{for} \quad 0 \le x \le 1 \text{ and } y \ge 0 \\ &= 0 \quad \text{otherwise} \end{aligned}$$

It can be shown that $f(x, y)$ is a valid joint pdf by calculating

$$\int_0^\infty \int_0^1 2xe^{-y} dx\,dy = 1$$

The marginal pdfs $f_X(x)$ and $f_Y(y)$ are calculated as follows:

$$f_X(x) = \int_0^\infty 2xe^{-y}\,dy = -2xe^{-y}\big|_0^\infty = 2x$$

$$f_Y(y) = \int_0^1 2xe^{-y}\,dx = x^2 e^{-y}\big|_0^1 = e^{-y}$$

Likewise, from the joint cdf, we can derive the marginal cdf of either X or Y separately.

Theorem 6.3.3 If X and Y are two jointly distributed random variables, with the joint *cumulative distribution function* $F_{X,Y}$, then the cdfs of X and Y are given, respectively, by

$$F_X(x) = \lim_{y \to \infty} F_{X,Y}(x,y)$$

$$F_Y(y) = \lim_{x \to \infty} F_{X,Y}(x,y)$$

Example 6.3.5: Let us consider the bivariate density function

$$\begin{aligned} f(x,y) &= \frac{3}{2}(x^2 + y^2) \quad \text{for} \quad 0 \le x \le 1 \text{ and } 0 \le y \le 1 \\ &= 0 \quad \text{otherwise} \end{aligned}$$

It can be verified that

$$\int_{-\infty}^\infty \int_{-\infty}^\infty \frac{3}{2}(x^2 + y^2)dx\,dy = \int_0^1 \int_0^1 \frac{3}{2}(x^2 + y^2)dx\,dy = 1$$

The joint cdf is given by

$$F_{X,Y} = \int_0^y \int_0^x \frac{3}{2}(x^2 + y^2)dx\,dy = \frac{1}{2}(x^3 y + xy^3)$$

From the joint cdf the marginal cdfs can be calculated as

$$F_X(x) = \frac{1}{2}(x^3 + x) \quad \text{and} \quad F_Y(y) = (y + y^3)$$

6.3.2 Generalized multivariate distribution

The concepts of multivariate distribution, as emerging from the discussion on the bivariate case, can now be generalized to n-dimensions. An n-dimensional ***random vector*** is considered an ordered list of n random variables. Evidently, the ***space*** of this random vector is the set of ordered n-tuples.

$$\mathbf{X} = [X_1, X_2, \ldots, X_n]$$

When the random variables $X_1, X_2, \ldots, X_n$ are discrete, we have the following definition of the *joint pmf.*

Definition 6.3.5 The ***joint probability mass function*** n random variables $X_1, X_2, \ldots, X_n$ is the function p: $\mathbb{R}^n \to [0, 1]$ defined by

$$p(x_1, x_2, \ldots, x_n) = P(X_1 = x_1, X_2 = x_2, \ldots, X_n = x_n) \quad \text{for} -\infty < x_1, x_2, \ldots, x_n < \infty$$

Further, we have

$$\sum_{\text{all } x_1, x_2, \ldots, x_n} \cdots \sum p(x_1, x_2, \ldots, x_n) = 1$$

In the continuous case, the probability of the joint event is expressed as

$$P(x_1 \leq X_1 \leq x_1 + dx_1, x_2 \leq X_2 \leq x_2 + dx_2, \ldots, x_n \leq X_n \leq x_n + dx_n)$$
$$= f(x_1, x_2, \ldots, x_n)\, dx_1\, dx_2 \ldots dx_n$$

where the function f satisfying the following properties

1. $f(x_1, x_2, \ldots, x_n) \geq 0$; and
2. $\int \cdots \int_S f(x_1, x_2, \ldots, x_n)\, dx_1\, dx_2 \ldots dx_n = 1$, S denotes entire parameter space

is called ***joint probability density function*** of the random vector $\mathbf{X}$. Consequently, we have the following definition.

Definition 6.3.6 If X_i takes values in $[a_i, b_i]$, $i = 1, 2, \ldots, n$, giving the joint probability density function $f(x_1, x_2, \ldots, x_n)$: $\mathbb{R}^n \to \mathbb{R}$ at $(x_1, x_2, \ldots, x_n)$, then

$$P(a_1 \leq X_1 \leq b_1, a_2 \leq X_2 \leq b_2, \ldots, a_n \leq X_n \leq b_n)$$
$$= \int_{a_n}^{b_n} \cdots \int_{a_2}^{b_2} \int_{a_1}^{b_1} f(x_1, x_2, \ldots, x_n)\, dx_1\, dx_2 \ldots dx_n$$

It is to be noted that a collection of random variables $X_1, X_2, \ldots, X_n$ are independent if their multivariate distribution can be expressed as a product of the univariate distributions for all individual random variables. In the discrete case

$$p(x_1, x_2, \ldots, x_n) = p(x_1)\, p(x_2) \cdots p(x_n) \tag{6.32}$$

whereas in the continuous case

$$f(x_1, x_2, \ldots, x_n) = f(x_1) f(x_2) \cdots f(x_n) \tag{6.33}$$

Now, we move on to the joint cdf which for n random variables is defined here.

Definition 6.3.7 For jointly distributed n random variables $X_1, X_2, \ldots, X_n$, the ***joint cdf*** $F_{\mathbf{X}}$: $\mathbb{R}^n \rightarrow [0, 1]$ is defined as

$$F_{\mathbf{X}}(x) = P(X_1 \leq x_1, X_2 \leq x_2, \ldots, X_n \leq x_n)$$

For discrete random variables $X_1, X_2, \ldots, X_n$ with joint pmf $p(x_1, x_2, \ldots, x_n)$, the joint cdf is given by

$$F_{\mathbf{X}}(\mathbf{x}) = \sum_{x_{1i} \leq x_1, x_{2i} \leq x_2, \ldots, x_{ni} \leq x_n} \sum \cdots \sum p(x_{1i}, x_{2i}, \ldots, x_{ni}) \tag{6.34}$$

If $X_1, X_2, \ldots, X_n$ are continuous random variables with joint probability density function $f(x_1, x_2, \ldots, x_n)$, then the joint cdf is given by

$$F_{\mathbf{X}}(\mathbf{x}) = \int_{-\infty}^{x_n} \cdots \int_{-\infty}^{x_2} \int_{-\infty}^{x_1} f(x_{1i}, x_{2i}, \ldots, x_{ni}) \, dx_{1i} \, dx_{2i} \ldots dx_{ni} \tag{6.35}$$

Further, for the continuous case

$$f(\mathbf{x}) = \frac{\partial^n}{\partial x_1 \, \partial x_2 \ldots \partial x_n} F_{\mathbf{X}}(\mathbf{x}) \tag{6.36}$$

Example 6.3.6: Let us consider the joint distribution of three random variable X_1, X_2, and X_3 with joint pdf

$$f(x_1, x_2, x_3) = \begin{cases} e^{-(x_1 + x_2 + x_3)} & \text{for } 0 < x_1, x_2, x_3 < \infty \\ 0 & \text{elsewhere} \end{cases}$$

and calculate the joint cdf.

$$\begin{aligned} F_{X_1, X_2, X_3}(x_1, x_2, x_3) &= \int_0^{x_3} \int_0^{x_2} \int_0^{x_1} e^{-(x_1 + x_2 + x_3)} dx_{1i} \, dx_{2i} \, dx_{3i} \\ &= (1 - e^{-x_1})(1 - e^{-x_2})(1 - e^{-x_3}) \text{for } 0 < x_1, x_2, x_3 < \infty \end{aligned}$$

6.4 Covariance and correlation

A distinguishable feature of protein dynamics is the collective nature of fluctuations involving atoms and residues. One approach to study the dynamics of a protein molecule has made use of an elastic network model where each "node" is the C_α atom of a residue and a "spring" represents the interaction between two nodes. The equilibrium position of a node, i, is denoted by the vector $\mathbf{R}_i^0$, while the instantaneous position by $\mathbf{R}_i$. The fluctuation or deformation from the mean position is, therefore, defined as $\Delta\mathbf{R}_i = \mathbf{R}_i - \mathbf{R}_i^0$. Typically, the primary goal is to analyze the fluctuations of a particular residue, i, or the "correlation" between the fluctuations between two different residues, i and j.

Earlier in this chapter (Section 6.2.4), we have discussed the basic concepts of expectation and variance for a single random variable. In this section, we shall see how the expectation of a combination of random variables can be computed from the joint multivariate distribution, and introduce the notions of covariance and correlation.

Using Definition 6.2.13 for the expectation of a random variable X or Y and Definition 6.3.1 for the joint pmf of two random variables X and Y, it can be shown that

$$E(X+Y) = E(X) + E(Y) \tag{6.37}$$

This linearity of expectations of random variables is formalized in the following theorem which holds for both discrete and continuous random variables.

Theorem 6.4.1 Linearity of expectations. For all r, s, and $t \in \mathbb{R}$, and random variables X and Y,

$$E(rX + sY + t) = rE(X) + sE(Y) + t \tag{6.38}$$

This can be generalized for n random variables and $s_0, s_1, s_2, \ldots, s_n \in \mathbb{R}$

$$E\,(s_0 + s_1X_1 + s_2X_2 + \ldots + s_nX_n) = s_0 + s_1E(X_1) + s_2E(X_2) + \ldots + s_nE(X_n) \tag{6.39}$$

In general, $E\,(XY) \neq E\,(X)\,E(Y)$. However, if X and Y are independent, we have the following theorem:

Theorem 6.4.2 If two random variables X and Y, discrete or continuous, are independent, then

$$E(XY) = E(X)E(Y) \tag{6.40}$$

In contrast, using Definition 6.2.15 for the variance of a random variable,

$$\mathrm{Var}(X+Y) = \mathrm{Var}(X) + \mathrm{Var}(Y) + 2E[(X - E(X))\,(Y - E(Y))] \tag{6.41}$$

That is, an additional term $2E\,[(X-E\,(X))\,(Y-E\,(Y))]$ is involved here.

Now, we move on to the definition of covariance which measures the relationship between two random variables.

Definition 6.4.1 The ***covariance*** of two random variables X and Y is defined as

$$\mathrm{Cov}(X, Y) = E\left(\left(X - \mu_X\right)\left(Y - \mu_Y\right)\right) \tag{6.42}$$

where $\mu_X = E\,(X)$ and $\mu_Y = E\,(Y)$.

From (6.42), we have

$$\mathrm{Cov}(X, Y) = E\left(XY - \mu_Y X - \mu_X Y + \mu_X \mu_Y\right)$$

Applying Theorem 6.4.1

$$\mathrm{Cov}(X, Y) = E(XY) - \mu_Y E(X) - \mu_X E(Y) + \mu_X \mu_Y$$

or,

$$\mathrm{Cov}(X, Y) = E(XY) - \mu_X \mu_Y \tag{6.43}$$

Eq. (6.43) is an alternative expression for the covariance of two random variables X and Y.

A positive covariance, Cov $(X, Y) > 0$, implies that if $X > E(X)$, it is likely that $Y > E(Y)$, and X and Y are said to be positively correlated (Fig. 6.4A). When Cov $(X, Y) < 0$, X and Y are negatively correlated (Fig. 6.4B). Cov $(X, Y) = 0$ indicates that X and Y are uncorrelated (Fig. 6.4C).

Applying Theorem 6.4.2 to (6.43), we have the following observation.

If two random variables X and Y are independent, then Cov $(X, Y) = 0$.

However, the converse is not necessarily true. In some cases, Cov $(X, Y) = 0$, although X and Y are not independent.

Although the covariance between random variables indicates how they may influence each other, one disadvantage is that it depends on the units in which the random variables are represented. Hence, Cov (X, Y) may not always be suitable to express the dependence between X and Y, and a standardized (unitless) version of the covariance is preferred. This version is called the *correlation coefficient* of X and Y.

Definition 6.4.2 Considering two jointly distributed random variables X and Y, with their means denoted, respectively, by μ_X and μ_Y, and their variances by σ_X^2 and σ_Y^2 ($\sigma_X, \sigma_Y > 0$), the ***correlation coefficient*** between X and Y is given by

$$\rho(X, Y) = \frac{E\left((X - \mu_X)(Y - \mu_Y)\right)}{\sigma_X \sigma_Y} = \frac{\mathrm{Cov}(X, Y)}{\sigma_X \sigma_Y} \tag{6.44}$$

Clearly, $\rho(X, Y)$ is dimensionless.

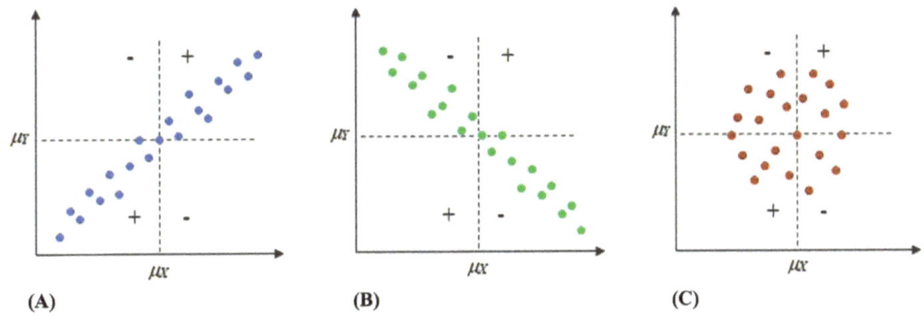

FIGURE 6.4

Figure cov (A) positive covariance, (B) negative covariance, and (C) covariance near zero.

Covariance matrix

Let $\mathbf{X}$ be a $p \times 1$ random vector representing a multivariate population under consideration.

$$\mathbf{X} = \begin{bmatrix} X_1 \\ \vdots \\ X_p \end{bmatrix}$$

We have the following definition for its expectation.

Definition 6.4.3 The expectation of a random vector $\mathbf{X}$ is the vector $\boldsymbol{\mu}$ of the expectations of its components.

$$\mathbf{E}(\mathbf{X}) = \begin{bmatrix} E(X_1) \\ \vdots \\ E(X_p) \end{bmatrix} = \begin{bmatrix} \mu_1 \\ \vdots \\ \mu_p \end{bmatrix} = \boldsymbol{\mu}$$

Based on Definition 6.2.3, the variance–covariance matrix can be defined.

Definition 6.4.4 For a p-dimensional random vector $\mathbf{X} = [X_1, \ldots, X_n]^T$, the ***variance–covariance matrix*** is defined as

$$\mathrm{Cov}(\mathbf{X}) = \mathbf{E}\left[(\mathbf{X} - \boldsymbol{\mu})(\mathbf{X} - \boldsymbol{\mu})^T\right] = \left[\sigma_{ij}\right]$$

where $\sigma_i^2 = \mathrm{Var}\,(X_i) < \infty$ and the off-diagonal (i, j)th entry is $\mathrm{Cov}\,(X_i, X_j)$.

The variance–covariance matrix can be denoted by $\boldsymbol{\Sigma}$ and written as

$$\sum = \mathrm{Cov}(\mathbf{X}) = \begin{bmatrix} \mathrm{Var}(X_1) & \mathrm{Cov}(X_1, X_2) & \ldots & \mathrm{Cov}(X_1, X_p) \\ \mathrm{Cov}(X_2, X_1) & \mathrm{Var}(X_2) & \ldots & \mathrm{Cov}(X_2, X_p) \\ \vdots & & & \vdots \\ \mathrm{Cov}(X_p, X_1) & \mathrm{Cov}(X_p, X_2) & \ldots & \mathrm{Var}(X_p) \end{bmatrix}$$

$$= \begin{bmatrix} \sigma_{11} & \sigma_{12} & \ldots & \sigma_{1p} \\ \sigma_{21} & \sigma_{22} & \ldots & \sigma_{2p} \\ \vdots & & & \vdots \\ \sigma_{p1} & \sigma_{p1} & \ldots & \sigma_{pp} \end{bmatrix}$$

It is to be noted that $\mathrm{Cov}\,(X_i, X_i) = \mathrm{Var}\,(X_i) = \sigma_{ii}$. Further, $\sigma_{ij} = \sigma_{ji}$, for all i, j so that the matrix is symmetric.

Example 6.4.1: Let us consider the bivariate density function

$$f(x, y) = x + \frac{3}{2}y^2 \quad \text{for} \quad 0 < x < 1 \text{ and } 0 < y < 1$$
$$= 0 \qquad \text{otherwise}$$

and define the random vector

$$\mathbf{Z} = \begin{bmatrix} X \\ Y \end{bmatrix}$$

and calculate the covariance matrix of **Z**.

First, we calculate the marginal pdfs of X and Y using Theorem 6.3.2

$$f_X(x) = \int_0^1 \left(x + \frac{3}{2}y^2 \right) dy = x + \frac{1}{2} \qquad \text{for } 0 < x < 1$$

$$f_Y(y) = \int_0^1 \left(x + \frac{3}{2}y^2 \right) dx = \frac{1}{2}\left(3y^2 + 1\right) \quad \text{for } 0 < y < 1$$

From the pdfs, we find

$E(X) = 7/12$
$E(X^2) = 5/12$
$E(Y) = 5/8$
$E(Y^2) = 7/15$

and

$$E(XY) = \int_0^1 \int_0^1 xy \left(x + \frac{3}{2}y^2 \right) dx\, dy = \frac{17}{48}$$

Then, we calculate

$$\text{Var}(X) = E\left(X^2\right) - E(X)^2 = 11/144$$

and

$$\text{Var}(Y) = E\left(Y^2\right) - E(Y)^2 = 73/960$$

so that

$$\text{Cov}(X, Y) = E(XY) - E(X)E(Y) = -1/96$$

The covariance matrix is given by

$$\sum\nolimits_{\mathbf{Z}} = \begin{bmatrix} \text{Var}(X) & \text{Cov}(X, Y) \\ \text{Cov}(X, Y) & \text{Var}(Y) \end{bmatrix} = \begin{bmatrix} \frac{11}{144} & -\frac{1}{96} \\ -\frac{1}{96} & \frac{73}{960} \end{bmatrix}$$

Multivariate normal distribution

Earlier, in (6.22), we have seen that for the univariate normal distribution, with mean μ and variance σ^2, the probability density function is

$$f_X(x) = \frac{1}{\sigma\sqrt{2\pi}} e^{-\frac{1}{2}\left(\frac{x-\mu}{\sigma}\right)^2}$$

We can write

$$\left(\frac{x-\mu}{\sigma}\right)^2 = (x-\mu)\left(\sigma^2\right)^{-1}(x-\mu) \tag{6.45}$$

Eq. (6.45) can be generalized to $p > 1$ variables as

$$(\mathbf{x}-\boldsymbol{\mu})^T \sum^{-1} (\mathbf{x}-\boldsymbol{\mu})$$

where

$$\mathbf{x} = \begin{bmatrix} x_1 \\ \vdots \\ x_p \end{bmatrix} \quad -\infty < x_i < \infty \quad \text{for } i = 1, \ldots, p$$

are the observations on the random vector

$$\mathbf{X} = \begin{bmatrix} X_1 \\ \vdots \\ X_p \end{bmatrix},$$

$$\boldsymbol{\mu} = \begin{bmatrix} \mu_1 \\ \vdots \\ \mu_p \end{bmatrix}$$

is the expectation, and

$$\sum = \begin{bmatrix} \sigma_{11} & \sigma_{12} & \cdots & \sigma_{1p} \\ \sigma_{21} & \sigma_{22} & \cdots & \sigma_{2p} \\ \vdots & \vdots & \vdots & \vdots \\ \sigma_{p1} & \sigma_{p1} & \cdots & \sigma_{pp} \end{bmatrix}$$

is the variance–covariance matrix of **X**. Accordingly, we have the following definition of the multivariate normal density function.

Definition 6.4.5 For the random vector $\mathbf{X} \in \mathbb{R}^p$ with normal distribution denoted by $N_p(\boldsymbol{\mu}, \boldsymbol{\Sigma})$ the probability density function is given by

$$f(\mathbf{x}) = \frac{1}{(2\pi)^{p/2}|\boldsymbol{\Sigma}|^{1/2}} e^{-(\mathbf{x}-\boldsymbol{\mu})^T \boldsymbol{\Sigma}^{-1} (\mathbf{x}-\boldsymbol{\mu})/2} \tag{6.46}$$

where $-\infty < x_i < \infty$ for $i = 1, \ldots, p$.

For $p = 1$, (6.46) reduces to (6.22), which is the univariate normal pdf.

For bivariate normal distribution, $p = 2$

$$\mathbf{x} = \begin{bmatrix} x_1 \\ x_2 \end{bmatrix} \boldsymbol{\mu} = \begin{bmatrix} \mu_1 \\ \mu_2 \end{bmatrix} \boldsymbol{\Sigma} = \begin{bmatrix} \sigma_{11} & \sigma_{12} \\ \sigma_{12} & \sigma_{22} \end{bmatrix}$$

and

$$\boldsymbol{\Sigma}^{-1} = \frac{1}{\sigma_{11}\sigma_{22} - \sigma_{12}^2} \begin{bmatrix} \sigma_{22} & -\sigma_{12} \\ -\sigma_{12} & \sigma_{11} \end{bmatrix}$$

It can be shown that for a random vector

$$\mathbf{X} = \begin{bmatrix} X_1 \\ X_2 \end{bmatrix}$$

following a bivariate normal distribution, the pdf is given by

$$f(\mathbf{x}) = \frac{1}{2\pi\sigma_1\sigma_2\sqrt{1-\rho^2}} e^{-q/2}, \ -\infty < x_1, x_2 < \infty$$

where

$$q = \frac{1}{1-\rho^2}\left[\left(\frac{x_1-\mu_1}{\sigma_1}\right)^2 - 2\rho\left(\frac{x_1-\mu_1}{\sigma_1}\right)\left(\frac{x_2-\mu_2}{\sigma_2}\right) + \left(\frac{x_2-\mu_2}{\sigma_2}\right)^2\right]$$

ρ is the correlation coefficient of X_1 and X_2 as in Definition 6.4.2, $\rho^2 < 1$.

It is to be noted that if X denotes a set of n equally likely values $x_1, x_2, \ldots, x_n$, then

$$\mathrm{Var}(X) = \frac{1}{n}\sum_{i=1}^{n}(x_i-\mu)^2 \tag{6.47}$$

where

$$\mu = \frac{1}{n}\sum_{i=1}^{n} x_i$$

For small n, say $n \le 20$, (6.47) is modified as

$$\mathrm{Var}(X) = \frac{1}{n-1}\sum_{i=1}^{n}(x_i-\mu)^2 \tag{6.48}$$

to correct for a small bias.

Similarly, for a vector

$$\mathbf{X} = \begin{bmatrix} X_1 \\ \vdots \\ X_n \end{bmatrix}$$

whose components are equally probable

$$\mathrm{Var}(\mathbf{X}) = \frac{1}{n-1}\sum_{i=1}^{n}(X_i-\overline{X})^2$$

where $\overline{X}$ is the mean. For a pair of n-dimensional vectors $\mathbf{X}$ and $\mathbf{Y}$

$$\mathrm{Cov}(\mathbf{X},\mathbf{Y}) = \frac{1}{n-1}\sum_{i=1}^{n}(X_i-\overline{X})(Y_i-\overline{Y})$$

Theorem 6.4.3 If $\mathbf{X} = (X_1, \ldots, X_n)$ is an n-dimensional random vector, such that $\sigma_i^2 = \text{Var}(X_i) < \infty$ and $\mathbf{A}$ is an m x n matrix of constants, then

$$\text{Var}(\mathbf{AX}) = \mathbf{A}\,\text{Cov}(\mathbf{X})\mathbf{A}^T \geq 0$$
$$\text{Var}(\mathbf{XA}) = \mathbf{A}^T\text{Cov}(\mathbf{X})\mathbf{A} \geq 0$$

6.5 Principal component analysis

Principal component analysis (PCA) is a multivariate statistical technique used in almost all of quantitative sciences. Its purpose is essentially to analyze a data table representing observations described by variables which may or may not be interrelated by extracting important information from the dataset and expressing this information as a set of new orthogonal variables. Let us begin with an example which may subsequently help us understand the fundamentals of PCA.

Example 6.5.1:

Considering a molecule containing six atoms numbered 1–6 (number of entities, $n = 6$), Table 6.2A shows the positions of each atom along a specific direction measured in an arbitrary unit at two different time points (number of variables, $p = 2$).

The variables X and Y are plotted in Fig. 6.5A. There seems to be a positive correlation between the two variables. For further processing the data are centered, that is, from all the values of each variable the corresponding mean is subtracted. The centered data summarized in Table 6.2B are plotted in Fig. 6.5B. Thus the datapoints are centered around the origin.

Table 6.2 Two-dimensional variables: (A) original, and (B) centered.

A.			B.		
Atom no.	Time = t_1 X	Time = t_2 Y	Atom no.	Time = t_1 ($X^* = X - \overline{X}$)	Time = t_2 ($Y^* = Y - \overline{Y}$)
1.	2	4	1.	−3	−4
2.	4	6	2.	−1	−2
3.	4	8	3.	−1	0
4.	6	8	4.	1	0
5.	6	10	5.	1	2
6.	8	12	6.	3	4

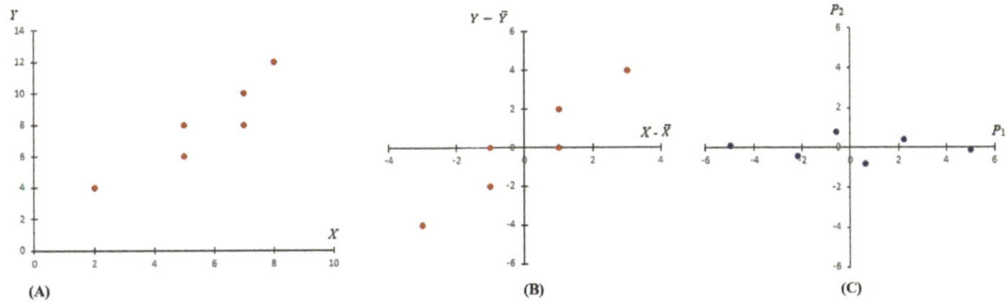

FIGURE 6.5

Graph of (A) sample data, (B) centered data and (C) transformed data.

From the centered datapoints, we have a centered 6 × 2 data matrix

$$\mathbf{D} = \begin{bmatrix} -3 & -4 \\ -1 & -2 \\ -1 & 0 \\ 1 & 0 \\ 1 & 2 \\ 3 & 4 \end{bmatrix} = \begin{bmatrix} \mathbf{X}^* & \mathbf{Y}^* \end{bmatrix} \text{(say)}$$

Let our objective be to obtain a linear combination of the columns of **D** with maximum variance as

$$\begin{bmatrix} -3 & -4 \\ -1 & -2 \\ -1 & 0 \\ 1 & 0 \\ 1 & 2 \\ 3 & 4 \end{bmatrix} \begin{bmatrix} a_1 \\ a_2 \end{bmatrix} = a_1 \begin{bmatrix} -3 \\ -1 \\ -1 \\ 1 \\ 1 \\ 3 \end{bmatrix} + a_2 \begin{bmatrix} -4 \\ -2 \\ 0 \\ 0 \\ 2 \\ 4 \end{bmatrix}$$

The variance and covariance can now be calculated from the centered data as

$$\text{Var}\left(\mathbf{X}^*\right) = \frac{1}{6-1}\sum_{i=1}^{6}\left(X_i - \overline{X}\right)^2 = \frac{22}{5} = 4.4$$

$$\text{Var}\left(\mathbf{Y}^*\right) = \frac{1}{6-1}\sum_{i=1}^{6}\left(Y_i - \overline{Y}\right)^2 = \frac{40}{5} = 8.0$$

$$\text{Cov}\left(\mathbf{X}^*, \mathbf{Y}^*\right) = \frac{1}{6-1}\sum_{i=1}^{6}\left(X_i - \overline{X}\right)\left(Y_i - \overline{Y}\right) = \frac{28}{5} = 5.6$$

Hence, the variance–covariance matrix is

$$\sum = \begin{bmatrix} 4.4 & 5.6 \\ 5.6 & 8.0 \end{bmatrix}$$

The eigenvalues of Σ can be calculated as

$$\lambda_1 = 12.08 \quad \text{and} \quad \lambda_2 = 0.32$$

from which we have the corresponding eigenvectors (normalized)

$$\begin{bmatrix} a_1 \\ a_2 \end{bmatrix} = \begin{bmatrix} 0.59 \\ 0.81 \end{bmatrix} \quad \text{and} \quad \begin{bmatrix} b_1 \\ b_2 \end{bmatrix} \begin{bmatrix} -0.81 \\ 0.59 \end{bmatrix}$$

giving us a transformation matrix

$$\mathbf{T} = \begin{bmatrix} a_1 & b_1 \\ a_2 & b_2 \end{bmatrix} = \begin{bmatrix} 0.59 & -0.81 \\ 0.81 & 0.59 \end{bmatrix}$$

$$\mathbf{DT} = \begin{bmatrix} -3 & -4 \\ -1 & -2 \\ -1 & 0 \\ 1 & 0 \\ 1 & 2 \\ 3 & 4 \end{bmatrix} \begin{bmatrix} 0.59 & -0.81 \\ 0.81 & 0.59 \end{bmatrix} = \begin{bmatrix} -5.0 & 0.1 \\ -2.2 & -0.4 \\ -0.6 & 0.8 \\ 0.6 & -0.8 \\ 2.2 & 0.4 \\ 5.0 & -0.1 \end{bmatrix} = \begin{bmatrix} \mathbf{P}_1 & \mathbf{P}_2 \end{bmatrix} \text{(say)}$$

So, we have two different linear combinations $\mathbf{P}_1$ and $\mathbf{P}_2$ of the columns of $\mathbf{D}$. The transformed variables are plotted in Fig. 6.5C. The variances of the linear combinations are given by

$$\text{Var}(\mathbf{P}_1) = (1/5)\left[(-5.0)^2 + (-2.2)^2 + (-0.6)^2 + (0.6)^2 + (2.2)^2 + (5.0)^2 = 60.4/5 = 12.08\right.$$

$$\text{Var}(\mathbf{P}_2) = (1/5)\left[(0.1)^2 + (-0.4)^2 + (0.8)^2 + (-0.8)^2 + (0.4)^2 + (-0.1)^2 = 1.62/5 = 0.32\right.$$

A few points are to be noted here:

1. Var ($\mathbf{X}^*$) + Var ($\mathbf{Y}^*$) = Var ($\mathbf{P}_1$) + Var ($\mathbf{P}_2$), that is, the total variance remains the same;
2. In the transformed data, the maximum, 97.4% of the total variance is captured by the $\mathbf{P}_1$-component;
3. Var ($\mathbf{P}_1$) = λ_1 and Var ($\mathbf{P}_2$) = λ_2.

It can also be shown that Cov ($\mathbf{P}_1$, $\mathbf{P}_2$) almost disappears indicating very little correlation between $\mathbf{P}_1$ and $\mathbf{P}_2$.

We now move on to the generalized problem of PCA involving a dataset with n individual or entities, and observations on p numerical variables for each entity. The dataset can be represented by a matrix $\mathbf{X} \in \mathbb{R}^{n \times p}$ which consists of column vectors as

$$\mathbf{X} = [\mathbf{x}_1, \ldots, \mathbf{x}_j, \ldots, \mathbf{x}_p]$$

A linear combination of the columns of matrix $\mathbf{X}$ can be written as

$$\mathbf{X}\mathbf{a} = \sum_{j=1}^{p} a_j \mathbf{x}_j \tag{6.49}$$

where

$$\mathbf{a} = \begin{bmatrix} a_1 \\ \vdots \\ a_p \end{bmatrix}$$

is a vector of constants. In PCA the objective is to find a linear combination with maximum variance. The variance of the linear combination (6.49) is

$$\mathrm{Var}(\mathbf{X}\mathbf{a}) = \mathbf{a}^T \mathrm{Cov}(\mathbf{X})\mathbf{a} = \mathbf{a}^T \mathbf{S}\mathbf{a}$$

The problem now reduces to obtaining a p-dimensional vector $\mathbf{a}$ by maximizing the quadratic form

$$f(\mathbf{a}) = \mathbf{a}^T \mathbf{S}\mathbf{a}$$

subject to a constraint that $\mathbf{a}$ is a unit-norm vector, that is, $\mathbf{a}^T\,\mathbf{a} = 1$.

Theorem 6.5.1 Lagrangian optimization

The maximum (or minimum) of a function $f(\mathbf{x})$ subject to a constraint $g(\mathbf{x}) = c$ (where c is a constant) can be obtained by maximizing (or minimizing) the Lagrangian function

$$\mathcal{L}(\mathbf{x}, \lambda) = f(\mathbf{x}) - \lambda(g(\mathbf{x}) - c)$$

where λ is a scalar called the Lagrange multiplier.

Hence, we must maximize the Lagrangian

$$\mathcal{L}(\mathbf{a}, \lambda) = \mathbf{a}^T \mathbf{S}\mathbf{a} - \lambda\left(\mathbf{a}^T\mathbf{a} - 1\right)$$

This requires

$$\frac{d}{d\mathbf{a}}\left[\mathbf{a}^T\mathbf{S}\mathbf{a} - \lambda\left(\mathbf{a}^T\mathbf{a} - 1\right)\right] = \mathbf{0}$$

which gives

$$\mathbf{S}\mathbf{a} - \lambda\mathbf{a} = \mathbf{0}$$

or,

$$\mathbf{S}\mathbf{a} = \lambda\mathbf{a} \tag{6.50}$$

Eq. (6.50) shows that λ is the eigenvalue and $\mathbf{a}$ is the corresponding (unit-norm) eigenvector of the variance–covariance matrix $\mathbf{S}$.

In Example 6.5.1, from a 6×2 data matrix, we calculated a 2×2 symmetric variance–covariance matrix which had two eigenvalues and two corresponding eigenvectors. For the generalized $n \times p$ data matrix $\mathbf{X}$, we have a $p \times p$ symmetric variance–covariance matrix $\mathbf{S}$ with exactly p real eigenvalues, λi ($i = 1, \ldots, p$). The corresponding eigenvectors form an orthonormal set of vectors, that is,

$$\begin{aligned} \mathbf{a}_i^T\mathbf{a}_j &= 1 \quad \text{if } i = j \\ &= 0 \quad \text{if } i \neq j \end{aligned}$$

If the eigenvalues are arranged in a decreasing order, that is,

$$\lambda_1 > \lambda_2 > \ldots > \lambda_p$$

then the linear combination defined by the eigenvector $\mathbf{a}_1$ will have the highest variance since

$$\mathrm{Var}(\mathbf{Xa}) = \mathbf{a}^T\mathbf{S}\,\mathbf{a} = \lambda\mathbf{a}^T\mathbf{a} = \lambda \tag{6.51}$$

Hence, the linear combination defined by the highest eigenvector bear the most and that corresponding to the lowest eigenvector the least information about the distribution of the data.

In general, as a part of preprocessing of the data table, the columns of $\mathbf{X}$ are "centered" so that the mean of each column is equal to 0. The centered $n \times p$ matrix, denoted by $\mathbf{X}^*$, consists of centered column vectors

$$\mathbf{X}^* = \mathbf{X} - \overline{\mathbf{X}} = \left[\mathbf{x}_1, \ldots, \mathbf{x}_p\right]$$

where all the elements of column j of $\overline{\mathbf{X}}$ are the mean values of the observations on variable $\mathbf{x}_j$. The centering does not affect the final result since the variance–covariance matrix with a set of uncentered or centered variables is the same. Therefore we have the following relation

$$(n-1)\mathbf{S} = \mathbf{X}^{*T}\mathbf{X}^* \tag{6.52}$$

Definition 6.5.1 Let $\mathbf{X}$ be an n x p data matrix and $\mathbf{S}$ the p x p variance–covariance matrix associated with the dataset. If $\mathbf{a}_i$ $(i = 1, \ldots, p)$ are orthonormal eigenvectors of the matrix $\mathbf{S}$, the linear combinations $\mathbf{Xa}_i$ $(i = 1, \ldots, p)$ are defined as the ***principal components*** (***PCs***) of the dataset. The elements of the eigenvectors $\mathbf{a}_i$ $(i = 1, \ldots, p)$ are called ***PC loadings*** and the elements of the linear combinations $\mathbf{Xa}_i$ $(i = 1, \ldots, p)$ are called ***PC scores***.

Eq. (6.52) relates PCA to singular value decomposition (SVD) as we shall see next.

Principal component analysis and singular value decomposition

We have seen in Section 3.3 that a matrix $\mathbf{Y} \in \mathbb{R}^{n\times p}$ of rank r ($r \le \min\{n, p\}$) can be decomposed as

$$\mathbf{Y} = \mathbf{U}\boldsymbol{\Sigma}\mathbf{A}^T \tag{6.53}$$

where $\mathbf{U}_{n\times n}$ and $\mathbf{A}_{p\times p}$ are orthogonal matrices and $\boldsymbol{\Sigma}_{p\times p}$ is a diagonal matrix. The diagonal elements of $\boldsymbol{\Sigma}_{p\times p}$, $\sigma_i \ge 0$, $i = 1, 2, \ldots, r$, are called the singular values of $\mathbf{Y}$ and are equal to the nonnegative square roots of the nonzero eigenvalues of either $\mathbf{Y}^T\mathbf{Y}$ or $\mathbf{Y}\mathbf{Y}^T$. Since $\mathbf{Y}$ is of rank r, both $\mathbf{Y}^T\mathbf{Y}$ and $\mathbf{Y}\mathbf{Y}^T$ have the same rank r and the same r nonzero eigenvalues. By convention $\sigma_1 \ge \sigma_2 \ge \ldots \sigma_r$. The columns of $\mathbf{A}$, $\mathbf{a}_i$ $(i = 1, 2, \ldots, r)$, are called the right singular vectors while the columns of $\mathbf{U}$, $\mathbf{u}_i$ $(i = 1, 2, \ldots, r)$, are called the left singular vectors of $\mathbf{Y}$. It should be noted that although the eigenvalues of $\mathbf{Y}^T\mathbf{Y}$ and $\mathbf{Y}\mathbf{Y}^T$ are the same, the vectors on the left- and right-hand sides are not. The columns of **A** are the eigenvectors of the p x p matrix $\mathbf{Y}^T\mathbf{Y}$ while that of **U** are the eigenvectors of the $n \times n$ matrix $\mathbf{Y}\mathbf{Y}^T$.

If we take $\mathbf{Y} = \mathbf{X}^*$, then the elements of the right singular vectors of $\mathbf{X}^*$ are the ***PC loadings***. **A** is called a ***loading matrix***. It can also be interpreted as a ***projection matrix*** since multiplication of $\mathbf{X}^*$ by A gives the projections of the observations on the PCs. Also,

$$\mathbf{X}^*\mathbf{A} = \mathbf{U}\boldsymbol{\Sigma}\mathbf{A}^T\mathbf{A} = \mathbf{U}\boldsymbol{\Sigma} \tag{6.54}$$

Hence, $\mathbf{U}\boldsymbol{\Sigma}$ are the PCs of $\mathbf{X}^*$. Combining (6.52) and (6.53), we have

$$(n-1)\mathbf{S} = \mathbf{X}^{*T}\mathbf{X}^* = \left(\mathbf{U}\boldsymbol{\Sigma}\mathbf{A}^T\right)^T\left(\mathbf{U}\boldsymbol{\Sigma}\mathbf{A}^T\right) = \mathbf{A}\boldsymbol{\Sigma}^T\mathbf{U}^T\mathbf{U}\boldsymbol{\Sigma}\mathbf{A}^T = \mathbf{A}\boldsymbol{\Sigma}^2\mathbf{A}^T \tag{6.55}$$

where $\boldsymbol{\Sigma}^2$ is the diagonal matrix with the squared singular values as diagonal entries (in decreasing order). Eq. (6.55) also shows that these square singular values are eigenvalues of the matrix $(n-1)$ **S** and establishes the relation between PCA and SVD.

An essential purpose of PCA is to extract important information from a complex dataset by reducing the dimensionality of the dataset. It increases interpretability but minimizes information loss by creating new uncorrelated variables with successive maximization of variance. We have seen earlier that corresponding to the order of the eigenvalues, the PCs are also obtained in the order of importance (variance).

It can be seen that the covariance between two linear combinations is given by

$$\mathbf{a}_i^T \cdot \mathbf{S}\mathbf{a}_j = \lambda_j \mathbf{a}_i^T \cdot \mathbf{a}_j = 0 \quad \text{if } i \neq j$$

Hence, two such linear combinations are uncorrelated.

The quality (importance) of a given PC is determined by the proportion of the total variance it accounts for,

$$q_i = \frac{\lambda_i}{\sum_{i=1}^{p} \lambda_i}$$

In Example 6.5.1 the qualities of the two PCs, $\mathbf{P}_1$ and $\mathbf{P}_2$, are, respectively,

$$q_1 = \frac{\lambda_1}{\sum_{i=1}^{2} \lambda_i} = \frac{12.08}{12.4} = 0.974$$

and

$$q_2 = \frac{\lambda_2}{\sum_{i=1}^{2} \lambda_i} = \frac{0.32}{12.4} = 0.026$$

The number r of PCs to be retained for subsequent analysis and interpretation depends on the amount of variance r retains

$$q_{1,\ldots,r} = \frac{\sum_{i=1}^{r} \lambda_i}{\sum_{i=1}^{p} \lambda_i}$$

Two additional problems of data processing that PCA resolves are noise and redundancy. As seen earlier, **S** is a $p \times p$ symmetric variance–covariance matrix. The diagonal elements of **S** are the variance of particular measurement types, while the off-diagonal entries are the covariance between measurement types. Evidently, the correlations between all possible pairs of measurements are captured by **S**. The extent of noise in the data is determined by the *signal-to-noise ratio* (SNR) as a ratio of the variances σ^2

$$\mathrm{SNR} = \frac{\sigma^2_{\text{signal}}}{\sigma^2_{\text{noise}}}$$

Hence, higher values of the diagonal terms correspond to higher precision of the data, while lower values indicate noise-contamination. On the other hand, large/small values of the off-diagonal elements correspond to high/low redundancy. Therefore an optimized variance–covariance matrix must be diagonal.

Exercise 6

1. We have two peptides—P1 and P2. P1 is made of three lysine residues and two arginine residues. P2 is made of four lysine residues and seven arginine residues. Suppose one of the two peptides is selected with a probability of half each. What is the probability that an arginine is present at the first position of the peptide selected? If we are given the information that an arginine is present at the first position of the peptide selected, what is the conditional probability that P2 is selected?
2. We have 10 different protein molecules and at least one of 4 amino acids $aa_{1,\ldots,4}$ is present in any of them. What is the probability that aa_2 is present in none of them? What is the probability that exactly one of the 10 protein molecules contains aa_3?
3. DNA contains four kinds of bases—A, T, G, and C. The GC-content of the genome of a particular organism is 59%. What is the probability that a sequence 10 bases long randomly selected from the genome will have exactly six A or T bases?
4. Let a function f be defined by

$$f(x) = \begin{cases} x^2 & \text{for } 0 < x < 1 \\ 0 & \text{otherwise} \end{cases}$$

   Check if f is a valid probability density function. If so, calculate the probability that X lies between 0.3 and 0.6.
5. Let us consider a random variable X whose pdf is given as follows:

$$f_X(x) = \begin{cases} 5 - x & \text{for } 0 < x < 3 \\ 0 & \text{otherwise} \end{cases}$$

and a transformation $Y = \sqrt{X}$

Find the pdf $f_Y(y)$

6. The pmf of a discrete random variable X is given as follows:

x	2	4	5	7
$p(x)$	0.2	0.1	0.3	0.4

Calculate $E(X)$.

7. The probability density function of a continuous random variable X is given by

$$f(x) = \begin{cases} 3x^4 & \text{for } 0 < x < 1 \\ 0 & \text{otherwise} \end{cases}$$

Calculate $E(X)$.

8. Let a function f be defined by

$$f_{X,Y}(x) = \begin{cases} x^2y + y^2x & \text{for } 0 \le x \le 1, 0 \le y \le 1 \\ 0 & \text{otherwise} \end{cases}$$

Check if f is a valid probability density function. In such case, calculate $P(0.5 \le X \le 0.7, 0.4 \le Y \le 0.6)$.

9. From the bivariate distribution function of continuous variables X and Y given by

$$f_{X,Y}(x) = \begin{cases} x^2y + y^2x & \text{for } 0 \le x \le 1, 0 \le y \le 1 \\ 0 & \text{otherwise} \end{cases}$$

calculate the marginal pdfs $f_X(x)$ and $f_Y(y)$.

10. Consider the bivariate density function

$$\begin{aligned} f(x, y) &= 2xe^{-y} && \text{for } 0 \le x \le 1 \text{ and } y \ge 0 \\ &= 0 && \text{otherwise} \end{aligned}$$

and compute the variance–covariance matrix.

11. The observations on three numerical variables each on four entities are presented in the form of a data matrix

$$\mathbf{X} = \begin{bmatrix} 2 & 2 & 2 \\ 4 & 4 & 5 \\ 6 & 4 & 7 \\ 8 & 6 & 10 \end{bmatrix}$$

Obtain the PCs and compute the proportion of the total variance that each of the components accounts for.

Further reading

Abdi, H., Williams, L.J., 2010. Principal component analysis. Wiley Interdiscip. Rev. 2, 433–459.

Dekking, F.M., Kraaikamp, C., Lopuhaä, Meester, L.E., 2005. A modern introduction to probability and statistics. Springer-Verlag, London, U.K.

Evans, M.J., Rosenthal, J.S., 2009. Probability and statistics: the science of uncertainty. Macmillan Higher Education.

Hogg, R.V., McKean, J.W., Craig, A.T., 2019. Introduction to mathematical statistics, 8th edition Pearson.

Jolliffe, I.T., Cadima, I., 2016. Principal component analysis: a review and recent developments. Phil. Trans. R. Soc. A 374, 20150202.

CHAPTER

X-ray crystallography

7

Ever since the first biomacromolecular structures were resolved, X-ray crystallography has been the most powerful tool in molecular structural biology. Recent progress in single particle cryo-electron microscopy (cryo-EM) has brought in (though not very meaningfully) a sense of rivalry between the two techniques. It is, however, more reasonable to appreciate that the two are complementary in nature.

Both X-ray crystallography and cryo-EM are capable of solving macromolecular structures at atomic resolution. Invariably, they are both dependent on rigorous mathematical and computational analyses. The mathematical theory behind X-ray crystallography will be discussed in this chapter and cryo-EM in the next.

7.1 X-ray scattering

In classical physics, electromagnetic radiation consists of electromagnetic waves, which are synchronized oscillations of electric and magnetic fields perpendicularly linked to each other. The oscillations of both fields are perpendicular to the direction of (wave) motion. Visible light constitutes a (small) part of the (wavelength- or frequency-based) electromagnetic spectrum. In quantum theory, electromagnetic radiation is the flow of photons (considered to be particulate in nature) through space.

X-rays are electromagnetic radiation just like visible light; however, their wavelength is closer to that of electrons. As such, they display both wave and particle properties and are used to image atomic-level structures based on a certain property of diffraction/scattering.

Considered as waves, X-rays have wavelengths and frequencies of the order of $\lambda = 1$ Å and $\omega = 2\pi\nu = 2\pi c/\lambda \approx 2 \times 10^{19}$/s respectively, where $c \approx 3 \times 10^{8}$ m/s is the speed of light (or, for that matter, any form of electromagnetic radiation). As particles, X-ray photons (corresponding to $\lambda = 1$ Å) have Planck-Einstein energies given by

$$E = h\,\nu = h\,c/\lambda = 1.988 \times 10^{-15}\ \text{J}$$

and de Broglie momentum

$$p = h/\lambda = 6.26 \times 10^{-24}\ \text{kg} \cdot \text{m} \cdot \text{s}^{-1}$$

where Planck's constant $h = 6.626 \times 10^{-34}\ \text{J} \cdot \text{s}$.

Mathematical Approaches to Molecular Structural Biology. DOI: https://doi.org/10.1016/B978-0-323-90397-4.00007-X
© 2023 Elsevier Inc. All rights reserved.

Diffraction and scattering are interrelated by wave-particle dualism. Diffraction is a wave phenomenon – when a wave encounters an obstacle there is a deviation in the direction of its propagation, accompanied by a change in its amplitude or phase, at the edge of the obstacle. This can be said as well for an electromagnetic wave interacting with an electron. The same phenomenon can also be seen as photon particles being deflected or scattered as a result of collision with an electron. Evidently, there are two types of X-ray scattering by electrons – (1) Thomson scattering as a result of electromagnetic wave–electron interaction and (2) Compton scattering due to photon–electron collision.

7.1.1 Electromagnetic waves

Electromagnetic waves are propagated by periodically varying electric and magnetic fields. The problem of electromagnetic wave propagation can therefore be addressed by the application of Maxwell's equations which are the basic equations of electromagnetism. Maxwell's equations describe (1) the relation between charge q and the electric field $\mathbf{E}$, (2) the nature of the magnetic field $\mathbf{B}$, (3) the effect on the electric field $\mathbf{E}$ of a changing magnetic field, and (4) the effect on the magnetic field $\mathbf{B}$ of a current or changing electric field.

It is known that a static electric charge produces electric field lines while a steady electric current, that is, a charge moving with a uniform speed, produces magnetic field lines. An energy density is associated with the electric and magnetic fields. However, in the steady state, there is no transport of energy or momentum and no propagation of electromagnetic waves.

For wave propagation, the charge needs to be accelerated. The oscillatory $\mathbf{E}$ and $\mathbf{B}$ thus produced are in phase and perpendicular to one another. From Maxwell's equations one can derive the wave equations for $\mathbf{E}$ and $\mathbf{B}$ as

$$\frac{\partial^2 \mathbf{E}}{\partial x^2} = \frac{1}{c^2}\left(\frac{\partial^2 \mathbf{E}}{\partial t^2}\right) \tag{7.1}$$

$$\frac{\partial^2 \mathbf{B}}{\partial x^2} = \frac{1}{c^2}\left(\frac{\partial^2 \mathbf{B}}{\partial t^2}\right) \tag{7.2}$$

where $c = 1/\sqrt{(\mu_0 \varepsilon_0)}$ is the speed of the waves; ε_0 is the electric constant and μ_0 is the magnetic constant. For a sinusoidal wave the solutions to (7.1) and (7.2) are

$$E(x,t) = E_m \sin(kx - \omega t) \tag{7.3}$$

$$B(x,t) = B_m \sin(kx - \omega t) \tag{7.4}$$

where $k = 2\pi/\lambda$ and $c = \omega/k$. E_m and B_m are the electric and magnetic field amplitudes. The intensity of the electromagnetic wave is

$$I = \left(\frac{1}{2}\varepsilon_0 c\right) E_m^2 \tag{7.5}$$

Eqs. (7.3) and (7.4) can also be written as a complex wave function

$$\bar{f}(x,t) = \bar{A}e^{i(kx-wt)} \tag{7.6}$$

where the complex amplitude includes the phase constant δ

$$\bar{A} \equiv Ae^{\delta}$$

The actual wave is the real part of $\bar{f}$

$$f(x,t) = \mathrm{Re}\left[\bar{f}(x,t)\right]$$

Exponentials are easier to manipulate than sines and cosines.

In (7.3) and (7.4), it has been assumed that **E** and **B** are functions of time (t) and one space coordinate (x) only. Furthermore, **E** and **B** are uniform throughout any plane perpendicular to the x-axis. Hence, such a wave, which is traveling parallel to the x-axis, is called a ***plane wave***. If only one frequency component is present in the wave, it is called a ***monochromatic*** plane wave.

Let us suppose that the wave is propagating in the z-direction. In such case, the electric field can oscillate in any direction that lies in the xy-plane. If the field is oscillating in a specific direction, it is said to be ***linearly polarized***. A wave polarized in the x-direction can be represented by

$$\bar{\mathbf{f}}_x(z,t) = \bar{A}e^{i(kz-wt)}\hat{\mathbf{x}}$$

and that in the y-direction by

$$\bar{\mathbf{f}}_y(z,t) = \bar{A}e^{i(kz-wt)}\hat{\mathbf{y}}$$

where $\hat{\mathbf{x}}$ and $\hat{\mathbf{y}}$ are unit vectors in the x-direction and y-direction, respectively.

As the wave travels, it transports energy. The energy flux density, that is, energy per unit area per unit time, transported by the electromagnetic field is given by the ***Poynting vector***

$$\mathbf{S} = \frac{1}{\mu_0}(\mathbf{E} \times \mathbf{B}) \tag{7.7}$$

Let us assume that the source of the wave, and hence the energy, is localized at the origin within a large spherical shell of radius r. The total power passing through the surface of the shell is given by the integral of Poynting vector:

$$P(r) = \oint \mathbf{S} \cdot d\mathbf{a} = \frac{1}{\mu_0}\oint (\mathbf{E} \times \mathbf{B}) \cdot d\mathbf{a} \tag{7.8}$$

The power radiated is then

$$P_{\mathrm{rad}} = \lim_{r \to \infty} P(r) \tag{7.9}$$

which is the energy per unit time transported to infinity.

If a point charge q, localized at the origin undergoes an acceleration $\mathbf{a}(t)$, its dipole moment is

$$\mathbf{p}(t) = q\,\mathbf{d}(t)$$

where $\mathbf{d}$ is the position of q with respect to the origin, and

$$\ddot{\mathbf{p}}\,(t) = q\,\mathbf{a}\,(t)$$

Using spherical coordinates, with the z-axis in the direction of $\ddot{\mathbf{p}}$, the field radiated can be represented by

$$\mathbf{E}\,(r,\,\theta,t) \cong \frac{\ddot{p}}{4\,\pi\epsilon_0\,c^2}\left(\frac{\sin\theta}{r}\right)\hat{\boldsymbol{\theta}} \tag{7.10a}$$

$$\mathbf{B}\,(r,\,\theta,t) \cong \frac{\ddot{p}}{4\,\pi\epsilon_0\,c^3}\left(\frac{\sin\theta}{r}\right)\hat{\boldsymbol{\phi}} \tag{7.10b}$$

where $\hat{\boldsymbol{\theta}}$ and $\hat{\boldsymbol{\phi}}$ are unit vectors. Hence, the radial Poynting flux is given by

$$\mathbf{S} \cong \frac{1}{\mu_0}\,(\mathbf{E}\times\mathbf{B}) = \frac{\ddot{p}^2}{16\,\pi^2\epsilon_0\,c^3}\left(\frac{\sin^2\theta}{r^2}\right)\hat{\mathbf{r}} \tag{7.11}$$

and the total radiated power is

$$P \cong \int \mathbf{S}\cdot d\mathbf{a} = \frac{\ddot{p}^2}{6\,\pi\epsilon_0\,c^3} = \frac{q^2a^2}{6\,\pi\epsilon_0\,c^3} \tag{7.12}$$

Eq. (7.12) is known as the ***Larmor formula***.

7.1.2 Thomson scattering

When an electromagnetic wave is incident on a charged particle moving with a velocity $\mathbf{v}$, the electric and magnetic components of the wave exert a force (known as ***Lorentz force***) on the particle given by

$$\mathbf{f} = q\,(\mathbf{E} + \mathbf{v}\times\mathbf{B}) \tag{7.13}$$

Since the wave is periodic in nature, the particle is also set into a periodic motion. The particle is accelerated and, as a consequence, emits radiation. Precisely, the energy absorbed by the particle from the incident wave is reemitted as electromagnetic radiation. Hence, the process can be considered as scattering of the electromagnetic wave by the particle.

Now, let us consider a linearly polarized monochromatic plane wave incident of a particle of charge q and mass m (initially at rest). The electric component of the wave can be expressed as

$$\mathbf{E} = \hat{\mathbf{e}}\,E_0 e^{i\,(\mathbf{k}\cdot\mathbf{r}-wt)} \tag{7.14}$$

where E_0 is the amplitude of the electric field, $\hat{\mathbf{e}}$ is the polarization vector, and $\mathbf{k}$ is the wave vector. (It is to be noted that $\hat{\mathbf{e}}\cdot\mathbf{k} = 0$). It is assumed that the particle undergoes an oscillation about an equilibrium position at the origin of the coordinate system and the amplitude of oscillation is small. Further, the velocity of the particle is considered to be subrelativistic and therefore the magnetic component of the force (7.13) is negligible.

The equation of motion of the charged particle is then

$$\mathbf{f} = q\,\mathbf{E} = m\,\mathbf{a} \tag{7.15}$$

where m is the mass of the particle and $\mathbf{a}$ is the acceleration. Using Larmor formula (7.12), we can find the time-averaged power radiated per unit solid angle by nonrelativistic accelerating charged particle as

$$\frac{dP}{d\Omega} = \frac{q^2 \langle a^2 \rangle}{16\,\pi^2 \epsilon_0\, c^3} \sin^2\theta \tag{7.16}$$

where $\langle a^2 \rangle$ denotes a time-average given by

$$\langle a^2 \rangle = \frac{q^2}{m^2} \langle E^2 \rangle = \frac{q^2 E_0^2}{2\,m^2} \tag{7.17}$$

Hence, from (7.16) we have

$$\frac{dP}{d\Omega} = \left(\frac{q^2}{4\,\pi\epsilon_0\, m\, c^2}\right)^2 \frac{\epsilon_0\, c E_0^2}{2} \sin^2\theta \tag{7.18}$$

The incident flux of the wave is given by the time-averaged Poynting vector $\langle \mathbf{S} \rangle$. Since the electric and magnetic field are perpendicular and have equal amplitudes, it can be found that

$$\langle \mathbf{S} \rangle = \frac{\epsilon_0\, c E_0^2}{2} \tag{7.19}$$

Defining the differential cross-section for scattering into θ as

$$\frac{d\sigma}{d\Omega} = \frac{dP/d\Omega}{\langle \mathbf{S} \rangle}$$

we find from (7.18) and (7.19)

$$\frac{d\sigma}{d\Omega} = \left(\frac{q^2}{4\,\pi\epsilon_0\, m\, c^2}\right)^2 \sin^2\theta \tag{7.20}$$

Integrating over the solid angle, the total scattering cross-section is

$$\sigma = \int_0^{\pi} \frac{d\sigma}{d\Omega}\, 2\,\pi \sin\theta\, d\,\theta = \frac{8\,\pi}{3} \left(\frac{q^2}{4\,\pi\epsilon_0\, m\, c^2}\right)^2 \tag{7.21}$$

If the charged particle under consideration is an electron, (7.21) becomes ***Thomson scattering cross-section***

$$\sigma_{\text{Thomson}} = \frac{8\,\pi}{3} r_0^2 \tag{7.22}$$

where

$$r_0 = \frac{e^2}{4\,\pi\epsilon_0\, m_e\, c^2}$$

is called the ***classical electron radius***. Thus, when scattering radiation, the electron acts as a solid sphere.

In (7.21), θ is the angle between the direction of acceleration of the particle and the direction of outgoing radiation. In arriving at (7.21), we have considered only scattering of linearly polarized radiation. For randomly polarized (or unpolarized) radiation, let ϕ be the angle between the direction of the incident radiation and that of the scattered radiation, and ψ the angle that specifies the orientation of the polarization vector in the plane perpendicular to the wave vector **k**. It can be shown mathematically that

$$\cos\theta = \cos\psi \sin\phi$$

or,

$$\sin^2\theta = 1 - \cos^2\psi \sin^2\phi$$

So, $\sin^2\theta$ in (7.20) has to be substituted by its average over all possible polarization of the incident wave, that is,

$$\overline{\sin^2\theta} = 1 - \overline{\cos^2\psi}\,\sin^2\phi = 1 - \frac{(\sin^2\phi)}{2} = \frac{1+\cos^2\phi}{2}$$

Hence, the differential scattering cross-section for unpolarized incident radiation is given by

$$\left(\frac{d\sigma}{d\Omega}\right)_{\text{unpolarized}} = \frac{1}{2}\left(\frac{q^2}{4\pi\epsilon_0 m c^2}\right)^2 (1+\cos^2\phi) \tag{7.23}$$

If the intensity of the incident radiation be I_0, the intensity of the scattered radiation, defined as the power per unit solid angle scattered through an angle ϕ, is given by

$$I_\phi = \frac{1}{2}\left(\frac{q^2}{4\pi\epsilon_0 m c^2}\right)^2 (1+\cos^2\phi) I_0 \tag{7.24}$$

In Thomson scattering, there exists a definite phase relationship between the incident and scattered radiation, hence, it is ***coherent***. Besides, Thomson scattering is ***elastic*** as the incident and scattered X-rays have the same wavelength (or frequency).

7.1.3 Compton scattering

As mentioned earlier, Compton scattering is due to the collision of a photon with a charged particle, usually an electron (Fig. 7.1). In Thomson scattering, the motion of the charge has been assumed to be subrelativistic. However, in Compton scattering, the energy of the incident photon is sufficiently high and a significant amount of its momentum is imparted to the charged particle. The energy of the photon is changed by an inelastic scattering process, which can then be explained only by relativistic dynamics.

Let us consider a photon, with energy and momentum, E_i and $\mathbf{p}_i$, respectively, colliding with a "relativistic" electron at rest. Evidently, before the collision the energy of the electron is entirely its "rest mass energy," m_0c^2. As a result of the

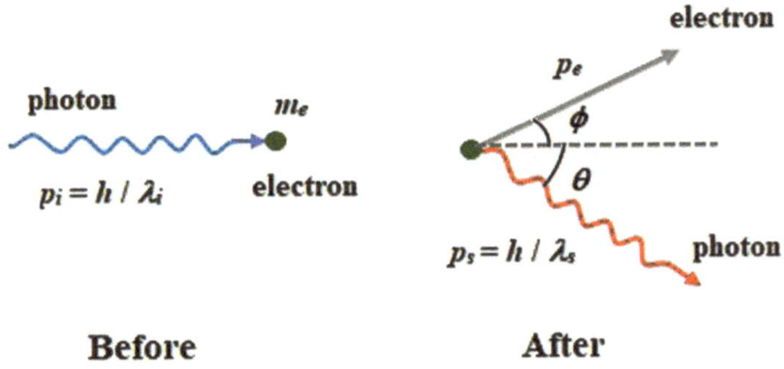

FIGURE 7.1

Scattering of X-rays considered as photon–electron collision.

collision, the photon is scattered at an angle θ, with energy and momentum E_s and $\mathbf{p}_s$, respectively, and the electron recoils at an angle ϕ with momentum $\mathbf{p}_e$ (Fig. 7.1). Relative expression of energy of the electron after the collision is

$$E = \sqrt{p_e^2 c^2 + (m_e c^2)^2} \tag{7.25}$$

The laws of conservation of momentum and energy can be applied to the collision process. The total linear momentum of the system (consisting of the photon and the electron) being conserved,

$$\mathbf{p}_i = \mathbf{p}_s + \mathbf{p}_e \tag{7.26}$$

where $p_i = h/\lambda_i$ and $p_s = h/\lambda_s$. Eq. (7.26) can be rearranged as

$$\mathbf{p}_e = \mathbf{p}_i - \mathbf{p}_s \tag{7.27}$$

With dot products on both sides

$$\mathbf{p}_e \cdot \mathbf{p}_e = (\mathbf{p}_i - \mathbf{p}_s) \cdot (\mathbf{p}_i - \mathbf{p}_s)$$

or,

$$p_e^2 = p_i^2 + p_s^2 - 2 p_i p_s \cos\theta \tag{7.28}$$

Applying the conservation of energy

$$E_i + m_e c^2 = E_s + E \tag{7.29}$$

where $E_i = p_i c$ and $E_s = p_s c$. From (7.29) we have

$$\left[(E_i - E_s) + m_e c^2\right]^2 = E^2 \tag{7.30}$$

Combining (7.25) and (7.30)

$$(E_i - E_s)^2 + 2 m_e c^2 (E_i - E_s) = p_e^2 c^2 \tag{7.31}$$

Dividing both sides of (7.31) by c^2 we can express the equation in terms of momenta as

$$(p_i - p_s)^2 + 2 m_e c (p_i - p_s) = p_e^2 \tag{7.32}$$

Substituting p_e^2 from (7.28) and (7.32) can be simplified to

$$\frac{1}{p_s} - \frac{1}{p_i} = \frac{1}{m_e c}(1 - \cos\theta)$$

which, in terms of the wavelengths, becomes

$$\lambda_s - \lambda_i = \frac{h}{m_e c}(1 - \cos\theta) \tag{7.33}$$

which is the ***Compton equation***. The quantity $h/m_e c$ has the dimension of length. It is called the ***Compton wavelength of the electron***. Compton scattering is an example of inelastic scattering since the scattered radiation has a longer wavelength (lower frequency) than that of the incident radiation.

7.2 Scattering by an atom

Elastic Thomson scattering accounts for most X-ray scattering by electrons, and it can be shown mathematically that under the conditions of diffraction of X-rays from crystals, inelastic Compton scattering may generally be ignored. Here, we shall consider elastic scattering of X-rays from electrons in which the incident radiation is absorbed by the scattering center and subsequently reemitted without any loss of energy.

Let the point O in Fig. 7.2 represents a scattering center (electron). When a monochromatic wave represented by the equation

$$y = A\exp(2\pi i\nu t) \tag{7.34}$$

where A is the amplitude and ν ($=c/\lambda$) the frequency, is incident on O in the direction as shown in the figure, it is scattered in all directions. The amplitude of the scattered wave at a point P (OP $= D$) is then given by

$$\eta(2\theta, D) = f_{2\theta}\frac{A}{D} \tag{7.35}$$

where 2θ is the scattering angle. $f_{2\theta}$ is a constant of proportionality referred to as the scattering length. The intensity (power per unit solid angle) of the scattered beam at P is then

$$I_{2\theta} = f_{2\theta}^2 I_0 \tag{7.36}$$

where I_0 is the intensity of the incident beam on the scatterer.

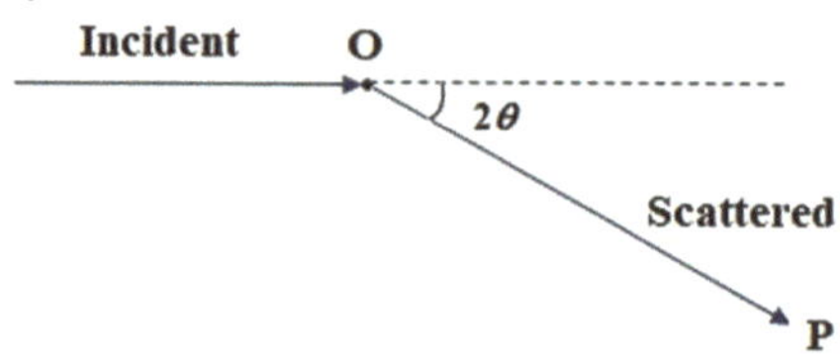

FIGURE 7.2

Scattering of X-rays by a point scatterer.

Now, let us consider scattering from a pair of electrons located at points O_1 and O_2 and a point P at a distance very large compared to O_1O_2 (Fig. 7.3); the phase difference between the wave arriving from O_2 with respect to that from O_1 is

$$\alpha_s = 2\pi \mathbf{r} \cdot \frac{(\mathbf{S} - \mathbf{S}_0)}{\lambda} \tag{7.37}$$

where $\mathbf{S}_0$ and $\mathbf{S}$ are two unit vectors along the directions of the incident and scattered waves respectively and $\mathbf{r}$ is the vector joining O_1 to O_2. Taking

$$\mathbf{s} = \frac{(\mathbf{S} - \mathbf{S}_0)}{\lambda} \tag{7.38}$$

we have

$$\alpha_s = 2\,\pi \mathbf{r} \cdot \mathbf{s} \tag{7.39}$$

The vectors $\mathbf{S}_0/\lambda$ and $\mathbf{S}/\lambda$, each is of magnitude $1/\lambda$. The vector $\mathbf{s}$, which is crucial in describing the scattering process, is geometrically perpendicular to the bisector of the angle between $\mathbf{S}_0$ and $\mathbf{S}$. The magnitude of $\mathbf{s}$ is given by

$$s = (2\sin\theta)/\lambda \tag{7.40}$$

If the positions of two identical scattering points O_1 and O_2 are given by vectors $\mathbf{r}_1$ and $\mathbf{r}_2$ with respect to the reference point O (an arbitrary point from which the radiation is scattered), the scattering amplitude can be expressed as

$$\eta_2(2\theta, D) = \eta(2\theta, D)\left\{\exp(2\pi i \mathbf{r}_1 \cdot \mathrm{s}) + \exp(2\pi i \mathbf{r}_2 \cdot \mathrm{s})\right\} \tag{7.41}$$

where η $(2\theta, D)$ is the amplitude of scattering from a single point.

Eq. (7.41) can be extended to the case of scattering from a general distribution of n identical scatterers (electrons) with $\mathbf{r}_j$ denoting the position of jth scatterer. The amplitude of scattering corresponding to a scattering vector s at a distance compared to the system of scatterers is then

$$\eta_n(2\theta, D) = \eta\,(2\theta, D)\sum_{j=1}^{n}\exp(2\pi i \mathbf{r}_j \cdot \mathbf{s}) \tag{7.42}$$

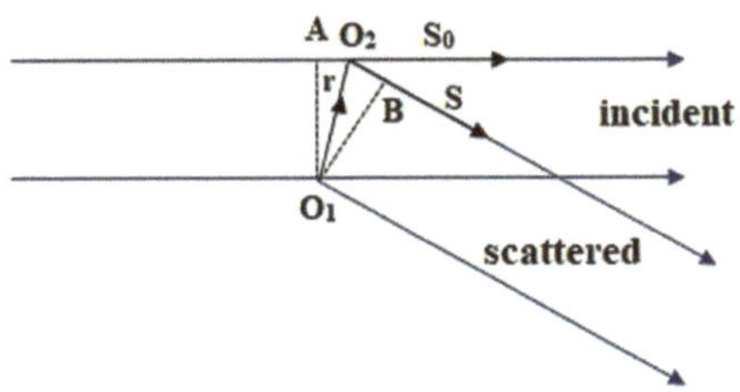

FIGURE 7.3

Scattering by a pair of point scatterers.

If the scatterers are nonequivalent, the above equation is modified as

$$\eta_n(2\theta, D) = \sum_{j=1}^{n} [\eta(2\theta, D)]_j \exp(2\pi i \mathbf{r}_j \cdot \mathbf{s})$$

$$= \frac{A}{D}\sum_{j=1}^{n} (f_{2\theta})_j \exp(2\pi i \mathbf{r}_j \cdot \mathbf{s}) \tag{7.43}$$

Till now we have considered free electrons as scatterers. However, in an atom, the electrons are bound in discrete energy states and X-rays can be scattered in two ways. In one case, Thomson scattering, there is no change in the energy of the electrons and therefore the incident radiation. The scattered wave is coherent with the incident wave. In the other case, Compton scattering, the electrons can gain or lose energy by moving from one energy state to another, leading to incoherent scattering. Furthermore, an atomic electron can be represented by a wave function Ψ and electronic charge is not located at a point but distributed in accordance with

$$\rho(r) = |\Psi|^2 \tag{7.44}$$

If the positional parameters are expressed in spherical polar coordinates (r, θ, ϕ), the charge associated with a small elemental volume (say $d\tau$) is

$$\rho(r, \theta, \phi)\, d\tau = \rho(r, \theta, \phi)\, r^2 \sin\theta \; dr \; d\theta \; d\phi \tag{7.45}$$

If the coordinate system is so chosen that the scattering vector $\mathbf{s}$ is parallel to the axis with respect to which θ is measured, then $\mathbf{r} \cdot \mathbf{s} = rs \cos\theta$. The total amplitude of the coherent scattered radiation can be obtained from (7.43) by replacing summation with integration

$$\eta_s = k_s \int_{r=0}^{\infty} \int_{\theta=0}^{\pi} \int_{\phi=0}^{2\pi} \rho(r, \theta, \phi)\, r^2 \exp(2\pi i r s \cos\theta) \sin\theta \; dr \; d\theta \; d\phi \tag{7.46}$$

where k_s is a constant dependent on $\mathbf{s}$. As a special case, let us take $|\Psi|$, and therefore ρ, to be spherically symmetric so that the electron density can be represented by $\rho(r)$. Since $\rho(r)$ is independent of θ and ϕ, and being spherically symmetric also centrosymmetric, (7.46) reduces to

$$\eta_s = 4\pi k_s \int_{r=0}^{\infty} \rho(r) r^2 \frac{\sin(2\pi rs)}{2\pi rs}\, dr \tag{7.47}$$

Now, to find the scattered amplitude from an electron at the origin, $(\eta_s)_0$, we can apply here the Dirac delta function that has been introduced in Chapter 5. The point electron has an electron density $\rho(r) = \delta(r)$, where

$$\delta(r) = 0, \quad \text{for } r \neq 0$$

$$\delta(r) = \infty, \quad \text{for } r = 0$$

and integrating over any finite volume of space around the origin

$$\int_V \delta(r)dv = \int_0^\infty \delta(r)\, 4\pi r^2\, dr = 1$$

That is, a delta function electron density has no contribution to (7.47) away from the origin. Furthermore,

$$\lim_{r\to 0} \frac{\sin(2\pi rs)}{2\pi rs} = 1$$

Hence, from (7.47), we obtain

$$(\eta_s)_0 = k_s \tag{7.48}$$

From (7.47) and (7.48), we have

$$p_s = \frac{\eta_s}{(\eta_s)_0} = 4\pi \int_{r=0}^{\infty} \rho(r) r^2 \frac{\sin(2\pi rs)}{2\pi rs} dr \tag{7.49}$$

which gives the fraction of the scattered amplitude from a point electron at the origin that is scattered from $\rho(r)$. If the atom contains Z electrons, the total electron density will be given by

$$\rho_a = \sum_{j=1}^{Z} \rho_j(r) \tag{7.50}$$

From (7.49) and (7.50), we obtain the ratio of the amplitude of coherent scattering (Thomson scattering) due to the total electron density of the atom to that from a single electron located at the atomic center as

$$f_a = 4\pi \int_{r=0}^{\infty} \rho_a(r) r^2 \frac{\sin(2\pi r s)}{2\pi r s} dr = \sum_{j=1}^{Z} (p_s)_j \tag{7.51}$$

which is defined as the ***atomic scattering factor***.

7.3 Diffraction from a crystal – Laue equations

Since X-rays can be scattered from atomic electrons, one could expect that by measuring the scattering from all the atoms in a molecule it should be possible to determine the structure of the molecule. However, it needs to be realized that X-ray scattering from a single molecule would be extremely weak and could hardly be detected above the noise level. To be detectable, the signal should be amplified by periodic arrays of molecules (and thereby atoms) such as a crystal or a fiber. A crystal is a periodic layout of a motif in a lattice – the motif can be a single atom, a small molecule, a macromolecule or a combination of two or more of these. A typical protein crystal is known to contain more than 10^{14} aligned molecules. A crystal can also be seen as consisting of billions of identical units – called the unit cells – closely packed in three-dimensions (Fig. 7.4). A unit cell may contain one or more molecules.

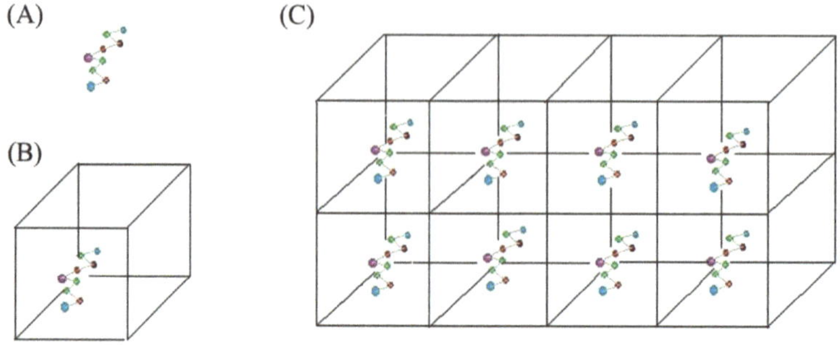

FIGURE 7.4

(A) Molecule, (B) unit cell, and (C) crystal.

From Pal, S. 2020. Fundamentals of Molecular Structural Biology, Academic Press.

The scattered radiation from all the atoms will be coherent with respect to each other. Since a crystal organizes an enormous number of atoms in the same orientation, the scattered waves add up in phase in certain directions while canceling out in others. As a result, the diffraction pattern is observed as an array of spots.

Let us first look at the diffraction from a one-dimensional array of n atoms, each atom being separated from its neighbor by a vector distance **a**. If we assume n to be odd and the atom at the center of symmetry as the coordinate origin, the amplitude of the scattered radiation corresponding to a scattering vector **s** at a distance large compared with the lattice dimension, is given by

$$A_n = f_a \sum_{j=-\frac{1}{2}(n-1)}^{\frac{1}{2}(n-1)} \cos(2\pi j \mathbf{a} \cdot \mathbf{s}) \tag{7.52}$$

where f_a is the atomic scattering factor. From (7.52), it can be shown that there will be significant diffracted intensity only when

$$\mathbf{a} \cdot \mathbf{s} = h \tag{7.53}$$

where h is an integer. Extending the concept to diffraction from a three-dimensional array of atoms which are separated by three vector distances **a**, **b**, and **c**, we can arrive at the conditions of diffraction as

$$\begin{aligned} \mathbf{a} \cdot \mathbf{s} &= h \\ \mathbf{b} \cdot \mathbf{s} &= k \\ \mathbf{c} \cdot \mathbf{s} &= l \end{aligned} \tag{7.54}$$

where h, k, and l are integers. These are known as ***Laue equations***.

As we proceed to find the scattering vector **s** that will satisfy the Laue equations for a given set of integers (hkl), it will be useful to have some basic notion of lattice and reciprocal lattice related to crystal structure.

7.3.1 Lattice and reciprocal lattice

Periodicities in three dimensions are a basic feature of crystals. A crystal is described by specifying the size, shape, and the contents of the simplest repeating unit and the manner in which the repeating units stack to form the crystal. The repeated pattern in space, such as the distribution of atoms in a crystal, is related to a ***space lattice*** of points. These space lattices, known as Bravais lattices, exhibit distinct translational symmetry. A Bravais lattice can be described in $\mathbb{R}^3$ by a linear combination of three linearly independent vectors $\mathbf{a}_1$, $\mathbf{a}_2$, and $\mathbf{a}_3$, known as repeat (translation) vectors, as

$$\mathbf{T}_{pqr} = p\,\mathbf{a}_1 + q\,\mathbf{a}_2 + r\,\mathbf{a}_3 \qquad p, q, \ r \in \mathbb{Z} \tag{7.55}$$

A translation by $\mathbf{T}_{pqr}$ leaves the lattice unchanged. All the lattice points in a Bravais lattice are equivalent, that is, each lattice point is in a similar environment.

The simplest repeating unit in a crystal is called a ***unit cell***. Each unit cell is defined in terms of lattice points. In $\mathbb{R}^3$, three repeating vectors **a**, **b**, and **c**, which separate the lattice points, together with the angles α, β, and γ, respectively between them, define a unit cell in the crystal lattice (Fig. 7.5).

The meaning of reciprocity is very straightforward. For example, if d denotes distance in Å, the reciprocal quantity is $d^* = 1/d$ expressed in 1/Å. Similarly, a unit cell in the direct (real) space with parameters **a**, **b**, **c**, α, β, and γ has a corresponding reciprocal unit cell (Fig. 7.6): **a***, **b***, **c***, α^*, β^*, and γ^*. There are exact mathematical relationships between the direct space and reciprocal unit cell parameters. The complete set of relationships between **a**, **b**, and **c** and **a***, **b***, and **c*** are as follows:

$$\begin{array}{lll} \mathbf{a}\cdot\mathbf{a}^* = 1 & \mathbf{a}\cdot\mathbf{b}^* = 0 & \mathbf{a}\cdot\mathbf{c}^* = 0 \\ \mathbf{b}\cdot\mathbf{a}^* = 0 & \mathbf{b}\cdot\mathbf{b}^* = 1 & \mathbf{b}\cdot\mathbf{c}^* = 0 \\ \mathbf{c}\cdot\mathbf{a}^* = 0 & \mathbf{c}\cdot\mathbf{b}^* = 0 & \mathbf{c}\cdot\mathbf{c}^* = 1 \end{array} \tag{7.56}$$

Evidently, **a*** is perpendicular to both **b** and **c**, and so on. The vectors **a***, **b***, and **c*** define what is known as the ***reciprocal lattice***. It can be shown that the scattering vector **s** that satisfies Laue equations is given by

$$\mathbf{s} = h\mathbf{a}^* + k\mathbf{b}^* + l\mathbf{c}^* \tag{7.57}$$

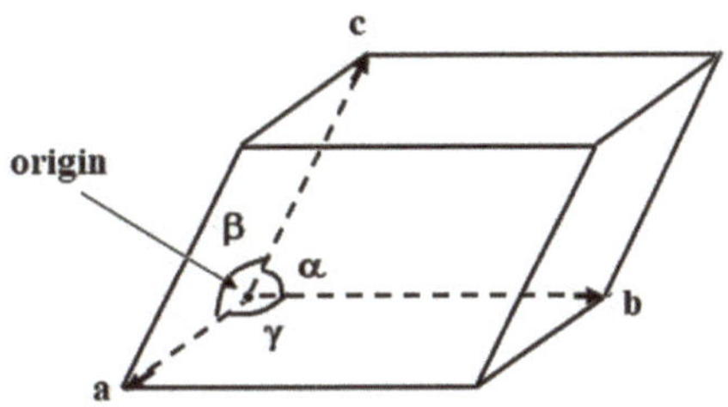

FIGURE 7.5

A unit cell defined by vectors **a**, **b**, and **c**, and angles α, β, and γ.

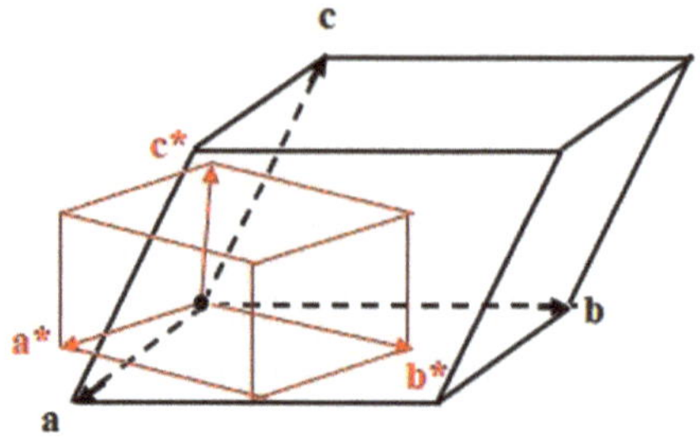

FIGURE 7.6

Unit cell in real space (black outline) and reciprocal space (red outline).

where h, k, and l are three integers. Evidently, the scattering vector is obtained from the reciprocal lattice. Since the reciprocal lattice is fixed relative to the crystal, so is the scattering vector.

There exists a reciprocal relationship also between the volume of a unit cell (in direct space) V and the volume of a reciprocal unit cell V^*.

$$V = \mathbf{a} \cdot \mathbf{b} \times \mathbf{c} \quad \text{and} \quad VV^* = 1$$

7.3.2 Structure factor

Now we need to determine the resultant effect of all atoms in the unit cell. For a crystal with N atoms in the unit cell, the total scattered amplitude is given by

$$F_{hkl} = \sum_{j=1}^{N} f_j \exp\left(2\pi i \mathbf{r}_j \cdot \mathbf{s}\right) \tag{7.58}$$

where f_j is the scattering factor of the jth atom. F_{hkl} is a function of the content of a unit cell and is called the ***structure factor***. If we separate (7.58) into real and imaginary parts

$$F_{hkl} = A_{hkl} + iB_{hkl} \tag{7.59}$$

where

$$A_{hkl} = \sum_{j=1}^{N} f_j \cos\left(2\pi \mathbf{r}_j \cdot \mathbf{s}\right) \tag{7.60}$$

and

$$B_{hkl} = \sum_{j=1}^{N} f_j \sin\left(2\pi \mathbf{r}_j \cdot \mathbf{s}\right) \tag{7.61}$$

we can see a phase associated with F_{hkl} given by

$$\tan \phi_{hkl} = \frac{B_{hkl}}{A_{hkl}} \tag{7.62}$$

Further, it can be shown that the intensity of the scattered radiation from a unit cell is

$$I_{hkl} = |F_{hkl}|^2 = A_{hkl}^2 + B_{hkl}^2 = \sum_{i=1}^{N}\sum_{j=1}^{N} f_i f_j \cos\left[2\pi\left(\mathbf{r}_i - \mathbf{r}_j\right)\cdot\mathbf{s}\right] \tag{7.63}$$

Eq. (7.63) clearly shows that the intensity is dependent on the interatomic distances and not on the actual atomic coordinates. It is to be noted that the intensity of the diffracted beam from the entire crystal which consists of a large number of unit cells (as mentioned earlier) is, nonetheless, proportional to I_{hkl}.

7.3.3 Bragg's law

The actual positions of the X-ray diffraction spots are given by Bragg's law where diffraction is considered as partial reflection of the incident radiation by sets of parallel planes. The phenomenon can be simplified as in Fig. 7.7 where scattering from two atoms A and A′ located on two parallel planes separated by a distance d produces a phase difference of

$$(\mathrm{BA}' + \mathrm{A}'\mathrm{C})/\lambda. = 2d\sin\theta/\lambda \tag{7.64}$$

Therefore the scattered waves from the two atoms will reinforce each other if they are in the same phase, that is, if

$$2d\sin\theta = n\lambda \tag{7.65}$$

where n is an integer. This is the mathematical form of Bragg's law. Eq. (7.64) gives the permitted angles of reflection θ, in terms of λ and d.

In three dimensions, the diffraction from the entire crystal can be viewed as partial reflection of X-rays from a number of sets of equispaced "atom-rich" parallel planes. These planes are defined by intercepts on the unit cell edges, a/h, b/k, and c/l, where h, k, and l are integers called the indices of X-ray reflection.

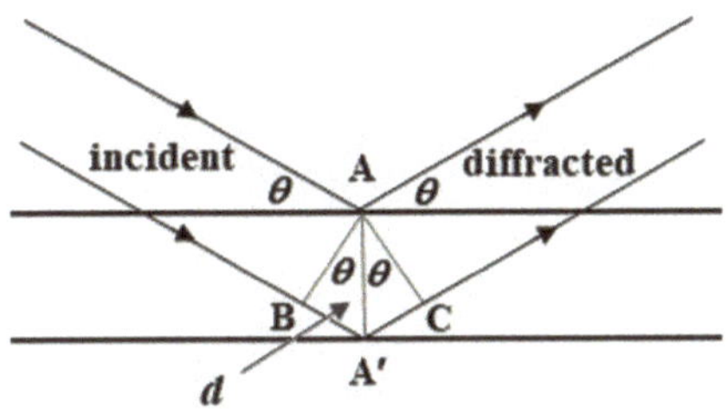

FIGURE 7.7

Parallel rays reflected from points on neighboring planes.

The spacing of the planes is denoted by d_{hkl}. From the geometry of the reflecting planes and Laue equation (7.54) it can be shown that

$$\frac{2\sin\theta}{\lambda} = \frac{1}{d_{hkl}} \tag{7.66}$$

which is a rearranged form of (7.65). $\mathbf{h}$ [= (hkl)] are also called ***Miller indices***. They specify the Braggs planes which intercept the unit cell edges h, k, and l times respectively. Also, $\mathbf{h}$ is a vector in the reciprocal space perpendicular to these planes.

7.4 Diffraction and Fourier transform

As mentioned in the beginning of Chapter 5, the diffraction by a lattice produces a lattice as a diffraction pattern (reciprocal lattice) which is in the reciprocal space. Nevertheless, the direct space of the crystal lattice and the reciprocal space of the diffraction pattern are related by Fourier transform.

In the same chapter, we have introduced the Fourier transform of a function $f(x)$. Here, we need to generalize the expression for a three-dimensional function $f(\mathbf{r})$, where $\mathbf{r} = (x, y, z)$, as

$$\bar{f}(\mathbf{s}) = \iiint f(\mathbf{r}) \exp(-2\pi i \mathbf{r} \cdot \mathbf{s})\, d\mathbf{r} \tag{7.67}$$

where $\mathbf{s} = (u, v, w)$ are three reciprocal variables. Similarly, the inverse Fourier transform is given by

$$f(\mathbf{r}) = \iiint \bar{f}(\mathbf{s}) \exp(2\pi i \mathbf{r} \cdot \mathbf{s})\, d\mathbf{s} \tag{7.68}$$

Now, let us consider a continuous distribution of electron density $\rho(\mathbf{r})$ in a volume v, at a distance $\mathbf{r}$ from the origin. An element of volume dv in the said space has an electron content of $\rho(\mathbf{r})dv$. Then the contribution of the element to the total scattering is $\rho(\mathbf{r})dv$ times as much as that from a single electron at the same position. Hence, the total scattered amplitude from the distribution of electron density $\rho(\mathbf{r})$, expressed as a fraction of that from a point electron at the origin, is given by

$$\mathbf{F}(\mathbf{s}) = \int_v \rho(\mathbf{r}) \exp(2\pi i \mathbf{r} \cdot \mathbf{s})\, dv \tag{7.69}$$

where the integration is over the entire volume under consideration. It is clear from (7.69) that $\mathbf{F}(\mathbf{s})$ is the Fourier transform of $\rho(\mathbf{r})$. The intensity of the transform is

$$I(\mathbf{s}) = \mathbf{F}(\mathbf{s})\mathbf{F}^*(\mathbf{s}) \tag{7.70}$$

Thus $\mathbf{F}(\mathbf{s})$, which is mathematically represented by (7.69), gives the diffraction pattern of the distribution of electron density $\rho(\mathbf{r})$.

The Fourier transform of a molecule with n atoms, each atom with a scattering factor $f_j(\mathbf{s})$, can be written as

$$\mathbf{F}(\mathbf{s}) = \sum_{j=1}^{n} f_j(\mathbf{s}) \exp\left(2\pi i \mathbf{r}_j \cdot \mathbf{s}\right) \tag{7.71}$$

Considering a unit cell whose edges are denoted by vectors **a**, **b**, and **c**, and jth atom by fractional coordinates x_j, y_j, and z_j, so that

$$\mathbf{r}_j = x_j\mathbf{a} + y_j\mathbf{b} + z_j\mathbf{c} \tag{7.72}$$

Eq. (7.58) can be written as

$$F_{hkl} = \sum_{j=1}^{N} f_j \exp\left[2\pi i\left(x_j\mathbf{a} + y_j\mathbf{b} + z_j\mathbf{c}\right) \cdot (h\mathbf{a}^* + k\mathbf{b}^* + l\mathbf{c}^*)\right]$$

$$= \sum_{j=1}^{N} f_j \exp\left[2\pi i\left(hx_j + ky_j + lz_j\right)\right] \tag{7.73}$$

Eq. (7.73) gives the Fourier transform for the unit cell. Thus, it can be said that, for a crystal, the structure factor equation is its Fourier transform defined at the reciprocal lattice points.

7.5 Convolution and diffraction

In Chapter 5, the convolution of two functions $f(x)$ and $g(x)$, denoted by $f * g$, over an interval $[-\infty, \infty]$, was defined as

$$(f * g)(t) = \int_{-\infty}^{\infty} f(x)g(t-x)dx \tag{7.74}$$

The integral is thus a "blending" of the two functions. According to the convolution theorem, the Fourier transform of a convolution $f * g$ is equal to the product of the Fourier transforms of f and g separately, multiplied by $\sqrt{2\pi}$, that is

$$FT\left[(f * g)(t)\right] = \sqrt{2\pi}\, FT\left[f(x)\right] \cdot FT\left[g(x)\right] \tag{7.75}$$

The convolution theorem can explain why diffraction from a lattice (in real space) produces another lattice (in reciprocal space). A crystal consists of a three-dimensional array of repeating unit cells – the periodic structure of the crystal is a convolution of a unit cell and a point lattice. Mathematically, this can be generated by convoluting the content of a unit cell by a three-dimensional lattice of δ-functions. Therefore the diffraction pattern is the product of the Fourier transform of the content of one unit cell and the Fourier transform of the (real) lattice. In other words, the diffraction pattern of a single unit cell is sampled at the reciprocal lattice points.

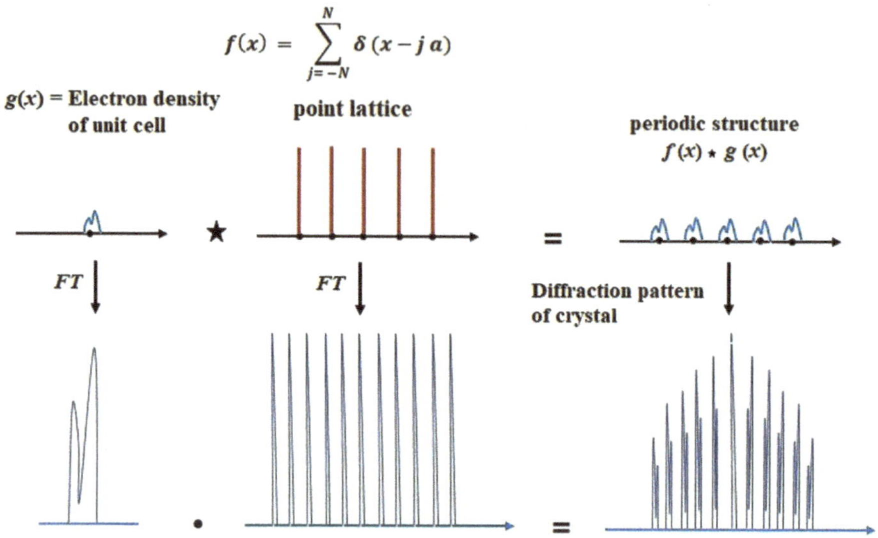

FIGURE 7.8

Diffraction from the convolution of the electron density in a unit cell and a one-dimensional lattice of δ-functions.

Let us consider a function $g(x)$ representing the electron density in a unit cell and a point lattice given by

$$f(x) = \sum_{j=-N}^{N} \delta(x - ja) \tag{7.76}$$

where a is the spacing between the lattice points. Then, the convolution

$$f(x) \star g(x)$$

generates a periodic structure (Fig. 7.8). As a consequence, the diffraction pattern from this lattice is generated by the product of the Fourier transforms of the two functions (Fig. 7.8).

7.6 The electron density equation

In (7.69) the scattered amplitude $\mathbf{F}(\mathbf{s})$ has been expressed as a Fourier transform of the electron density $\rho(\mathbf{r})$. On the basis of Definition 5.1.2 in Chapter 5, we can regenerate the electron density with an inverse Fourier transform.

$$\rho(\mathbf{r}) = \int_{v^*} \mathbf{F}(\mathbf{s}) \exp(-2\pi i \mathbf{r} \cdot \mathbf{s})\, dv^* \tag{7.77}$$

where v^* is the corresponding volume in reciprocal space. (7.77) is called the electron density equation. In general, the equation would involve an integral like the forward Fourier transform. However, for a crystal, **F** (s) exists only at reciprocal lattice points; the weight of **F** (s) at each of these points is (1/*V*) F_{hkl}. Hence, the integral must be replaced by a summation.

$$\rho(\mathbf{r}) = \frac{1}{V}\sum_{\mathbf{s}} F(\mathbf{s}) \exp(-2\pi i \mathbf{r} \cdot \mathbf{s}) \tag{7.78}$$

The electron density is completely defined in one unit cell of volume *V*. Therefore substituting

$$\mathbf{r} = x\mathbf{a} + y\mathbf{b} + z\mathbf{c} \text{ and } \mathbf{s} = h\mathbf{a}^* + k\mathbf{b}^* + l\mathbf{c}^*$$

we have

$$\rho(x, y, z) = \frac{1}{V}\sum_h\sum_k\sum_l F_{hkl} \exp\{-2\pi i(hx + ky + lz)\} \tag{7.79}$$

x, *y*, and *z* being the fractional coordinates of a point in the unit cell.

7.6.1 Phase problem and the Patterson function

(7.79) can be modified by including the relative phase α_{hkl} and written as

$$\begin{aligned}\rho(\mathbf{r}) &= \frac{1}{V}\sum_{\mathbf{s}} \mathbf{F}(\mathbf{s}) \exp(-2\pi i \mathbf{r} \cdot \mathbf{s}) \\ &= \frac{1}{V}\sum_h\sum_k\sum_l |F_{hkl}| \exp(i\alpha_{hkl}) \exp\{-2\pi i(hx + ky + lz)\}\end{aligned} \tag{7.80}$$

where *V* is the unit cell volume. *h*, *k*, and *l* can be obtained from the position of the diffraction spot. The amplitude $|F_{hkl}|$ can be calculated from the measured intensity of the diffracted wave at the corresponding spot. However, the measurements are unable to provide any information on the relative phase α_{hkl}. This is precisely known as the "phase problem."

In crystallography, the phase problem has been addressed by the ***Patterson function*** *P* (**u**), which is defined as the "autocorrelation" function of the electron density. In other words, *P* (**u**) is the convolution of $\rho(\mathbf{r})$ with $\rho(-\mathbf{r})$:

$$P(\mathbf{u}) = \rho(\mathbf{r}) * \rho(-\mathbf{r}) = \int_V \rho(\mathbf{r})\rho(\mathbf{u} + \mathbf{r})\, d\mathbf{r} \tag{7.81}$$

From the property of convolution, we have

$$FT\{P(\mathbf{u})\} = |F(\mathbf{s})|^2 \text{ so that } P(\mathbf{u}) = FT^{-1}\{|F(\mathbf{s})|^2\} \tag{7.82}$$

and we can calculatte

$$P(\mathbf{u}) = \frac{1}{V}\sum_{\mathbf{s}} |F(\mathbf{s})|^2 \exp(-2\pi i \mathbf{u} \cdot \mathbf{s}) \tag{7.83}$$

(7.83) indicates that a map calculated using the Patterson function does not require the pahse information. It is also to be noted that $P(\mathbf{u})$ is maximum when **u** is equal to an interatomic distance. Hence, the Patterson map has peaks not at the absolute atomic positions but at the interatomic vectors.

For molecules with relatively small number of atoms it is possible to determine the original positions of the atoms from the observed Patterson peaks. Now, if there are N atoms in a unit cell, there will be N electron density peaks but $N\ (N-1)$ interatomic vector peaks. So, for macromolecules (with $N > 1000$) the Patterson peaks will not be resolved from each other. Nonetheless, the Patterson function is still useful as a part of other methods, such as the isomorphous replacement method, to solve macromolecular structures.

7.6.2 Isomorphous replacement

In this approach, heavy atoms are incorporated into protein crystals such that the shape and size of the unit cell and orientation of the protein in the cell remain unchanged. The heavy atom will enhance the scattered intensity significantly. The difference in the scattered intensities between that produced by the native crystal and the derivative crystal can be used to compute a Patterson map. Since there will be only a few heavy atoms, the Patterson map will be relatively easy to deconvolute. The contribution of the heavy atom to the structure factor can then be calculated. The structure factor of the derivative crystal F_{PH} is related to that for the native crystal F_P by the equation

$$F_{PH} = F_p + F_H \tag{7.84}$$

where F_H is the heavy atom structure factor. The relation, however, gives two possible values for F_P. The twofold phase ambiguity can be resolved by more than one heavy atom derivative in multiple isomorphous replacement (MIR).

7.6.3 Electron density sharpening

The atoms in a crystal vibrate about their mean positions with amplitude that increases with temperature. However, the frequencies of vibrations are very low in the time-scale of measurements. Hence, the diffraction pattern is produced from a "frozen" crystal where the atoms are stationary but displaced from their mean positions. Consequently, the time-average of the pattern effectively smears an atom from a point to a sphere or ellipsoid of electron density. This is accounted for by including the ***temperature- or B-factor*** in (7.73). For the jth atom, the B-factor is given by

$$B_j = 8\pi^2 \mu\, j^2 \tag{7.85}$$

where μ is the mean square displacement in the three principal lattice directions. The structure factor can then be calculated as

$$F_{hkl,calc} = \sum_{j=1}^{N} f_j \exp\left[2\pi i\left(hx_j + ky_j + lz_j\right)\right]\exp\left(-\frac{B_j \sin^2\theta}{\lambda^2}\right) \tag{7.86}$$

It is to be noted that besides the thermal motion of the atoms and molecules, other factors such as static defects in crystal packing, and nonideal X-ray beam and detector responses also contribute to the B-factor. Whatever the causes, the result is a smearing effect which is usually modeled as a convolution of the ideal electron density with a "blurring" Gaussian function:

$$F_{obs} = F_{ideal} * e^{-B_j\left(\frac{\sin\theta}{\lambda}\right)^2} \tag{7.87}$$

where $B_j > 0$. This affects the diffraction intensity leading to a loss of resolution.

The blurring effect can be corrected by a process known as electron density sharpening which is essentially a deconvolution. Since the blurring effect is modeled by a Gaussian function with a positive B-factor (7.87), the deconvolution can be accomplished by applying a similar Gaussian function with a negative sharpening factor $B^{(-)}$.

$$F_{sharpened} = F_{obs} * e^{-B_j^{(-)}\left(\frac{\sin\theta}{\lambda}\right)^2} = F_{ideal} * e^{-\left(B_j + B_j^{(-)}\right)\left(\frac{\sin\theta}{\lambda}\right)^2} \tag{7.88}$$

where $B_j^{(-)} < 0$.

In general, sharpening based on (7.88), sometimes with necessary modifications, has been found to be a powerful approach for restoring the precision of electron density maps that are affected by the blurring effect.

Exercise 7

1. When electromagnetic radiation is incident on a free electron and the scattered radiation is observed at point **r** from the electron, the scattered-to-incident amplitude ratio is given by r_e/r, where r_e is the radius of the electron. This suggests that only a fraction of the incident radiation is scattered. Yet, the scattered radiation is measurable – explain.
2. How would the scattering of electromagnetic radiation be affected if an electron is replaced by a proton as the scatterer? – explain.
3. Electromagnetic radiation of wavelength λ is incident on a row of four identical scatterers in a direction normal to the row. 2λ is the space between neighboring scatterers. Determine the scattered amplitude and intensity (as a fraction of that from all four scatterers located at one point) for a scattering angle 60 degrees.
4. A unit cell has the following parameters:

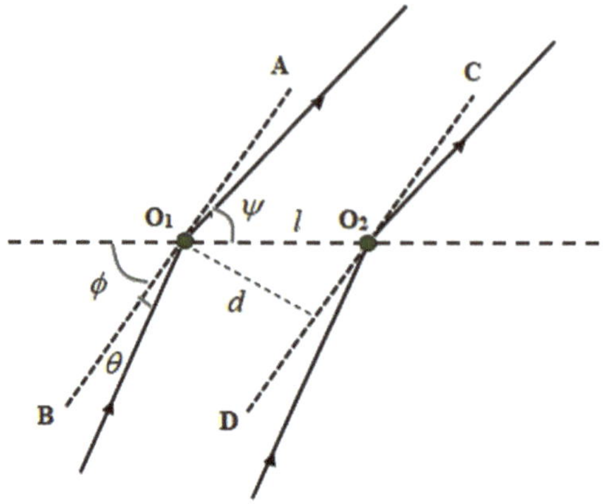

FIGURE 7.9

Diagram for exercise 5.

$a = 10$ Å, $b = 10$ Å, $c = 20$ Å, $\alpha = 90$ degrees, $\beta = 90$ degrees, $\gamma = 120$ degrees

a. Determine the parameters of the reciprocal cells.
b. Calculate the volume V of the real cell and volume V^* of the reciprocal cell, and show their reciprocity, $VV^* = 1$.

5. shows that X-ray is diffracted by scattering centers O_1 and O_2 which are at a distance l apart. Equivalently, this can also be considered as reflection from planes AB and CD with spacing d between them. The relevant angles are labeled. On the basis of the figure, prove the equivalence between the Laue equations and Bragg's law (Fig. 7.9).
6. Consider that the electron density of an atom in a crystal is represented by a Gaussian

$$f_a = \sqrt{\frac{a}{2\pi}} e^{-\frac{ax^2}{2}}$$

It is convolved with a blurring Gaussian function

$$f_b = \sqrt{\frac{b}{2\pi}} e^{-\frac{bx^2}{2}}$$

Show quantitatively what would be the effect of the convolution on the diffraction pattern.

Further reading

Drenth, J., 2007. Principles of Protein X-Ray Crystallography, third ed. Springer.
Griffiths, D.J., 2013. Introduction to Electrodynamics, fourth ed. Pearson.
Hendrickson, W.A., 1995. X-ray in molecular biophysics. Phys. Today 48 (11), 42–48.

Ladd, M., Palmer, R., 2013. Structure Determination by X-ray Crystallography, fifth ed. Springer.

Liu, C., Xiong, Y., 2014. Electron density sharpening as a general technique in crystallographic studies. J. Mol. Biol. 426, 980–993.

Tipler, P.A., Mosca, 2008. Physics for Scientists and Engineers, sixth ed. W. H. Freeman & Company.

Woolfson, M.M., 1997. An Introduction to X-Ray Crystallography, second ed. Cambridge University Press.

CHAPTER

Cryo-electron microscopy 8

The term "cryo-electron microscopy" (cryo-EM) encompasses several experimental techniques, each of which is based on the principle of imaging radiosensitive objects in a transmission electron microscope under cryogenic conditions. Just as the light microscope extends the resolving power of the human eye to visualize objects, the electron microscope extends the resolving power of the light microscope. A light microscope and an electron microscope, though significantly different in their resolving power, are, nonetheless, based on analogous principles of imaging.

In light microscopy, visible electromagnetic waves or photons are made to pass through the object and then refracted through optical lenses fabricated with glass to form an image. In an electron microscope, electrons are emitted by a source in high vacuum and accelerated down the microscope column toward the sample by electrical potential differences in the range of 100–300 keV. As waves, they interact with the atoms and are diffracted (scattered) in the course of passing through the sample. Thereafter, the electrons are focused to produce an image by the electromagnetic lenses of the microscope—being (negatively) charged particles, they are prone to be deflected by the electromagnetic field. Evidently and interestingly, here the electron reveals its dual nature—wave as well as particle. Hence, a brief review on wave–particle duality and quantum physics should be useful for our later discussion.

8.1 Quantum physics

8.1.1 Wave–particle duality

Making a radical departure from the classical concept, Max Planck introduced the concept of ***quantized energy*** in its exchange between matter and radiation. This concept was taken forward by Albert Einstein to the description of light (electromagnetic wave) as discrete bundles of energy or photons. The energy of a photon is given as

$$E = h\nu = hc/\lambda \tag{8.1}$$

where ν and λ are, respectively, the frequency and the wavelength of the light, and

$$h = 6.626 \times 10^{-34}\ \mathrm{J \cdot s}$$

Mathematical Approaches to Molecular Structural Biology. DOI: https://doi.org/10.1016/B978-0-323-90397-4.00008-1
© 2023 Elsevier Inc. All rights reserved.

is known as the ***Planck's constant***. Further, behaving as a particle the photon is assigned a linear momentum with magnitude

$$p = h/\lambda \tag{8.2}$$

Now, it is quite logical to think that if light can behave as both wave and particle, matter, such as an electron, should display wave nature and be assigned a wavelength and a frequency. Based on this symmetry argument, Louis de Broglie proposed that an electron with an energy E and linear momentum p can be conceptualized as a matter wave with wavelength and frequency given by

$$\lambda = h/p \tag{8.3}$$

and

$$v = E/h \tag{8.4}$$

The simplest way to describe the matter wave related to a particle with a defined momentum p_0 (and, therefore, defined wavelength λ_0 or wave number $k_0 = 2\pi/\lambda_0$) would be a perfect sinewave moving along the x-axis which at a time $t = 0$ can be represented as

$$\psi(x) = \psi_0 \sin k_0 x \tag{8.5}$$

8.1.2 Schrödinger equation

A mathematical theory of the matter wave was developed by Erwin Schrödinger. This is known as **quantum theory**, **quantum mechanics**, or **wave mechanics**. Schrödinger introduced the **wave function Ψ** that would obey a fundamental equation of quantum mechanics known as **Schrödinger equation** which in one dimension is written as

$$\mathrm{i}\hbar \frac{\partial \Psi}{\partial t} = -\frac{\hbar^2}{2m}\frac{\partial^2 \Psi}{\partial x^2} + V(x)\Psi \tag{8.6}$$

where $\hbar = h/2\pi$ and potential $V(x)$ is assumed to be time-independent. The equation is known as **time-dependent Schrödinger** equation. For a particle moving in the x-direction, it can be shown that

$$\Psi(x,t) = \Psi_0 \, \mathrm{e}^{\mathrm{i}(kx-\omega t)} \tag{8.7}$$

is a solution to the Schrödinger equation and, hence, the wave function. In (8.7), k $(=2\pi/\lambda)$ is the wave number and ω $(=2\pi\,\nu)$ is the angular frequency. It is not possible to know the precise location of a quantum particle. Therefore Max Born introduced the concept of ***probability density*** expressed as

$$P(x) = \Psi\Psi^* \tag{8.8}$$

so that the probability of finding the particle between x and $x + dx$ is given by $\Psi\Psi^* dx$. Separating the time and space variations of the wave function, we can derive from (8.6) the **time-independent Schrödinger** equation

$$-\frac{\hbar^2}{2m}\frac{d^2\psi(x)}{dx^2} + V(x)\psi(x) = E\psi(x) \tag{8.9}$$

where E $(=\hbar\,\omega = h\,\nu)$ is the energy of the particle. For a particle, not under the influence of any potential $[V(x) = 0]$, the total energy is entirely kinetic, that is, $E = K = p^2/2\,m$. In such case, (8.9) becomes

$$\frac{\hbar^2}{p^2}\frac{d^2\psi(x)}{dx^2} + \psi(x) = 0 \tag{8.10}$$

The solution to this equation is

$$\psi(x) = \psi_0\, e^{ikx} \tag{8.11}$$

where $k = 2/\lambda$. This is the wave function for a free particle traveling in the x-direction.

For the motion of a particle in three dimensions, the time-independent Schrödinger equation (8.9) can be written as

$$-\frac{\hbar^2}{2m}\left(\frac{\partial^2\psi(x,y,z)}{\partial x^2} + \frac{\partial^2\psi(x,y,z)}{\partial y^2} + \frac{\partial^2\psi(x,y,z)}{\partial z^2}\right) + V(x,y,z)\psi(x,y,z) = E\psi(x,y,z) \tag{8.12}$$

8.1.3 Hamiltonian

The Schrödinger equation can be written in another form based on the concepts of observables and operators. An ***observable*** is defined as any dynamical variable that can be measured. In classical mechanics, observables are represented by functions, whereas in quantum mechanics, they are represented by mathematical operators. An ***operator*** is a mathematical entity that transforms one function into another. It can be as simple as $x\times$, which is essentially a multiplication by x.

$$x \times f(x) = x f(x) \tag{8.13}$$

The function $x\,f(x)$ is different from $f(x)$. Often operators are represented by the symbol Ω. If Ω is a linear operator, then for functions f and g,

$$\Omega(af + bg) = a\,\Omega f + b\,\Omega g \tag{8.14}$$

where a and b are constants. Operators in quantum mechanics, representing different observables, are all linear.

In certain cases, when an operator operates on a function, the outcome is the same function multiplied by a constant. Such a function is called an

Table 8.1 Some quantum mechanical operators.

	Operators	
Observables	**In one dimension**	**In three dimensions**
Position	$x \rightarrow x \times$	$\mathbf{r} \rightarrow \mathbf{r} \times$
Momentum	$p_x \rightarrow -i\hbar \frac{\partial}{\partial x}$	$\mathbf{p} \rightarrow -i\hbar\left(\hat{\mathbf{i}}\frac{\partial}{\partial x} + \hat{\mathbf{j}}\frac{\partial}{\partial y} + \hat{\mathbf{k}}\frac{\partial}{\partial z}\right)$
Kinetic energy	$T \rightarrow -\frac{\hbar^2}{2m}\frac{\partial^2}{\partial x^2}$	$T \rightarrow -\frac{\hbar^2}{2m}\left(\frac{\partial^2}{\partial x^2} + \frac{\partial^2}{\partial y^2} + \frac{\partial^2}{\partial z^2}\right) = -\frac{\hbar^2}{2m}\nabla^2$
Potential energy	$V(x) \rightarrow V(x) \times$	$V(\mathbf{r}) \rightarrow V(\mathbf{r}) \times$
Total energy	$E = (T + V) \rightarrow H = (T + V)$	$E = (T + V) \rightarrow \hat{H} = (T + V)$

∇^2: read as "del squared," called the Laplacian.

eigenfunction and the constant is called an ***eigenvalue*** of the operator. In the eigenvalue equation

$$\Omega f = \lambda f \tag{8.15}$$

f is the eigenfunction and λ is the eigenvalue of the operator Ω. The observables and their corresponding quantum mechanical operators that are relevant to our discussion in this chapter are given in Table 8.1.

In classical mechanics, the total energy (kinetic and potential) of a particle moving in one dimension is called the ***Hamiltonian***

$$H(x, p) = T + V = \frac{p^2}{2m} + V(x) \tag{8.16}$$

In quantum mechanics the corresponding ***Hamiltonian operator*** is obtained by substituting

$$p \rightarrow -i\hbar\frac{d}{dx}$$

in (8.16).

$$\hat{H} = -\frac{\hbar^2}{2m}\frac{d^2}{dx^2} + V(x) \tag{8.17}$$

So, the time-dependent Schrödinger equation (8.6) can be written in the form

$$i\hbar\frac{\partial \Psi}{\partial t} = \hat{H}\Psi \tag{8.18}$$

and the time-independent Schrödinger equation (8.9) can be written as

$$\hat{H}\psi(x) = E\psi(x) \tag{8.19}$$

In three dimensions, (8.19) becomes

$$\hat{H}\psi(\mathbf{r}) = E\psi(\mathbf{r}) \tag{8.20}$$

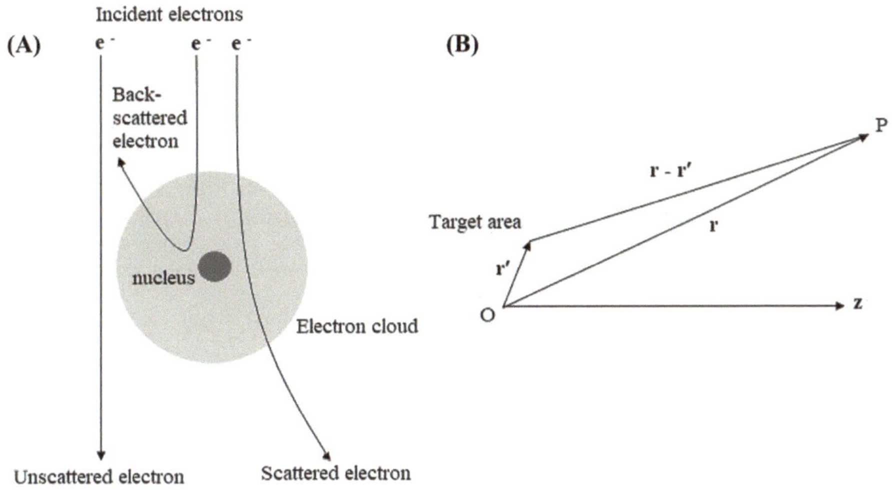

FIGURE 8.1

Physical (A) and geometric (B) representations of scattering of an electron by an atom.

8.2 Wave optics of electrons—scattering

The scattering of an electron (mass μ) by an atomic potential, $V(\mathbf{r})$, (Fig. 8.1) can be represented by (8.20). Substituting

$$\hat{H} = -\frac{\hbar^2}{2\mu}\nabla^2 + V(\mathbf{r}) \tag{8.21}$$

in (8.20)

$$-\frac{\hbar^2}{2\mu}\nabla^2\psi(\mathbf{r}) + V(\mathbf{r})\psi(\mathbf{r}) = E\psi(\mathbf{r}) \tag{8.22}$$

which can be rewritten as

$$(\nabla^2 + k^2)\psi(\mathbf{r}) = \frac{2\mu}{\hbar^2}V(\mathbf{r})\psi(\mathbf{r}) \tag{8.23}$$

where

$$k^2 = \frac{2\mu E}{\hbar^2}$$

Green's function

Green's function is used to solve a differential equation containing an inhomogeneous term, which is called a ***source term***, related to an integral operator containing the source.

Let us consider a linear, inhomogeneous equation of the form

$$\mathcal{L}h(x) = f(x) \tag{8.24}$$

(Continued)

Green's function (Continued)

where h and f are two functions in the domain Ω, and $\mathcal{L}$ is a linear operator. If $\mathcal{L}$ has an inverse that is an integral operator, there may be a solution to (8.24) in the form

$$h(x) = \int_{\Omega} G(x,t) f(t)\, dt \tag{8.25}$$

In such case, the kernel of the integral operator $G(x, y)$ is called the ***Green's function***. Let $f(x)$ in (8.24) be a point source and the equation be valid on the range $a \le x \le b$ subject to homogeneous boundary conditions at $x = a$ and $x = b$. Then, the Green's function in (8.25) must satisfy the boundary conditions and

$$\mathcal{L} G(x,t) = \delta(x - t) \tag{8.26}$$

so that the solution to (8.24) can be written as

$$h(x) = \int_a^b G(x,\ t) f(t)\, dt \tag{8.27}$$

Using (8.24), it can be verified that

$$\mathcal{L} h(x) = \int_a^b \mathcal{L} G(x,t) f(t)\, dt = \int_a^b \delta(x - t) f(t)\, dt = f(x) \tag{8.28}$$

The general solution of (8.23) consists of two components:

(1) general solution to the homogeneous equation

$$(\nabla^2 + k^2)\psi_0(\mathbf{r}) = 0 \tag{8.29}$$

as

$$\psi_0(\mathbf{r}) = \frac{1}{(2\pi)^{3/2}} \exp(\mathrm{i}\,\mathbf{k}\cdot\mathbf{r}) \tag{8.30}$$

which is the incident plane wave, and

(2) a particular solution of (8.23) with the interaction potential

Assuming a point source, the general solution of (8.23) can be expressed as

$$\psi(\mathbf{r}) = \psi_0(\mathbf{r}) + \frac{2\mu}{\hbar^2} \int G(\mathbf{r} - \mathbf{r}') V(\mathbf{r}') \psi(\mathbf{r}')\, \mathrm{d}\mathbf{r}' \tag{8.31}$$

where $G(\mathbf{r} - \mathbf{r}')$ is the Green's function obtained by solving the point source equation

$$(\nabla^2 + k^2)\, G(\mathbf{r} - \mathbf{r}') = \delta(\mathbf{r} - \mathbf{r}') \tag{8.32}$$

by the method of contour integration (not covered in the present book—interested readers may refer to books on complex analysis).

$$G^{\pm}(\mathbf{r} - \mathbf{r}') = -\frac{1}{4\pi} \frac{\exp(\mathrm{i}\,k|\mathbf{r} - \mathbf{r}'|)}{|\mathbf{r} - \mathbf{r}'|} \tag{8.33}$$

For a stationary scattering wave, one may choose $G(\mathbf{r}-\mathbf{r}') = G^{\pm}(\mathbf{r}-\mathbf{r}')$, so that

$$\psi(\mathbf{r}) = \psi_0(\mathbf{r}) - \frac{\mu}{2\pi\hbar^2}\int \frac{\exp(\mathrm{i}k|\mathbf{r}-\mathbf{r}'|)}{|\mathbf{r}-\mathbf{r}'|} V(\mathbf{r}')\psi(\mathbf{r}')\,\mathrm{d}\mathbf{r}' \tag{8.34}$$

At this point, to solve (8.34), two assumptions are made as follows.

1. As shown in Fig. 8.1B, the target is localized around the origin O on which the wave is incident in the direction **z** and scattering is observed at P. The observation point and the integration point are, respectively, represented by **r** and **r**′. If we calculate the wave function far away from the scattering center, it can be assumed that

$$\mathbf{r} >> \mathbf{r}'$$

In that case,

$$\exp\left(\mathrm{i}k\left|\mathbf{r}-\mathbf{r}'\right|\right) = \mathrm{e}^{\mathrm{i}kr}\mathrm{e}^{-\mathrm{i}\mathbf{k}'\cdot\mathbf{r}'}$$

and

$$\frac{1}{|\mathbf{r}-\mathbf{r}'|} \approx \frac{1}{r}$$

2. **Born approximation**. The complexity of the integral in (8.34) makes its solution quite formidable. However, it can be solved by a series of successive or iterative approximations known as the Born series. Born approximation considers only the first term of the series, and hence the method is called the ***first Born approximation***. Further, if the potential $V(\mathbf{r})$ is weak, it does not significantly affect the wave function. Then,

$$\Psi(\mathbf{r}') \approx \Psi_0(\mathbf{r}') = \frac{1}{(2\pi)^{3/2}}\mathrm{e}^{\mathrm{i}\mathbf{k}\cdot\mathbf{r}'} \tag{8.35}$$

Based on these assumptions, we can rewrite (8.34) as

$$\Psi(\mathbf{r}) = \Psi_0(\mathbf{r}) - \frac{\mu}{2\pi\hbar^2}\frac{1}{(2\pi)^{3/2}}\frac{\mathrm{e}^{\mathrm{i}kr}}{r}\int \mathrm{e}^{-\mathrm{i}(\mathbf{k}'-\mathbf{k})\cdot\mathbf{r}}V(\mathbf{r}')\,\mathrm{d}\mathbf{r}' \tag{8.36}$$

which can also be written as

$$\Psi(\mathbf{r}) = \frac{1}{(2\pi)^{3/2}}\left[\mathrm{e}^{\mathrm{i}\mathbf{k}\cdot\mathbf{r}} + \frac{\mathrm{e}^{\mathrm{i}kr}}{r}f(\mathbf{k}',\mathbf{k})\right] \tag{8.37}$$

where

$$f(\mathbf{k}',\mathbf{k}) = -\frac{\mu}{2\pi\hbar^2}\int e^{-\mathrm{i}(\mathbf{k}'-\mathbf{k})\cdot\mathbf{r}'}V(\mathbf{r}')\,\mathrm{d}\mathbf{r}' \tag{8.38}$$

The first term in (8.37) denotes the incident plane wave in the propagation direction **k**, while the second term denotes the outgoing wave with amplitude

$f(\mathbf{k}', \mathbf{k})$. Clearly, the scattering amplitude is the Fourier transform of the potential energy with the wave vector $(\mathbf{k}' - \mathbf{k})$.

It can be shown by quantum mechanical calculations that in the first Born approximation, the differential cross section is given by

$$\frac{d\sigma}{d\Omega} = \left| f(\mathbf{k}', \mathbf{k}) \right|^2 = \frac{\mu^2}{2\pi^2 \hbar^4} \left| \int e^{-i(\mathbf{k}'-\mathbf{k}) \cdot \mathbf{r}'} V(\mathbf{r}') d\mathbf{r}' \right|^2 \tag{8.39}$$

8.3 Theory of image formation

8.3.1 Electrodynamics of lens system

The electron microscope lens system, shown in Fig. 8.2, is based on the principle of deflection of electrons, which are negatively charged particles of small mass, by an electromagnetic field. Condenser lenses transform the diverging electron beam into a parallel one incident on the sample. The objective lens is responsible for primary magnification while the image is further magnified by projector lenses before the electrons reach the detector.

Empirically, it is found that a particle with mass m and electric charge q experiences a force in an electric field given by

$$\mathbf{F}_E = q\mathbf{E} \tag{8.40}$$

If q is positive (as for a proton), $\mathbf{F}_E$ is in the direction of $\mathbf{E}$. For an electron, which is negatively charged, $\mathbf{F}_E$ is antiparallel, that is, in the direction opposite to $\mathbf{E}$. According to Newton's second law,

$$\mathbf{F}_E = q\mathbf{E} = m\mathbf{a}$$

or,

$$\mathbf{a} = (q/m)\mathbf{E} \tag{8.41}$$

where $\mathbf{a}$ is the acceleration of the electron caused by the field. For high energy electrons, a relativistic correction is necessary.

$$\mathbf{F}_E = q\mathbf{E} = m\left(1 - \frac{v^2}{c^2}\right)^{-3/2} \mathbf{a} \tag{8.42}$$

where $\mathbf{v}$ is the velocity of the electron and c is the speed of light.

Experimentally, it is also observed that an electric charge q moving with velocity $\mathbf{v}$ in a magnetic field $\mathbf{B}$ experiences a force given by

$$\mathbf{F}_M = q\mathbf{v} \times \mathbf{B} \tag{8.43}$$

The direction of $\mathbf{F}_M$, perpendicular to both $\mathbf{v}$ and $\mathbf{B}$, is given by the right-hand rule for positive charge. For an electron (negative charge), it is just the opposite.

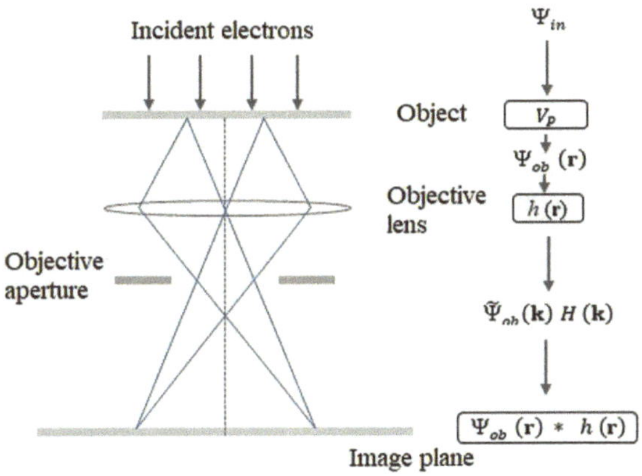

FIGURE 8.2

Schematic representation of an electron microscopic imaging system.

Combining (8.40) and (8.43), it can be said that a charge q moving with velocity $\mathbf{v}$ in a region of space where both an electric field $\mathbf{E}$ and a magnetic field $\mathbf{B}$ exist, experiences an electromagnetic force given by

$$\mathbf{F} = \mathbf{F}_E + \mathbf{F}_M = q(\mathbf{E} + \mathbf{v} \times \mathbf{B}) \tag{8.44}$$

It is called the ***Lorentz force***. It is evident from (8.44) that $\mathbf{F}_E$ changes the magnitude while $\mathbf{F}_M$ changes the direction of the velocity.

8.3.2 Image formation

The interaction between an incident electron wave and a macromolecule embedded in the sample can be considered due to two interaction potential components:

$$V_{\text{int}}(\mathbf{r}) = V_{\text{atom}}(\mathbf{r}) + V_{\text{bond}}(\mathbf{r}) \tag{8.45}$$

where V_{atom} is the "atom" contribution from a set of isolated atoms, and V_{bond} is the "bond" contribution representing the charge redistribution due to the solvent, ions and other molecular interactions. In simpler models to explain high-energy electron diffraction and microscopy, the second component is often ignored.

As a result of the interaction, some of the incident electrons are scattered. The images are generated by phase contrast caused by the interference between the unscattered and scattered parts of the exit wave. The sample is the phase object and the exit wave function is

$$\Psi_{ob}(\mathbf{r}) = \Psi_{in} \exp\left[\mathrm{i}\,\sigma V_P(\mathbf{r})\right] \tag{8.46}$$

where Ψ_{in} is the incident wave, m_e is the electron mass, λ is the electron wavelength, V_p is the obect potential and

$$\sigma = \frac{m_e \lambda}{2\pi\hbar^2}$$

In general, biological specimens are weak electron scatterers. That is, the phase shift introduced by the sample is small. So, as a "weak phase object approximation," from (8.46), we have

$$\Psi_{ob}(\mathbf{r}) = \Psi_{in}[1 + \mathrm{i}\,\sigma V_P(\mathbf{r})] \tag{8.47}$$

Clearly, the exit wave consists of two parts—(1) the unscatttered wave and (2) the scattered wave which is proportional to the object potential. Being an imaginary part, the second term has a phase shift of 90°. Sometimes, to make calculations simpler, it is assumed that the incident wave $\Psi_{\text{in}} = 1$ and (8.47) becomes

$$\Psi_{ob}(\mathbf{r}) \approx 1 + \mathrm{i}\sigma V_P(\mathbf{r}) \tag{8.48}$$

Then, the intensity distribution observed in the image plane is

$$I_i(\mathbf{r}) = \Psi_{ob}(\mathbf{r})\Psi^*_{ob}(\mathbf{r}) \approx 1 + [\sigma V_P(\mathbf{r})]^2 \tag{8.49}$$

Since $[\sigma\ V_p\ (\mathbf{r})]^2 << 1$, the image has very little contrast. To increase the contrast and make the image sufficiently observable, the lens system of the microscope changes the phase of the scattered wave by 90° so that

$$\Psi_{ob}(\mathbf{r}) \approx 1 - \sigma V_P(\mathbf{r}) \tag{8.50}$$

and, therefore,

$$I_i(\mathbf{r}) \approx 1 - 2\sigma V_P(\mathbf{r}) \tag{8.51}$$

The intensity now becomes proportional to the object potential. The objective lens executes this change in the phase by introducing some "aberrations" to the scattered wave. Such aberrations, mainly spherical and defocus, are represented by a function $H(\mathbf{r})$. Its influence on the scattered wave can be mathematically represented by

$$\tilde{\Psi}_{ob}(\mathbf{k})\,H(\mathbf{k})$$

where $\mathbf{k}$ is the spatial frequency vector (in Fourier space) and $\tilde{\Psi}_{ob}(\mathbf{k})$ is the Fourier transform of $\Psi_{ob}(\mathbf{r})$. Hence, the image wave function is given by the inverse Fourier transform

$$\Psi_{\text{image}}(\mathbf{r}) = FT^{-1}\left[\tilde{\Psi}_{ob}(\mathbf{k})H(\mathbf{k})\right] = \Psi_{ob}(\mathbf{r}) * h(\mathbf{r}) \tag{8.52}$$

where $h(\mathbf{r})$, called the ***point spread function*** (PSF) or the **transfer function**, is the inverse Fourier transform of $H(\mathbf{k})$. (8.52) shows that the observed image is the convolution of the scattered wave with the PSF. The observed intensity is then

$$I_{image} = \left[\tilde{\Psi}_{ob}(\mathbf{k})H(\mathbf{k})\right]^2 \tag{8.53}$$

The aberration function is described by

$$H(\mathbf{k}) = \exp[-i\chi(\mathbf{k})] \tag{8.54}$$

The phase shift arising from spherical aberration and defocus is described by

$$\chi(\mathbf{k}) = \frac{2\pi}{\lambda}\left[\frac{1}{4}C_s\lambda^4k^4 - \frac{1}{2}\Delta f\,\lambda^2 k^2\right] \tag{8.55}$$

where C_s is the coefficient of spherical aberration and Δf is the defocus. (8.55) indicates that the image contrast of biological samples could be enhanced by the combined effects of spherical aberration and defocus. The image plane is moved away from the exact focus leading to a phase shift between scattered an unscattered electrons.

8.4 Image processing by multivariate statistical analysis—principal component analysis

Initially, the three-dimensional structure of macromolecules by electron microscopy was based essentially on visual recognition and interpretation of two-dimensional images. The approach, therefore, lacked objectivity and reproducibility. Advances in electron microscopy, amply supported by efficient algorithm and powerful computation, have remarkably improved the precision of macromolecular structure analysis.

In general, two problems are associated with electron microscopy of biological macromolecules. First, inside the cell the biomacromolecules are intrinsically flexible and exist in multiple conformations. Besides, the conditions of recording lead to relatively high level of noise in the raw electron images. To address the problems of heterogeneity and poor signal-to-noise ratio, single-particle cryo-EM works with enormously large datasets and averages similar images prior to further processing and interpretation.

Averaging similar images from a mixed population is necessary in single-particle processing. Evidently, there is a need for a robust similarity measure such as correlation or distance between the images. Further, to make the process computationally efficient, the acquired dataset must be compressed to a manageable size. The answer to the need are multivariate statistical analysis and eigenvector–eigenvalue-based data compression by principal component analysis (PCA).

8.4.1 Hyperspace and data cloud

Present-day cryo-EM involves the automatic collection of huge datasets—tens of thousands images of the sample are recorded. This is essential to establish that the subtle structural differences observed within the sample are of statistical significance. In a statistical dataset, an image is basically a "measurement" comprising a "collection of numbers." Each number in the collection represents the density of one pixel of the image. For example, an image of the size 400×400 pixels consists of $400 \times 400 = 160{,}000$ density values starting from the top left, that is, pixel (1, 1) and ending at the bottom right [pixel (400, 400)].

In a typical cryo-EM experiment, one may be working with, say, 200,000 images. For true comparison, all the 200,000 images must be of the same size, with all the pixels arranged in the same order. Hence, mathematically each of the measurements can be represented by a point (or, more correctly, as a vector) in a multidimensional vector space (***hyperspace***), where its coordinates are defined by the "gray" values of the pixels. In the present example, the collection of 200,000 points in hyperspace constitutes a ***data cloud***. So, the problem of image comparison boils down to computation of the distance between images in the data cloud—similar images are expected to be separated by a small distance or, in other words, have a high degree of correlation.

8.4.2 Distance metrics

Euclidean metric

There are different metrics to determine the distance and associated correlation. Of these, the classical Euclidean metric is the one most widely used. In Chapter 2 the distance between two vectors **u** and **v** has been defined as the inner product (Definition 2.7.4)

$$d(\mathbf{u},\mathbf{v}) = \|\mathbf{u}-\mathbf{v}\| = \sqrt{(\mathbf{u}-\mathbf{v},\ \mathbf{u}-\mathbf{v})} \tag{8.56}$$

Accordingly, the Euclidean distance between two measurements (images) represented by vectors **X** and **Y**, each containing p pixels can be written as

$$d^2_{\mathbf{XY}} = \sum_{i=1}^{p} (X_i - Y_i)^2 \tag{8.57}$$

In Section 6.4, the concepts of covariance and correlation concerning two vectors have been introduced. Considering two vectors **X** and **Y** as standardized, so that each has a mean = 0 and standard deviation = 1, the correlation can be written as

$$\mathrm{Cov}(\mathbf{X},\mathbf{Y}) = c_{\mathbf{XY}} = \frac{1}{p}\sum_{i=1}^{p} X_i Y_i \tag{8.58}$$

Then, from (8.57), we have

$$\begin{aligned} d^2_{\mathbf{XY}} &= \sum_{i=1}^{p} (X_i - Y_i)^2 = \sum_{i=1}^{p} X_i^2 + \sum_{i=1}^{p} Y_i^2 - 2\sum_{i=1}^{p} X_i Y_i \\ &= p\,\mathrm{Var}(\mathbf{X}) + p\,\mathrm{Var}(\mathbf{Y}) - 2p\,\mathrm{Cov}(\mathbf{X},\mathbf{Y}) \end{aligned} \tag{8.59}$$

As Var (**X**) = Var (**Y**) = 1, (8.59) reduces to

$$c_{\mathbf{XY}} = 1 - \frac{d^2_{\mathbf{XY}}}{2p} \tag{8.60}$$

(8.60) shows that the correlation, $c_{\mathbf{XY}}$, and Euclidean distance is related in a very simple way—the shorter the distance between **X** and **Y**, the higher the correlation between them.

Chi-square metric (χ^2–metric)

In certain cases, it may be disadvantageous of use the Euclidean metric—the χ^2–metric provides a useful alternative. It is based on relative counts, and not on the original ones. Further, it standardizes the measurements (vectors) by the mean and not by the variance. This pretreatment of the data produces sets of proportions called ***profiles***.

The chi-square distance between two profiles (vectors) $\mathbf{X} = [\, X_1 X_2 \ldots X_p]$ and $\mathbf{Y} = [Y_1 Y_2 \ldots Y_p]$ is defined as

$$\chi^2_{\mathbf{XY}} = \sum_{i=1}^{p} \left(\frac{X_i}{\overline{X}} - \frac{Y_i}{\overline{Y}} \right)^2 \tag{8.61}$$

where $\overline{X}$ and $\overline{Y}$ are the averages. If the average is subtracted from the original measurements and the relative counts are used in the expression of distance, we have

$$\chi^2_{\mathbf{XY}} = \sum_{i=1}^{p} \left(\frac{X_i - \overline{X}}{\overline{X}} - \frac{Y_i - \overline{Y}}{\overline{Y}} \right)^2 \tag{8.62}$$

which can be simplified to yield the same distance as in (8.61).

Modulation metric

The χ^2–metric was found to have some fundamental problems (not discussed here) when applied to electron microscopic data. To circumvent the problems, the "modulation" metric was introduced. In this the measurements are divided by their standard deviations, which for vectors $\mathbf{X} = [X_1 X_2 \ldots X_p]$ and $\mathbf{Y} = [Y_1\ Y_2 \ldots\ Y_p]$, are given by

$$\sigma_{\mathbf{X}} = \sqrt{\frac{1}{p} \sum_{i=1}^{p} X_i^2}$$

and

$$\sigma_{\mathbf{Y}} = \sqrt{\frac{1}{p} \sum_{i=1}^{p} Y_i^2}$$

Hence, with modulation metric, the squared distance between **X** and **Y** is given by

$$d^2_{m\mathbf{XY}} = \sum_{i=1}^{p} \left(\frac{X_i}{\sigma_{\mathbf{X}}} - \frac{Y_i}{\sigma_{\mathbf{Y}}} \right)^2 \tag{8.63}$$

8.4.3 Data compression

The basic concept of PCA has been introduced in Chapter 6. In Example 6.5.1 (Fig. 6.5), we have seen that by capturing 97.4% of the total variance of the data, the first principal component reduces the dimension of the problem from two to just one. Further, it has also been explained how the basic methodology can be extended to higher dimensions.

Looking from another angle, PCA aims to optimize the orthogonal coordinate system of the hyperspace to best fit the shape of the data cloud, the first principal axis corresponding to the greatest elongation. Importantly, the reduction in dimensionality of the space leads to a compression of the dataset, with insignificant loss of information. Typically, 50–200 orthogonal directions are found to be sufficient to describe the most significant variations in the dataset of 200,000 images, each image comprising 400 × 400 pixels. In the compressed data, each image is no longer described by 160,000 density values, but by 50–200 coordinates corresponding to the principal directions of variations.

For PCA, we need a matrix formulation of the cryo-EM dataset. Let the entire dataset be represented by a matrix $\mathbf{X} \in \mathbb{R}^{n \times p}$. The matrix contains n (=200,000) rows, one for each measured image, while each row contains p (=160,000) pixel densities of that image. Evidently, the dataset can be seen as a data cloud of n points in a p-dimensional image-space or, alternatively, that of p points in an n-dimensional pixel-space. In other words, two hyperspaces are associated with the matrix **X**: (1) the p-dimensional image-space in which there are n points (each representing an image) and (2) the n-dimensional pixel-space with a set of p points (each representing a pixel-vector).

The matrix **X** can be considered consisting of p column vectors:

$$\mathbf{X} = [\mathbf{x}_1 \ldots \mathbf{x}_j \ldots \mathbf{x}_p]$$

and, as in (6.46) of Chapter 6, a linear combination of the columns of **X** can be written as

$$\mathbf{X}\mathbf{u} = \sum_{j=1}^{p} u_j \mathbf{x}_j \tag{8.64}$$

where

$$\mathbf{u} = \begin{bmatrix} u_1 \\ \cdot \\ \cdot \\ \cdot \\ u_p \end{bmatrix}$$

is a vector of constants. In PCA the objective is to find a linear combination with maximum variance. The variance of the linear combination (8.64) is

$$\mathrm{Var}(\mathbf{Xu}) = \mathbf{u}^T \mathrm{Cov}(\mathbf{X})\,\mathbf{u} = \mathbf{u}^T \mathbf{S}_p\,\mathbf{u} \tag{8.65}$$

Hence, the problem reduces to obtaining a p-dimensional vector $\mathbf{u}$ by maximizing the quadratic form

$$f(\mathbf{u}) = \mathbf{u}^T \mathbf{S}_p\,\mathbf{u} \tag{8.66}$$

subject to the normalization condition

$$\mathbf{u}^T\mathbf{u} = 1 \tag{8.67}$$

Lagrangian optimization of (8.66) gives the eigenvector–eigenvalue equation

$$\mathbf{S}_p\,\mathbf{u} = \lambda\,\mathbf{u} \tag{8.68}$$

where λ is the eigenvalue describing the variance and $\mathbf{u}$ is the variance maximizing eigenvector of the variance-covariance matrix

$$\mathbf{S}_p = \mathbf{X}^T\mathbf{X} \tag{8.69}$$

which is a $p \times p$ symmetric matrix with exactly p real eigenvalues, λi ($i = 1,\ldots, p$). The corresponding eigenvectors form an orthonormal set of vectors, that is,

$$\begin{aligned} \mathbf{u}_i^T\mathbf{u}_j &= 1 \quad \text{if } i = j \\ &= 0 \quad \text{if } i \neq j \end{aligned}$$

If the eigenvalues are arranged in a decreasing order, that is,

$$\lambda_1 > \lambda_2 > > \lambda_p$$

then the linear combination defined by the eigenvector $\mathbf{u}_1$ will have the highest variance since

$$\mathrm{Var}(\mathbf{Xu}) = \mathbf{u}^T \mathbf{S}_p\,\mathbf{u} = \lambda\,\mathbf{u}^T\mathbf{u} = \lambda \tag{8.70}$$

Hence, the linear combination defined by the highest eigenvector bears the greatest and that corresponding to the lowest eigenvector the least information about the distribution of the data.

It can be recollected that the aim of our PCA has also been to optimize (rotate) the orthogonal coordinate system such that the rotated coordinate system best fits the largest elongation of the data cloud. For that, a rotation matrix $\mathbf{U}$ can be constructed with the orthonormal eigenvectors obtained from (8.68) as its columns. $\mathbf{U}$ can have a maximum of p columns spanning the p-dimensional space. However, depending on the proportion of the total variance that needs to be retained for subsequent analysis, the number of columns of $\mathbf{U}$ is restricted to $q \leq p$ (in the descending order of the eigenvalues). To obtain the coordinate of the images in the rotated system, the image vectors stored as rows in the data matrix

$\mathbf{X} \in \mathbb{R}^{n \times p}$ are to be projected onto the unit vectors stored as columns of $\mathbf{U} \in \mathbb{R}^{p \times q}$

$$\mathbf{C}_{\text{image}} = \mathbf{X}\mathbf{U} \tag{8.71}$$

Clearly, each row of $\mathbf{C}_{\text{image}} \in \mathbb{R}^{n \times q}$ represents a full input image (***eigen-image***), but in the rotated coordinated system.

Similar optimization can be carried out in the pixel-space. The pixel-vectors are represented by the rows of the matrix $\mathbf{X}^T \in \mathbb{R}^{p \times n}$, and the variance-covariance matrix is an $n \times n$ matrix

$$\mathbf{S}_n = \mathbf{X}\mathbf{X}^T \tag{8.72}$$

The eigenvalues of $\mathbf{X}^T\ \mathbf{X}$ and $\mathbf{X}\ \mathbf{X}^T$ are the same; hence the total variance described in the image-space and the pixel-space is the same. However, the corresponding set of orthonormal eigenvectors is different. If $r \leq p < n$ orthonormal eigenvectors of $\mathbf{S}_n$ (in the descending order of the corresponding eigenvalues) are collected as columns in a matrix $\mathbf{V} \in \mathbb{R}^{n \times r}$, the coordinates in the transformed (rotated) pixel-space is given by

$$\mathbf{C}_{\text{pixel}} = \mathbf{X}^T\,\mathbf{V} \tag{8.73}$$

Each row of $\mathbf{C}_{\text{pixel}} \in \mathbb{R}^{p \times r}$ represents a pixel-vector in the rotated coordinate system.

8.5 Clustering

The eigenvector–eigenvalue-based data compression must be followed by classification in the abridged eigenvector space. The classification enables selection of images that are then averaged to improve the signal-to-noise ratio. The principal components of the data cloud are used to classify the images into groups by a clustering procedure. The clustering procedures are based on the distance between the images, or their similarity.

8.5.1 Hierarchical clustering

The essential aim of hierarchical clustering is to minimize intraclass variance while maximizing interclass variance calculated between cluster centers. Choice of the starting point separates hierarchical clustering into two different approaches—agglomerative and divisive. In the agglomerative (also called ascendant) approach, each point in the data cloud is initially considered a group or class. This is followed by merging of the closest points into a stipulated number of clusters. The divisive approach initially puts all the datapoints in one cluster, and then separates them into smaller groups based on their dissimilarity.

Agglomerative hierarchical clustering is a "bottom-up" strategy. In this method, each point forms its own cluster and iteratively merges smaller clusters into larger clusters. The iteration continues until all the points are in a single cluster or specific termination conditions are satisfied. The merger is based on some similarity measure and two clusters are merged per iteration. With n datapoints the agglomerative method requires at most n iterations. The ultimate single cluster becomes the hierarchy's root.

Divisive hierarchical clustering adopts a "top-down" approach. Initially, all datapoints are placed in one cluster, which is the root of the hierarchy. Subsequently, the root cluster is divided into smaller subslusters by iterations. The process continues until the lowermost clusters are stable, with each cluster containing either a single point or multiple points sufficiently close to each other.

Ward's method sets a criterion for hierarchical clustering in relation to the distance between two clusters A and B

$$\begin{aligned}\Delta(A,B) &= \sum_{i\in A\cup B}\left\|\mathbf{r}_i-\mathbf{c}_{A\cup B}\right\|^2-\sum_{i\in A}\left\|\mathbf{r}_i-\mathbf{c}_A\right\|^2-\sum_{i\in B}\left\|\mathbf{r}_i-\mathbf{c}_B\right\|^2\\ &= \frac{n_A n_B}{n_A+n_B}\left\|\mathbf{c}_A-\mathbf{c}_B\right\|^2\end{aligned} \tag{8.74}$$

where $\mathbf{r}$ denotes the point in question, and $\mathbf{c}$ denotes the center of a cluster with n points in it. In the clustering process the sum of the squares starts at zero, since every point is in its own cluster. $\Delta(A,B)$ grows as the clusters merge. The method keeps the growth to a minimum.

8.5.2 *K*-means

The most fundamental variety of cluster analysis, which organizes a set of objects into multiple exclusive groups or clusters, is partitioning. Centroid (mean of a group of points)-based K-means clustering is the most commonly used partitioning method. It is based on the notion of existence of a central vector, which may not necessarily be a member of the dataset. Nevertheless, it is this notional central vector around which the subdata can be grouped.

The basic algorithm begins by randomly choosing K initial centroids, where K is a user-specified parameter. The optimization process then assigns each point to the closest centroid, so that each collection of points assigned to a centroid forms a cluster. Based on each assignment, the centroid of every cluster is updated. The iterative assignment and updating continues until no point changes the clusters or, in other words, the centroids remain the same. Thus, if there are N images (I_1, I_2, ..., I_N) where each image is a p-dimensional vector, K-means clustering separates the N images into K subsets S_K, where $K << N$ and $I_j \in \{S_1, S_2, \ldots, S_N\}$.

Hence, we can quantitatively formulate the K-means optimization process as follows. Let us consider a dataset ($\mathbf{x}_1$, $\mathbf{x}_2$, ..., $\mathbf{x}_N$) consisting of N observations of a random p-dimensional variable $\mathbf{x}$. The objective is to partition the dataset into K clusters. Initial centroids chosen are represented by p-dimensional vectors $\boldsymbol{\mu}_k$ ($k = 1, 2, \ldots, K$) so that $\boldsymbol{\mu}_k$ is a prototype associated with the kth cluster.

Evidently, μ_ks represent the center of the clusters. Let us now introduce a binary indicator variable $r_{nj} = \{0. 1\}$ so that if $\mathbf{x}_n$ is assigned to cluster k, then

$$\begin{aligned} r_{nj} &= 1, \ \text{for } j = k \\ &= 0, \ \text{for } j \neq k \end{aligned} \tag{8.75}$$

and we can define a "distortion measure"

$$J = \sum_{n=1}^{N} \sum_{k=1}^{K} r_{nk} \left\| \mathbf{x}_n - \mu_k \right\|^2 \tag{8.76}$$

(8.76) represents the sum of squares of the distances of the datapoints from the kth centroid. The iterative two-phase optimization, assignment and updating, can now proceed. In the first phase, optimization is carried out for each n separately to obtain a value of k that satisfies the following:

$$r_{nk} = \begin{cases} 1 & \text{if } k = \arg\min_j \left\| \mathbf{x}_n - \mu_j \right\|^2 \\ 0 & \text{otherwise} \end{cases} \tag{8.77}$$

So, the nth datapoint is assigned to the closest cluster center (centroid). In the second phase, optimization is carried out for μ_k, that is by setting the derivative of J with respect to μ_k to zero:

$$2 \sum_{n=1}^{N} r_{nk} \left(\mathbf{x}_n - \mu_k \right) = 0$$

to obtain

$$\mu_k = \frac{\sum_n r_{nk} \, \mathbf{x}_n}{\sum_n r_{nk}} \tag{8.78}$$

Clearly, the denominator in (8.78) is equal to the number of points assigned to cluster k and, hence, μ_k is the (new) mean of all the datapoints assigned to this cluster—the kth centroid is thus updated. The iterative datapoint assignment and centroid recalculation continue up to the step that produces no changes in the clusters.

8.6 Maximum likelihood

Statistical image processing in cryo-EM seeks to optimize a "likelihood function," that is, given the observed data, the aim is to find the model that has the highest probability of being the correct one. The maximum likelihood (ML) estimation is essentially the maximization of a (conditional) probability function. Denoting the observed dataset by $\mathbf{X}$ and the parameter set of the target model by Θ, the task is to maximize the probability $P(\Theta | \mathbf{X})$ related to the correctness of the model.

Direct computation of $P(\Theta | \mathbf{X})$ has been found to be difficult. It is relatively easier to compute the probability of observing $\mathbf{X}$, given Θ. Hence, the likelihood function is defined as

$$\mathcal{L}(\Theta) = P(\mathbf{X}|\Theta) \tag{8.79}$$

Clearly, $\mathcal{L}$ is a function of the model Θ, and not the data $\mathbf{X}$.

From Bayes' theorem (Theorem 6.1.2), we have

$$P(\Theta|\mathbf{X}) = P(\mathbf{X}|\Theta)\frac{P(\Theta)}{P(\mathbf{X})} \tag{8.80}$$

$P(\mathbf{X})$ does not depend on Θ and, for all practical purpose, can be treated as a constant. $P(\Theta)$ is based on prior information in the absence of any data—it is called the ***prior probability***. The product $P(\mathbf{X} | \Theta)\ P(\Theta)$ is called the ***posterior probability***. If we consider that all possible models are equally likely, then $P(\Theta)$ is also a constant. In such case, maximization of $P(\Theta | \mathbf{X})$ is equivalent of maximization of the likelihood $\mathcal{L}(\Theta)$. In ML the goal is to find a set of parameters that maximizes $\mathcal{L}(\Theta)$. Hence, the maximum likelihood estimator (MLE) is

$$\Theta^{\text{MLE}} = \arg\max_{\Theta} \mathcal{L}(\Theta) \tag{8.81}$$

Now, if our dataset contains N-independent measurements $\mathbf{X} = (\mathbf{X}_1, \mathbf{X}_2, \ldots, \mathbf{X}_N)$, the likelihood can be expressed as a product of the probability density functions

$$\mathcal{L}(\Theta) = \prod_{i=1}^{N} P(\mathbf{X}_i|\Theta) \tag{8.82}$$

For computational convenience, (8.82) can be written as a sum of likelihood logarithm

$$L(\Theta) = \ln(\mathcal{L}(\Theta)) = \sum_{i=1}^{N} \ln P(\mathbf{X}_i|\Theta) \tag{8.83}$$

The difficulty in computation of $L(\Theta)$ depends on the form of $P(\mathbf{X} | \Theta)$. In a relatively simple problem, a direct analytic calculation is possible; in other cases, more elaborate algorithms are needed.

By now it is clear that the images obtained by cryo-EM contain convoluted structural information of the 3D object as projected on a 2D plane. Hence, once the image dataset is classified into homogeneous subsets, it is possible to recreate a model of the object. There are several methods for the reconstruction of a 3D object from its 2D projections. Methods such as "back-projection" perform the reconstruction in real space. In contrast, there are computational methods in which the reconstruction is carried out in Fourier space. The Fourier approach in single-particle cryo-EM is somewhat similar to that in X-ray crystallography—the data are collected in Fourier space so that the inverse Fourier transform gives a 3D map of the object in real space. The mathematics of 3D reconstruction is not discussed here. Interested readers may refer to (Orlova and Saibil, 2011) for an elaborate account.

Exercise 8

1. The wave function for the ground state of the hydrogen atom is given by

$$\psi(r) = \frac{1}{\sqrt{\pi a_0^3}} \, e^{-r/a_0}$$

   Calculate the electron probability density $P(r)$ for the ground state of the hydrogen atom. Where does the maximum value of $P(r)$ occur?
2. Show if the following operations are linear or nonlinear.
   (a) differentiation (first and second order), (b) integration, (c) extraction of a square root, and (d) inversion (replacement of x by $-x$).
3. Are the following functions, eigenfunctions of (1) the momentum operator, and (2) the kinetic energy operator? If found to be an eigenfunction, give the corresponding eigenvalue.
   (a) $a\,e^{bx}$, (b) $a\,e^{bx^2}$, (c) $a + b\,x$, (d) $\cos x$
4. Based on the mathematical expression for the scattered wave and the concept of Fourier transform, explain how is the scattering amplitude dependent on the atomic composition of the sample.
5. The observed image in an electron microscope is the convolution of the wave scattered by the object with the transfer function of the objective lens—explain.
6. Explain quantitatively the basic principles of multivariate statistical eigenvector–eigenvalue data compression as applied to single-particle electron microscopic image analysis.
7. Consider the data points given in the following table.

	X_1	X_2
A	2	4
B	8	2
C	9	3
D	1	5

Choose two initial centroids AB and CD, and using K-means algorithm partition the data points into two optimized clusters, $K = 2$.

Reference

Orlova, E.V., Saibil, H.R., 2011. Structural analysis of macromolecular assemblies by electron microscopy. Chem. Rev. 111, 7710–7748.

Further reading

Abdi, H., Williams, L.J., 2010. Principal component analysis. WIREs: Wiley Interdisciplinary Reviews 2, 433–459.

Atkins, P., Friedman, R., 2011. Molecular Quantum Mechanics. Oxford University Press.

Fitzpatrick, R., 2015. Quantum Mechanics. World Scientific Publishing.

Griffiths, D.J., 2013. Introduction to Electrodynamics, 4th edition Pearson.

Tipler, P.A., Mosca, 2008. Physics for Scientists and Engineers, 6th edition W. H. Freeman & Company.

van Heel, M., Portugal, R.V., Schatz, M., 2016. Multivariate statistical analysis of large datasets: single particle electron microscopy. Open J. Stat. 6, 701–739.

White, H.E., et al., 2017. Structural study of heterogeneous biological samples by cryoelectron microscopy and image processing. BioMed Res. Int. Available from: https://doi.org/10.1155/2017/1032432.

CHAPTER

Biomolecular structure and dynamics

9

Needless to reiterate, biomolecular structures play a paramount role in molecular biology. In the previous two chapters, the mathematical basis for two most powerful and widely used tools utilized in biomolecular structure determination has been discussed. However, not only the "average" or "equilibrium" structure of a protein molecule but also the dynamics around this structure is important for its function. This chapter highlights the essential mathematics concerning some selected and widely used computational techniques to investigate biomolecular structure and dynamics.

9.1 Comparison of biomolecular structures

In molecular structural biology, protein structure alignment is a crucial strategy. This is based mostly on the comparison of protein structures determined experimentally and/or predicted computationally. One of the most useful features for protein structure comparison is the ***root mean square deviation*** (RMSD) which is a metric that indicates the similarity between molecular conformations or, in general, between two ordered set of points.

Alignment of two conformations seeks to identify a correspondence between equivalent atoms constituting them. Understandably, this involves the translation and rotation of one of the two structures so that weighted sum of the squares of the distances between equivalent atoms is minimized. Considering the atoms as point masses, the optimal translation superimposes the centroids or barycenters of the two sets of points. This step is succeeded by the optimal rotation about the common barycenter.

9.1.1 Definition of the problem

For the purpose of computing the RMSD, a molecule containing N atoms is defined by a set of points (vectors) $A = \{\mathbf{a}_i\}_N$ and associated weights $w = \{w_i\}_N$, $1 \le i \le N$. The weight factor w_i emphasizes the importance of different parts of the protein structure, such as the backbone or side chains. Accordingly, let us consider the two molecules under comparison as represented by two ordered sets of vectors $X = \{\mathbf{x}_i\}_N$ (model) and $Y = \{\mathbf{y}_i\}_N$ (target), $1 \le i \le N$, both having

Mathematical Approaches to Molecular Structural Biology. DOI: https://doi.org/10.1016/B978-0-323-90397-4.00001-9
© 2023 Elsevier Inc. All rights reserved.

barycenters at the origin. The residual, which is the square of the RMSD, between the two point-sets is then defined as

$$E = \frac{1}{N} \sum_{i=1}^{N} w_i \left| \mathcal{R}\, \mathbf{x}_i - \mathbf{y}_i \right|^2 \tag{9.1}$$

where $\mathcal{R}$ is an orthogonal transformation. For the sake of simplicity, the weight factor w_i is set to 1, so that (9.1) becomes

$$E = \frac{1}{N} \sum_{i=1}^{N} \left| \mathcal{R}\, \mathbf{x}_i - \mathbf{y}_i \right|^2 \tag{9.2}$$

Hence, the problem is to find an orthogonal transformation $\mathcal{R}$ such that the residual E is minimized.

There are two approaches to address the problem – one is based on singular value decomposition (SVD), while the second, based on what are known as quaternions. SVD has been introduced in Chapter 3. Here, a basic introduction to quaternions and their use in three-dimensional rotation applications is given.

9.1.2 Quaternions

In Chapter 2, we have seen that rotation in $\mathbb{R}^3$ about an axis through the origin can be represented by a 3×3 orthogonal matrix. Interestingly, as we shall see, rotation can be more concisely represented by a ***quaternion***. It is a 4-tuple which can be considered as a point in four-dimensional Euclidean space

$$q = (q_0, \mathbf{q}) = (q_0, q_1, q_2, q_3) \tag{9.3}$$

In (9.3), the first term, q_0, plays the role of a real number, while the last three terms together constitute a generalized imaginary number. A quaternion can also be written as

$$q = q_0 + \mathbf{i}\, q_1 + \mathbf{j}\, q_2 + \mathbf{k}\, q_3 \tag{9.4}$$

where q_0, q_1, q_2, and q_3 are real numbers, and **i**, **j**, and **k** are mutually orthogonal imaginary unit vectors

$$\mathbf{i} = (1, 0, 0),\ \mathbf{j} = (0, 1, 0),\ \mathbf{k} = (0, 0, 1)$$

with the property

$$\mathbf{i}^2 = \mathbf{j}^2 = \mathbf{k}^2 = \mathbf{i}\mathbf{j}\mathbf{k} = -1$$

Here are some algebraic properties of quaternions:

1. Addition of two quaternions p and q componentwise:

$$p + q = (p_0 + q_0) + (p_1 + q_1)\mathbf{i} + (p_2 + q_2)\mathbf{j} + (p_3 + q_3)\mathbf{k} \tag{9.5}$$

2. The componentwise breakup of the product of p and q is too long; it can be written in a concise form as

$$pq = p_0 q_0 - \mathbf{p} \cdot \mathbf{q} + p_0 \mathbf{q} + q_0 \mathbf{p} + \mathbf{p} \times \mathbf{q} \tag{9.6}$$

(9.5) and (9.6) show that the set of quaternions is closed under both addition and multiplication. Furthermore, multiplication can be seen to be associative, $p\,(qr) = (pq)\,r$, though not commutative, $pq \neq qp$.

3. The complex conjugate of a quaternion q, denoted by q^*, is defined as

$$q^* = q_0 - \mathbf{q} = q_0 - q_1\,\mathbf{i} - q_2\,\mathbf{j} - q_3\,\mathbf{k} \tag{9.7}$$

From the definition it can be shown that

$$\begin{aligned} (q^*)^* &= q \\ q + q^* &= 2\,q_0 \\ q^*\,q &= q\,q^* \end{aligned} \tag{9.8}$$

It can also be shown that for two quaternions p and q

$$(p\,q)^* = q^*\,p^*$$

4. The norm of a quaternion q is given by the scalar

$$|q| = \sqrt{q^* q} \tag{9.9}$$

For a unit quaternion, the norm is 1. It can be shown that the norm of the product of two quaternions p and q is equal to the product of the individual norms, that is,

$$|p\,q| = |p|\,|q| \tag{9.10}$$

5. The inverse of a quaternion q is given by

$$q^{-1} = \frac{q^*}{|q|^2} \tag{9.11}$$

and it can be verified that $q^{-1}q = qq^{-1} = 1$.

9.1.3 Quaternion rotation operator

A quaternion $q \in \mathbb{R}^4$; in general, it should not be able to operate on a vector $\mathbf{v} \in \mathbb{R}^3$. However, a three-dimensional vector can be "promoted" to a "pure" quaternion. Pure quaternions constitute an important class of quaternions having the form $q = (0, \mathbf{q})$. That is, a vector $\mathbf{v} \in \mathbb{R}^3$ is a pure quaternion whose real part is zero.

Any rotation in three-dimension can be specified by two parameters – a unit 3-vector, say, $\hat{\mathbf{r}} = (\hat{\mathbf{x}}, \hat{\mathbf{y}}, \hat{\mathbf{z}})$, that defines an axis of rotation, and an angle θ which describes the magnitude of the rotation about $\hat{\mathbf{r}}$. Hence, an axis-angle rotation can be represented by $(\theta, \hat{\mathbf{x}}, \hat{\mathbf{y}}, \hat{\mathbf{z}})$.

To find a rotation quaternion similar to the axis-angle representation, let us consider a unit quaternion $q = q_0 + \mathbf{q}$ so that

$$q_0^2 + \|\mathbf{q}\|^2 = 1 \tag{9.12}$$

(9.12) implies that there must be an angle θ such that

$$\cos^2\theta = q_0^2$$
$$\sin^2\theta = \|\mathbf{q}\|^2$$

The unit quaternion can then be written as

$$q = \cos\theta + \hat{\mathbf{r}} \sin\theta \tag{9.13}$$

where the unit vector $\hat{\mathbf{r}} = \mathbf{q}/\ \|\mathbf{q}\|$.

Now, using the unit quaternion q, (9.13), an operator on a vector $\mathbf{v} \in \mathbb{R}^3$ can be defined as

$$\begin{aligned} \mathcal{R}_q(\mathbf{v}) &= q\,\mathbf{v}\,q^* \\ &= \left(q_0^2 - \|\mathbf{q}\|^2\right)\mathbf{v} + 2(\mathbf{q}\cdot\mathbf{v})\,\mathbf{q} + 2\,q_0\,(\mathbf{q}\times\mathbf{v}) \end{aligned} \tag{9.14}$$

From (9.14) and the basic algebraic properties of a quaternion, it can be shown that

$$\|\mathcal{R}_q(\mathbf{v})\| = \|q\,\mathbf{v}\,q^*\| = \|\mathbf{v}\| \tag{9.15}$$

Furthermore, if $\mathbf{v} = k\,\mathbf{q}$, it can be found that

$$q\,\mathbf{v}\,q^* = q(k\,\mathbf{q})\,q^* = k\,\mathbf{q} \tag{9.16}$$

(9.15) and (9.16) show that the quaternion operator $\mathcal{R}_q(\mathbf{v})$ neither changes the length of the vector $\mathbf{v}$, nor does it change the direction of $\mathbf{v}$, if along $\mathbf{q}$. Hence, $\mathcal{R}_q(\mathbf{v})$ acts like a rotation about $\mathbf{q}$. It can be shown that for any unit quaternion

$$q = q_0 + \mathbf{q} = \cos\left(\frac{\theta}{2}\right) + \hat{\mathbf{r}} \sin\left(\frac{\theta}{2}\right) \tag{9.17}$$

the action of the operator

$$\mathcal{R}_q(\mathbf{v}) = q\,\mathbf{v}\,q^*$$

on any vector $\mathbf{v} \in \mathbb{R}^3$ is equivalent to a rotation of the vector through an angle θ about $\hat{\mathbf{r}}$ as the axis of rotation. The axis-angle representation $(\theta, \hat{\mathbf{x}}, \hat{\mathbf{y}}, \hat{\mathbf{z}})$ is related to the quaternion $q = (q_0, q_1, q_2, q_3)$ as follows:

$$q_0 = \cos\left(\frac{\theta}{2}\right),\quad q_1 = \hat{\mathbf{x}} \sin\left(\frac{\theta}{2}\right),\quad q_2 = \hat{\mathbf{y}} \sin\left(\frac{\theta}{2}\right),\quad q_3 = \hat{\mathbf{z}} \sin\left(\frac{\theta}{2}\right) \tag{9.18}$$

Since the quaternion representation is less intuitive, its utility in preference to the axis-angle representation may be questioned. Both representations require trigonometric operations to be performed to rotate a set of points in the three-dimensional space. Performing them beforehand allows most quaternion operations to restrict themselves to addition, subtraction, multiplication, and division only, thus economizing computation time.

9.1.4 Minimization of residual

We now proceed to minimize E as expressed in (9.2). With this goal, let us promote $\mathbf{x}_i$ and $\mathbf{y}_i$ to pure quaternions so that

$$x_i : = (0, \mathbf{x}_i) \text{ and } y_i : = (0, \mathbf{y}_i)$$

The rotation $\mathcal{R}(q)$ on $\mathbf{x}_i$ is then written as

$$(0, \mathcal{R}(q)\mathbf{x}_i) = q x_i q^*$$

Then, (9.2) can be written in terms of quaternions as

$$E_q = \frac{1}{N}\sum_{i=1}^{N}(q x_i q^* - y_i)\ (q x_i q^* - y_i)^* \tag{9.19}$$

Expanding (9.19), we have

$$N E_q = \sum_{i=1}^{N}(x_i x_i^* + y_i y_i^* + (q x_i q^*) y_i + y_i (q x_i q^*)) \tag{9.20}$$

Applying the properties of pure quaternions, the last two terms of (9.20) can be combined as

$$(q x_i q^*) y_i + y_i (q x_i q^*) = 2\left([y_i (q x_i q^*)]_0, \mathbf{0}\right) \tag{9.21}$$

To rewrite the nonzero component in (9.21) as matrix products, we define the correlation matrix

$$\mathbf{C} = \mathbf{X}\mathbf{Y}^T = \sum_{i=1}^{N}\mathbf{x}_i \mathbf{y}_i^T \tag{9.22}$$

whose components are given by

$$C_{jk} = \sum_{i=1}^{N} x_{ji} y_{ki}, \quad k, j = 1, 2, 3 \tag{9.23}$$

The explicit form of the product matrix

$$\text{C} = \begin{bmatrix} C_{11} + C_{22} + C_{33} & C_{23} - C_{32} & C_{31} - C_{13} & C_{12} - C_{21} \\ C_{23} - C_{32} & C_{11} - C_{22} - C_{33} & C_{12} + C_{21} & C_{13} + C_{31} \\ C_{31} - C_{13} & C_{12} + C_{21} & -C_{11} + C_{22} - C_{33} & C_{23} + C_{32} \\ C_{12} - C_{21} & C_{13} + C_{31} & C_{23} + C_{32} & -C_{11} - C_{22} + C_{33} \end{bmatrix}$$

Hence, it can be shown that (9.2) reduces to

$$NE_q = \sum_{i=1}^{N} \left(|\mathbf{x}_i|^2 + |\mathbf{y}_i|^2 \right) - 2\mathbf{Q}^T\mathbf{C}\mathbf{Q} \tag{9.24}$$

where $\mathbf{Q}$ is the 4-vector corresponding to the quaternion q. Now, as in Section 6.5, the problem is to find the extreme of the quadratic form $\mathbf{Q}^T\mathbf{C}\,\mathbf{Q}$, that is to maximize

$$f(\mathbf{Q}) = \mathbf{Q}^T\mathbf{C}\mathbf{Q}$$

subject to the constraint, $\mathbf{Q}^T\,\mathbf{Q} = 1$. As before, Lagrangian optimization leads to the eigenproblem

$$\mathbf{C}\mathbf{Q} = \lambda\mathbf{Q} \tag{9.25}$$

where λ is the eighenvalue of the 4×4 symmetric matrix $\mathbf{C}$. Therefore the unit quaternion q that maximizes $\mathbf{Q}^T\mathbf{C}\,\mathbf{Q}$ is the eigenvector corresponding to the largest eigenvalue $\lambda_{\max}$ of the matrix $\mathbf{C}$. It describes the optimal rotation for (9.2) such that

$$E_{q_{min}} = \sqrt{\frac{\sum_{i=1}^{N} \left(|\mathbf{x}_i|^2 + |\mathbf{y}_i|^2 \right) - 2\lambda_{max}}{N}} \tag{9.26}$$

and

$$\mathbf{R}(q) = \begin{bmatrix} q_0^2 + q_1^2 - q_2^2 - q_3^2 & 2(q_1q_2 - q_0q_3) & 2(q_1q_3 + q_0q_2) \\ 2(q_1q_2 + q_0q_3) & q_0^2 - q_1^2 + q_2^2 - q_3^2 & 2(q_2q_3 - q_0q_1) \\ 2(q_1q_3 - q_0q_2) & 2(q_2q_3 + q_0q_1) & q_0^2 - q_1^2 - q_2^2 + q_3^2 \end{bmatrix} \tag{9.27}$$

It is to be noted that $\mathbf{R}(q)$ is essentially a two-to-one mapping from quaternion space $q \in \mathbb{R}^4$ to the three-dimensional rotation space ($\mathbb{R}^3$) since changing the sign of the quaternion preserves the rotation matrix, that is,

$$\mathbf{R}(q) = \mathbf{R}(-q)$$

9.2 Conformational optimization

The potential energy of a complex biomolecular system, such as a protein, is calculated as a function of its atomic coordinates. The multidimensional energy surface thus obtained forms the basis for simulation strategies of such systems. The energy surface can be explored to find the configuration for which the energy is a minimum. Many simulation strategies require prior minimization of energy and identification of the stable conformation.

Minimization is a specific case of a more general problem – ***optimization***. Mathematical optimization seeks to find the best available values of an objective function within a given domain and subject to given constraints. Simply, an optimization problem is to find the input to a given function that yields the maximum or minimum output from the function. Here we discuss the basic concepts underlying some optimization procedures.

9.2.1 Born–Oppenheimer approximation

In quantum mechanics, an atomic or a molecular system is described by a Hamiltonian (introduced in Section 8.1.3)

$$H(x,p) = T + V = \frac{p^2}{2m} + V(x) \tag{9.28}$$

and the electronic structure of the system is determined by solving the time-independent Schrödinger equation which, in the Hamiltonian form, is written as

$$\hat{H}\psi(x) = E\psi(x) \tag{9.29}$$

For a hydrogen atom, with the position of the nucleus (proton) fixed, the Schrödinger equation takes a relatively simple form and can easily be solved.

In contrast, even for the simplest molecule H_2^+, the Hamiltonian takes a more complex form

$$\hat{H} = T_e + T_N + V \tag{9.30}$$

where the three terms on the right are respectively the electron kinetic energy, the nuclear kinetic energy and the potential energy of the system. As a result, the Schrödinger equation cannot be solved analytically and approximations are required.

Born–Oppenheimer approximation is based on the great difference in masses of electrons and nuclei (a proton is ~2000 times heavier than an electron). So, as an approximation, the nuclei are considered to be nearly fixed with respect to electron motion. In such case, there is no need to solve the Schrödinger equation for all the particles simultaneously. Now, the solution (electronic wave function) depends "parametrically" on the positions of the nuclei. A set of solutions obtained with various arrangements of the nuclei can be used to construct the molecular potential energy curve for a diatomic molecule (Fig. 9.1A) and, in general, a potential energy surface (Fig. 9.1B) for a polyatomic molecule. The equilibrium conformation of the molecule corresponds to the lowest point on the curve (or surface).

The structure of a molecule can be specified by giving the locations (coordinates) of its constituent atoms. Evidently, the potential energy surface of the molecule is a function of its structure. The process of finding the coordinates that minimize the energy (corresponding to the lowest point on the potential energy surface) is known as ***geometry optimization***.

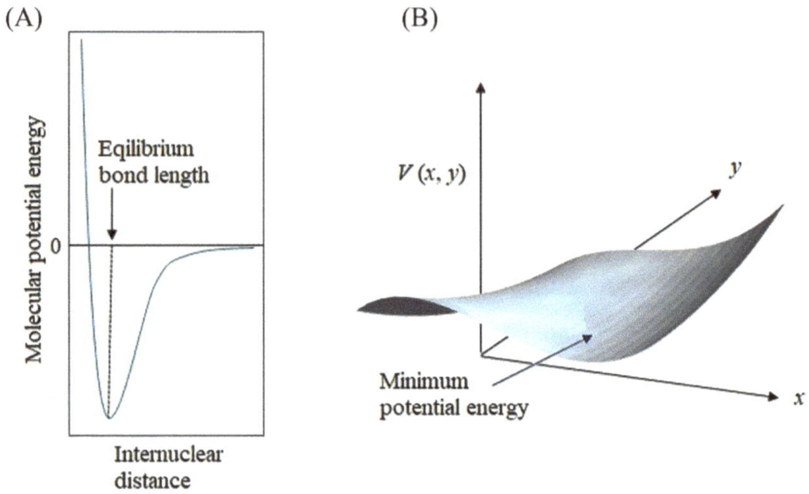

FIGURE 9.1

Univariate potential energy curve (A) and bivariate potential energy surface (B).

9.2.2 Biomolecular geometry optimization

Let us consider a biomolecular system consisting of N atoms located at positions $\mathbf{r} = (\mathbf{r}_1, \ldots, \mathbf{r}_j, \ldots, \mathbf{r}_N)$, where $\mathbf{r}_j = (x_1, y_1, z_1)$ are the Cartesian coordinates of the jth atom. The potential energy surface of the system is defined by the function $V(\mathbf{r})$. The maxima, minima, and saddles are the stationary or critical points on the potential energy surface. At all such points, the gradient of the potential energy function vanishes, that is,

$$\mathbf{g} = \nabla_{\mathbf{r}} V(\mathbf{r}) = 0 \tag{9.31}$$

Now, the potential energy surface can be approximated at a point $\mathbf{r}_0$, close to a critical point $\mathbf{r}$, by a Taylor expansion (4.35) about $\mathbf{r}_0$

$$V(\mathbf{r}) = V(\mathbf{r}_0) + \mathbf{g}_0^T \Delta\mathbf{r} + \frac{1}{2} \Delta\mathbf{r}^T \mathbf{H}_0 \Delta\mathbf{r} \tag{9.32}$$

where $\Delta\mathbf{r} = \mathbf{r} - \mathbf{r}_0$, g_0 is the gradient at $\mathbf{r}_0$ and

$$\mathbf{H}_0 = \frac{\partial^2 V}{\partial \mathbf{r}_i \, \partial \mathbf{r}_j}$$

is the Hessian at $\mathbf{r}_0$.

Among different coordinate systems, the Cartesian system has been the most widely used in geometry optimization. However, the Cartesian coordinates have their limitations. Internal coordinates, such as bond lengths and valence angles, provide a more useful alternative. However, the transformation from the internal coordinates to the Cartesians requires a time-consuming iterative process and a generalized inverse of the transformation matrix.

Newton–Raphson method

Optimization can be carried out based on a local quadratic approximation of the potential energy surface. Newton and quasi-Newton methods are some of the most efficient optimization procedures. Differentiating (9.32) with respect to the coordinates and applying condition (9.31) for the critical point, we have

$$\Delta \mathbf{r} = -\mathbf{H}_0^{-1}\mathbf{g}_0 \tag{9.33}$$

(9.33) defines what is known as the Newton–Raphson step, which takes us from the current point to the stationary point, provided $V(\mathbf{r})$ is truly quadratic over the region $[\mathbf{r}_o, \mathbf{r}]$. If not, an iterative procedure is required to be adopted. At each step, an approximate Hessian is calculated. For minimization, the Hessian matrix must be positive definite (with all positive eigenvalues).

Quasi-Newton methods start from an approximate Hessian $\mathbf{H}_{old}$ and updates the Hessian as

$$\mathbf{H}_{new} = \mathbf{H}_{old} + \Delta \mathbf{H} \tag{9.34}$$

where $\Delta\mathbf{H}$ is the difference between the calculated change in the gradient and the change predicted using $\mathbf{H}_{old}$. The updated Hessian must satisfy the Newton condition

$$\Delta \mathbf{g} = \mathbf{H}_{new}\,\Delta \mathbf{r} \tag{9.35}$$

where $\Delta \mathbf{g} = \mathbf{g}(\mathbf{r}_{new}) - \mathbf{g}(\mathbf{r}_{old})$ and $\Delta \mathbf{r} = \mathbf{r}_{new} - \mathbf{r}_{old}$. The new point is taken to be a stationary point if $|\mathbf{g}(\mathbf{r}_{new})| <$ a preset threshold ε (which is a very small number). If not, the procedure is repeated.

There are multiple ways to update the Hessian, that is, to calculate $\Delta\mathbf{H}$. However, each algorithm has its limitations. For example, the symmetric rank one (SR1) update, also called the Murtagh-Sargent update, uses the following scheme:

$$\Delta \mathbf{H}_{SR1} = \frac{(\Delta \mathbf{g} - \mathbf{H}_{old}\,\Delta \mathbf{r})\;(\Delta \mathbf{g} - \mathbf{H}_{old}\,\Delta \mathbf{r})^T}{((\Delta \mathbf{g} - \mathbf{H}_{old}\,\Delta \mathbf{r}))^T\,\Delta \mathbf{r}} \tag{9.36}$$

The SR1 algorithm confronts numerical problems when $|\Delta \mathbf{g} - \mathbf{H}_{old}\,\Delta \mathbf{r}|$ is small. Such problems are averted in the Broyden–Fletcher–Goldfarb–Shanno (BFGS) update

$$\Delta \mathbf{H}_{BFGS} = \frac{\Delta \mathbf{g}\,\Delta \mathbf{g}^T}{\Delta \mathbf{g}^T\,\Delta \mathbf{r}} - \frac{\mathbf{H}_{old}\,\Delta \mathbf{r}\,\Delta \mathbf{r}^T\,\mathbf{H}_{old}}{\Delta \mathbf{r}^T\,\mathbf{H}_{old}\,\Delta \mathbf{r}} \tag{9.37}$$

which ensures that the Hessian update is positive definite.

As mentioned earlier, Quasi-Newton methods start from an initial approximate Hessian. This can be obtained from the knowledge of the structure and connectivity of the molecule. In this regard, empirical estimates of stretching, bending and torsional force constants are found to be useful.

Conjugate gradient method

The conjugate gradient (CG) method was originally formulated to minimize a quadratic function

$$\min_{\mathbf{x}} q(\mathbf{x}) = \frac{1}{2}(\mathbf{x}, \mathbf{Qx}) - (\mathbf{b}, \mathbf{x}) \tag{9.38}$$

where $\mathbf{b}, \mathbf{x} \in \mathbb{R}^m$ ($\mathbf{b} \neq 0$), and $\mathbf{Q}$ is an $m \times m$ symmetric positive definite matrix. Two nonzero vectors $\mathbf{d}_i$ and $\mathbf{d}_j$ are said to be $\mathbf{Q}$-orthogonal, or conjugate with respect to $\mathbf{Q}$, if

$$(\mathbf{d}_{i,} \mathrm{Q}\,\mathbf{d}_j) = 0 \quad \text{for } i \neq j$$

CG is an iterative method – at each iteration it chooses a new search direction that lowers the energy while remaining at or very close to the minimum in the previous search direction. The search directions are conjugate so that

$$\Delta \mathbf{r}_{\text{new}} \mathbf{H} \Delta \mathbf{r}_{\text{old}} = 0$$

The number of iterations needed are equal to or less than the number of different eigenvalues of $\mathbf{H}$. Fletcher–Reeves and Polak–Ribiere are two most widely used CG methods. In Fletcher–Reeves method, the search direction in a particular step is decided by

$$\mathbf{d}_i = -\mathbf{g}_i + \frac{\mathbf{g}_i^T \mathbf{g}_i}{\mathbf{g}_{i-1}^T \mathbf{g}_{i-1}} \mathbf{d}_{i-1} \tag{9.39}$$

and in Polak–Ribiere by

$$\mathbf{d}_i = -\mathbf{g}_i + \frac{(\mathbf{g}_i - \mathbf{g}_{i-1})^T \mathbf{g}_i}{\mathbf{g}_{i-1}^T \mathbf{g}_{i-1}} \mathbf{d}_{i-1} \tag{9.40}$$

where $\mathbf{d}_i$ and $\mathbf{d}_{i-1}$ are respectively the current and preceding search directions, while $\mathbf{g}_i$ and $\mathbf{g}_{i-1}$ are respectively the current and previous gradients.

9.3 Molecular dynamics

It is universally acknowledged that for the functionality of a molecule, its structure and motion are both important. The motion of a protein molecule can be characterized at different levels – local, medium scale, large scale, and global. Localized motions, characteristic of atomic fluctuations, are responsible for ligand-docking flexibility of a protein. Medium-scale motions in the protein loops and terminal arms are often correlated with active-site conformational adaptation. Allosteric transitions of protein molecules are caused by large-scale motion of the domains while helix-coil transition and protein-folding are examples of global motion. However, the different types of motion are interdependent and coupled to each other – the atomic-scale movement being the most crucial.

In a biomolecule, the atoms are in perpetual motion. Molecular dynamics (MD) simulations predict the movement of each atom in a molecular system, such as a protein. Based on physical laws governing interatomic interaction, the technique has been widely used in physics, chemistry and biological sciences. It can be considered a "computational microscope" that facilitates detailed atomic-scale modeling. The technique can meticulously analyze both static and dynamic properties of a molecule and correlate them with its function.

9.3.1 Basic theory

Molecules are quantum mechanical entities; hence, the dynamic behavior of a molecule is best described by the quantum mechanical equation of motion, that is, the time-dependent Schrödinger Equation (8.6). However, this equation is extremely difficult to solve for a large molecule. For this reason, MD simulations of the atomic motions of a biomolecule (say, protein) are based on classical mechanics and utilizes Newton's equation of motion.

The state of a biomolecular system containing N atoms can be defined by $6N$ values – $3N$ coordinates consisting of position vectors $\mathbf{r}_i = (x_i, y_i, z_i)$ and $3N$ momenta $\mathbf{p}_i = (p_{xi}, p_{yi}, p_{zi})$, $i = 1, 2, \ldots, N$. Each combination of coordinates and momenta define a point in the $6N$-dimensional ***phase space*** $\boldsymbol{\Omega}$. In Chapter 8, the Hamiltonian of a particle moving in one-dimension has been defined (8.16). Here, we extend the definition to the system of N particles (atoms).

$$H(\mathbf{r}, \mathbf{p}) = T(\mathbf{p}) + V(\mathbf{r}) \tag{9.41}$$

where the kinetic energy

$$T(\mathbf{p}) = \sum_{i=1}^{N} \frac{\mathbf{p}_i^2}{2m_i} \tag{9.42}$$

and $V(\mathbf{r})$ is the potential energy function. From (9.41) and (9.42) we find that the motion of the system is governed by

$$\dot{\mathbf{r}}_i = \frac{\partial H}{\partial \mathbf{p}_i} = \frac{\mathbf{p}_i}{m_i} \tag{9.43}$$

$$\dot{\mathbf{p}}_i = \frac{\partial H}{\partial \mathbf{r}_i} = \mathbf{F}_i \tag{9.44}$$

where $\mathbf{F}_i$ is the force acting on atom i of mass m_i. (9.43) and (9.44) are the Hamiltonian equations of motion. (9.44) can also be written as

$$\mathbf{F}_i = m_i \mathbf{a}_i = m_i \frac{d^2 \mathbf{r}_i(t)}{dt^2} \tag{9.45}$$

which is the Newton's equation of motion. The force $\mathbf{F}_i$ is given by the negative gradient of the potential energy function $V(\mathbf{r_i})$ at position $\mathbf{r}_i$

$$\mathbf{F}_i = -\nabla V(\mathbf{r}_i) \tag{9.46}$$

Evidently, the solution of (9.46) requires information on $\mathbf{F}_i$ as well as the initial position, velocity and acceleration of the atom.

MD simulations require a potential energy function that describes the interactions of all the atoms in a molecular system. The potential energy is defined by what in computational chemistry is known as molecular mechanics force field (MM FF). Mathematically, a force field is an expression that describes the dependence of the energy of a system on the coordinates of its constituent particles. For a biomolecular system containing N atoms, it is an analytical form of the interatomic potential energy $V(\mathbf{r}_1, \mathbf{r}_2, \ldots, \mathbf{r}_N)$ along with a set of relevant parameters.

In MD simulations, an atom is considered as the basic unit of a model so that only interatomic interactions are required to be parameterized. The MD atom is a "hard ball" without directional properties and internal degrees of freedom; sometimes, it is considered to be a charged point. Depending on the conditions, an "atom" can be anything from a simple proton to an entire amino acid. Bonding interactions and bond angles are modeled as springs (harmonic oscillators).

The potential energy, which is a function of the three-dimensional structure of the system (let us denote it as R), consists of bonding (covalent) and nonbonding (noncovalent) components.

$$V(R) = \sum V_{\text{bonded}}(\mathbf{r}, \theta, \phi) + \sum V_{\text{nonbonded}}(\mathbf{r}) \tag{9.47}$$

The bonding component consists of terms which arise out of the geometry of a chain molecule as shown in Fig. 9.2. It involves two-, three-, and four-body interactions of covalently bonded atoms and can be broken up into three parts:

$$V_{\text{bonded}}(\mathbf{r}, \theta, \phi) = V_{\text{bonds}}(\mathbf{r}) + V_{\text{angle}}(\theta) + V_{\text{tors}}(\phi) \tag{9.48}$$

where the variables (oscillatory) $\mathbf{r}$ = bond length; θ = valence or "bend" angle; and ϕ = torsion or dihedral angle (Fig. 9.2).

The first term, a two-body spring bond potential, represents the motion of a simple (linear) harmonic oscillator formed by a (j, k) pair of covalently bonded atoms.

$$V_{\text{bonds}}(\mathbf{r}) = k_{r,jk}\left(r_{jk} - r_{o,jk}\right)^2 \tag{9.49}$$

where $r_{jk} = |\mathbf{r}_k - \mathbf{r}_j|$ is the distance between the atoms j and k (bond length), $r_{o,jk}$ is the equilibrium distance and $k_{r,jk}$ is the spring constant.

The angular vibrational motion occurring between a (j, k, l) triple of covalently bonded atoms is described by the three-body angular bond potential,

$$V_{\text{angle}}(\theta) = k_{\theta,jkl}\left(\theta_{jkl} - \theta_{o,jkl}\right)^2 \tag{9.50}$$

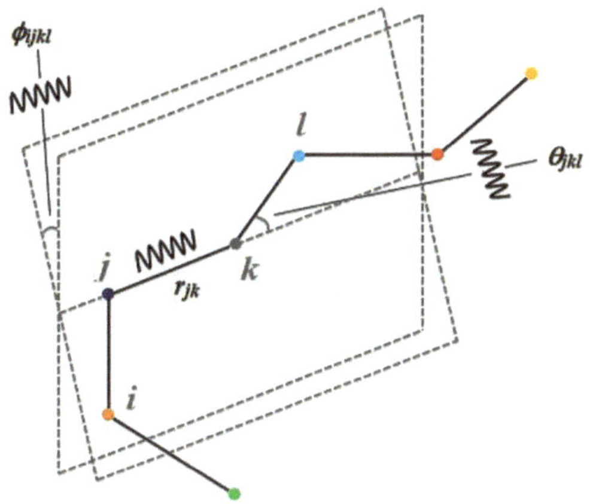

FIGURE 9.2

Geometric representation of a simple chain molecule. The oscillatory nature of **r**, θ, and ϕ has been indicated by springs.

where θ_{jkl} is the angle between successive bond vectors $\mathbf{r}_{jk} = \mathbf{r}_k - \mathbf{r}_j$ and $\mathbf{r}_{kl} = \mathbf{r}_l - \mathbf{r}_k$, $\theta_{o,jkl}$ is the equilibrium angle and $k_{\theta,jkl}$ is the spring constant.

The angle between the planes formed by the first three and the last three atoms of a consecutively bonded (i, j, k, l) quadruple of atoms is the torsion or dihedral angle. The four-body torsion angle potential can be expressed as

$$V_{\text{tors}}(\phi) = \sum_n k_{\phi,n,ijkl}\left[1 + \cos\left(n\phi_{ijkl} - \psi_n\right)\right] \tag{9.51}$$

where ϕ_{ijkl} is the angle between the (i, j, k) plane and the (j, k, l) plane, n is non-negative and indicator of periodicity, ψ is the phase angle and $k_{\phi n,ijkl}$ is a multiplicative constant. (9.48) can now be rewritten as

$$\begin{aligned} V_{\text{bonded}}(\mathbf{r}, \theta, \phi) = {} & k_{r,jk}\left(r_{jk} - r_{o,jk}\right)^2 \\ & + k_{\theta,jkl}\left(\theta_{jkl} - \theta_{o,jkl}\right)^2 \\ & + \sum_n k_{\phi,n,ijkl}\left[1 + \cos\left(n\phi_{ijkl} + \psi_n\right)\right] \end{aligned} \tag{9.52}$$

It is to be noted that (9.52) is somewhat oversimplification. Depending on the desired accuracy and sophistication of the simulation, cross-terms (such as the Urey-Bradley term used to describe a noncovalent spring between j and l atoms) are included in the MM FF.

The nonbonded or noncovalent component in (9.47) arises from interatomic ion–ion interaction (el) and Leonard–Jones (LJ) potential.

$$V_{\text{nonbonded}}(\mathbf{r}) = V_{\text{el}} + V_{\text{LJ}} \tag{9.53}$$

The ionic interaction is guided by Coulomb's law. For two ions with charges q_1 and q_2, the interaction can be expressed in terms of the potential energy

$$V_{\mathrm{el}} = \frac{q_1 q_2}{4\pi\varepsilon_0\varepsilon_r r} \tag{9.54}$$

where ε_r is the relative permittivity or dielectric constant of the medium.

The LJ potential arises from both attractive as well as repulsive forces. Three types of interaction, all electrostatic in nature, are collectively called van der Waals interactions – permanent dipole–dipole, permanent dipole-induced dipole, and induced dipole-induced dipole. These attractive forces are effective between two atoms when they are sufficiently close to each other. The interactions drive the atoms closer until they actually contact one another. Beyond this point, if the atoms try to move closer into each other and occupy the same space, a repulsive force comes into play between their electron clouds. This energy of repulsion, when added to van der Waals attractive interactional energy, gives the LJ potential:

$$V_{\mathrm{LJ}} = -\frac{A}{r^6} + \frac{B}{r^{12}} \tag{9.55}$$

which is the total interaction energy of two closely interacting atoms. Hence, (9.53) takes the form

$$V\text{nonbonded } (\mathbf{r}) = \left[\frac{q_1 q_2}{4\pi\varepsilon_0\varepsilon_r r} + \varepsilon_{ij}\left\{\left(\frac{r_{\min,ij}}{r_{ij}}\right)^{12} - 2\left(\frac{r_{\min,ij}}{r_{ij}}\right)^{6}\right\}\right] \tag{9.56}$$

where ε_{ij} is the minimum potential observed. The nonbonded terms represent interactions between all (i, j) pairs of atoms; the pairs already involved in a bonded term are usually excluded. Putting together all the components, (9.47) can now be written as

$$\begin{aligned} V(R) = {} & \sum_{\text{bonds}} k_{r,jk}\left(r_{jk} - r_{0,jk}\right)^2 \\ & + \sum_{\text{bond angles}} k_{\theta,jkl}\left(\theta_{jkl} - \theta_{0,jkl}\right)^2 \\ & + \sum_{\text{torsion angles}} \sum_{n} k_{\phi,n,ijkl}\left[1 + \cos\left(n\phi_{ijkl} - \psi_n\right)\right] \\ & + \sum_{\text{nonbonded atom pairs}} \left[\frac{q_1 q_2}{4\pi\varepsilon_0\varepsilon_r r} + \varepsilon_{ij}\left\{\left(\frac{r_{\min,ij}}{r_{ij}}\right)^{12} - 2\left(\frac{r_{\min,ij}}{r_{ij}}\right)^{6}\right\}\right] \end{aligned} \tag{9.57}$$

Evidently, the energy function contains terms associated with the three-dimensional structure of a molecule as well as certain specific parameters. The values of r_{ij}, θ_{jkl}, and ϕ_{ijkl} are all obtained from experimentally determined (by X-ray crystallography, NMR, etc.) or homology modeled structures. The parameters, such as the spring constants $k_{r,jk}$, $k_{\theta,jkl}$, and $k_{\phi,n,ijkl}$, are associated respectively with a pair, a triple and a quadruple of atoms. Besides, each atom is assigned a charge and LJ parameters. The precise formulation of the potential energy function, together with selection of suitable parameters, defines the force field.

The development of a force field is a nontrivial task. Experiments with small molecules (which are usually the building blocks of macromolecules of interest) and, most importantly, quantum mechanical calculations initially generate the parameters. The force field is then optimized with respect to the structural, dynamic and thermodynamic properties of previously characterized molecules. Currently, the most widely used force fields in MD are AMBER, CHARMM, and GROMOS.

After a molecular system has been parameterized, optimization of potential encrgy function is carried out to compute its stable configuration. Such optimization is necessary for reasons mentioned earlier. Often, a multidimensional potential energy surface is used to describe the energy of a molecule in terms of its atomic positions. The stable state of the molecule corresponds to the global and local minima on this surface. In a mathematically executed energy minimization/ optimization procedure, the atomic coordinates are varied to reach the lowest energy configuration (optimal structure) of the molecule.

9.3.2 Computation of molecular dynamics trajectory

Computation of the MD trajectory requires the solution of Newton's equation of motion (9.45) for all the atoms in the force field at each selected timestep. The force acing on atom i is given by (9.46). Essentially, $\mathbf{r}_i(t)$ is required to be determined. A standard procedure for solving ordinary differential equations, such as (9.45), is the finite-difference approach. Here, given the coordinates and velocities of the atoms at time t, those at a time $t + \Delta t$ can be obtained with a reasonable degree of accuracy. The equations are solved on a step-by-step basis using the Taylor expansion (Section 1.5.1) expressed as

$$\mathbf{r}\,(t+\Delta t) = \mathbf{r}\,(t) + \dot{\mathbf{r}}\,(t)\Delta t + \frac{1}{2}\ddot{\mathbf{r}}\,(t)\Delta t^2 + \cdots\cdots\cdots \tag{9.58}$$

Alternatively,

$$\mathbf{r}(t+\Delta t) = \mathbf{r}\,(t) + \mathbf{v}\,(t)\Delta t + \frac{1}{2}\mathbf{a}\,(t)\Delta t^2 + \cdots\cdots\cdots \tag{9.59}$$

where $\mathbf{v}(t)$ is the velocity and $\mathbf{a}\,(t)$ is the acceleration. The selection of time interval Δt depends on the properties of the molecular system and the nature of the question to be addressed by the simulation; in any case, it is significantly smaller than the simulation time scale.

(9.59) forms the basis of an integration algorithm. The initial positions $\{\mathbf{r}(0)\}$ are taken from the optimized structure of the molecule. The initial velocities $\{\mathbf{v}(0)\}$ are usually randomly assigned based on Maxwell–Boltzmann distribution of molecular speed at a temperature T (described in the box below). The initial acceleration can be calculated from the force field in accordance with

$$\{\mathbf{a}(0)\} = -\left(\frac{1}{m}\right)\nabla V\{\mathbf{r}(0)\} \tag{9.60}$$

Maxwell–Boltzmann distribution

In kinetic theory of gases, the constituent molecules of an ideal gas are considered as independent, randomly moving particles obeying Newton's laws of motion. The distribution of molecular speed v, in a volume containing N molecules at temperature T, is given by

$$N(v) = 4\pi N\left(\frac{m}{2\pi kT}\right)^{\frac{3}{2}} v^2 e^{-\frac{mv^2}{2kT}}$$

where m is the mass of each molecule and k is the Boltzmann constant. This is called the Maxwell-Boltzmann distribution of molecular speed.

The described algorithm is very simplistic and does not produce high quality results. Other algorithms with greater accuracy, such as the Verlet algorithm, have been developed. However, in most cases the same kind of reasoning is applied.

Let us consider rewriting (9.59) "backward in time," that is, $t - \Delta t$, as

$$\mathbf{r}(t-\Delta t) = \mathbf{r}(t) - \mathbf{v}(t)\,\Delta t + \frac{1}{2}\mathbf{a}(t)\,\Delta t^2 + \cdots\cdots\cdots\cdots \tag{9.61}$$

Adding (9.59) and (9.61) and assuming that the third- and higher-order terms are negligible, we have

$$\mathbf{r}\ (t+\Delta t) = 2\mathbf{r}(t) - \mathbf{r}\ (t-\Delta t) + \mathbf{a}(t)\,\Delta t^2 \tag{9.62}$$

(9.62) is the basic form of ***Verlet algorithm***. The velocity does not explicitly appear in the expression, but it can be calculated as

$$\mathbf{v}(t) = \frac{\mathbf{r}\ (t+\Delta t) - \mathbf{r}\ (t-\Delta t)}{2\Delta t} \tag{9.63}$$

A variant to (9.62) is the velocity-Verlet algorithm given by

$$\mathbf{r}(t+\Delta t) = \mathbf{r}(t) + \mathbf{v}(t)\Delta t + \frac{1}{2}\mathbf{a}(t)\,\Delta t^2 \tag{9.64}$$

and

$$\mathbf{v}(t+\Delta t) = \mathbf{v}(t) + \frac{1}{2}[\mathbf{a}(t) + \mathbf{a}(t+\Delta t)]\,\Delta t \tag{9.65}$$

The Verlet algorithm or its velocity-variant is an efficient integrator for MD simulation.

9.4 Normal mode analysis

It is beyond any doubt that MD has been the most commonly used computational technique to study the dynamical properties of biomolecules. Nonetheless, as a consequence of the computational complexity of all-atom MD simulations, its application is restricted to relatively short time scales. In many circumstances, a more feasible alternative is normal mode analysis (NMA), albeit with an acceptable compromise. MD uses a realistic force field to solve the equations of motion approximately. In contrast, NMA finds the exact solutions but uses a simplified force field.

NMA is a harmonic analysis – it provides information on the equilibrium modes available to a system, based on the assumption that the system is stabilized by harmonic potentials. Furthermore, it assumes that the potential energy of the system varies about a single well-defined energy minimum. The solutions resulting from the analysis are a set of orthogonal displacement vectors which are called the ***normal modes***.

As it appears, the complex dynamical behavior of a biomolecule can be approximated by considering its constituent atoms as a set of simple harmonic oscillators vibrating around a given equilibrium conformation. Hence, it will be useful here to review some underlying mathematics of such oscillation.

9.4.1 Oscillatory systems

The motion of a simple oscillatory system, that is, a particle subject to a "restoring" force, can be represented by a linear differential equation

$$\ddot{x} + \omega^2 x = 0 \tag{9.66}$$

where ω is the angular frequency. The solution to (9.66) is given as

$$x(t) = A\cos(\omega t + \phi) \tag{9.67}$$

where A and ϕ are constants determined from initial conditions. For the ease of calculation, the solution to (9.66) is sometimes tried in the form

$$x = A\,\mathrm{e}^{\mathrm{i}\alpha t}$$

Then, (9.66) is satisfied when $\alpha = \pm\omega$, so that the general solution can be written as

$$x = A_1\,\mathrm{e}^{\mathrm{i}\omega t} + A_2\,\mathrm{e}^{-\mathrm{i}\omega t} \tag{9.68}$$

If x describes the position of the particle, it must be real. Hence, A_1 and A_2 must be complex conjugates. Defining

$$A_1 = A_2^* = \left(\frac{A}{2}\right) e^{i\phi}$$

where A and ϕ are real, we have (9.67).

Let us now consider a case where we have two "coupled" differential equations;

$$2\ddot{x}_1 + \omega^2 (5x_1 - 3x_2) = 0$$

$$2\ddot{x}_2 + \omega^2 (5x_2 - 3x_1) = 0 \tag{9.69}$$

(9.69) can be written in a matrix form as

$$\ddot{\mathbf{x}} = -\omega^2 \mathbf{A}\,\mathbf{x} \tag{9.70}$$

where vector $\mathbf{x} = \begin{bmatrix} x_1 \\ x_2 \end{bmatrix}$ and the coefficient matrix $\mathbf{A} = \begin{bmatrix} \frac{5}{2} & -\frac{3}{2} \\ -\frac{3}{2} & \frac{5}{2} \end{bmatrix}$

To solve (9.70), let us consider a vector $\mathbf{q}$ linearly related to $\mathbf{x}$ by

$$\mathbf{x} = \begin{bmatrix} x_1 \\ x_2 \end{bmatrix} = \mathbf{P}\,\mathbf{q} = \mathbf{P}\begin{bmatrix} q_1 \\ q_2 \end{bmatrix} \tag{9.71}$$

where $\mathbf{P}$ is an invertible orthonormal transformation matrix that diagonalizes $\mathbf{A}$. Hence, from (9.70) we have

$$\mathbf{P}\ddot{\mathbf{q}} = -\omega^2 \mathbf{A}\,(\mathbf{P}\mathbf{q})$$

or

$$\ddot{\mathbf{q}} = -\omega^2 (\mathbf{P}^{-1}\mathbf{A}\,\mathbf{P})\mathbf{q} \tag{9.72}$$

Since $\mathbf{P}$ diagonalizes $\mathbf{A}$, (9.72) takes the form

$$\ddot{\mathbf{q}} = -\omega^2 \mathbf{D}\,\mathbf{q} \tag{9.73}$$

where $\mathbf{D}$ is a diagonal matrix. Orthogonal diagonalization has been discussed in Chapter 3 (Section 3.2). On calculation, we can find

$$\mathbf{P} = \begin{bmatrix} -\frac{1}{\sqrt{2}} & \frac{1}{\sqrt{2}} \\ \frac{1}{\sqrt{2}} & \frac{1}{\sqrt{2}} \end{bmatrix} \qquad \mathbf{P}^{-1} = \frac{1}{2}\begin{bmatrix} -\sqrt{2} & \sqrt{2} \\ \sqrt{2} & \sqrt{2} \end{bmatrix}$$

and

$$\mathbf{D} = \begin{bmatrix} \lambda_1 & 0 \\ 0 & \lambda_2 \end{bmatrix} = \begin{bmatrix} 4 & 0 \\ 0 & 1 \end{bmatrix}$$

where λ_1 and λ_2 are the eigenvalues of $\mathbf{A}$. So, we have from (9.73)

$$\ddot{\mathbf{q}} = \begin{bmatrix} -4\omega^2 & 0 \\ 0 & -\omega^2 \end{bmatrix} \mathbf{q}$$

which can be decomposed into two "uncoupled" differential equations

$$\begin{aligned} \ddot{q}_1 &= -4\omega^2 q_1 \\ \ddot{q}_2 &= -\omega^2 q_1 \end{aligned} \tag{9.74}$$

The solutions to (9.74) are given by

$$q_1 = B_1 \cos\left(2\omega t + \phi_1\right)$$

and

$$q_2 = B_2 \cos\left(\omega t + \phi_2\right) \tag{9.75}$$

where B_1, B_2, ϕ_1, and ϕ_2 are all constants. Evidently, the transformed coordinates $\mathbf{q} = (q_1, q_2)$ oscillate with the pure frequencies 2ω and ω. These are called the ***normal coordinates***. The two frequency modes are called the ***normal modes***. The original (Cartesian) coordinates can be obtained from the normal coordinates using (9.71).

$$\begin{bmatrix} x_1 \\ x_2 \end{bmatrix} = \begin{bmatrix} -\dfrac{1}{\sqrt{2}} & \dfrac{1}{\sqrt{2}} \\ \dfrac{1}{\sqrt{2}} & \dfrac{1}{\sqrt{2}} \end{bmatrix} \begin{bmatrix} q_1 \\ q_2 \end{bmatrix} = \frac{1}{\sqrt{2}} \begin{bmatrix} -q_1 + q_2 \\ q_1 + q_2 \end{bmatrix} \tag{9.76}$$

so that, the solutions to (9.69) in terms of coordinates $\mathbf{x} = (x_1, x_2)$ can be written as

$$x_1 = C_1 \cos\left(\omega t + \phi_2\right) - C_2 \cos\left(2\omega t + \phi_1\right)$$

and

$$x_2 = C_1 \cos\left(\omega t + \phi_2\right) + C_2 \cos\left(2\omega t + \phi_1\right) \tag{9.77}$$

where C_1 and C_2 are constants. None of x_1 and x_2 oscillates with a pure frequency.

9.4.2 Normal mode analysis theory

The macromolecule/biomolecule under investigation is considered to be a mechanical system consisting of N atoms under a given force field. The atoms are located at positions $\mathbf{r} = (\mathbf{r}_1, \ldots, \mathbf{r}_j, \ldots, \mathbf{r}_N)$, where $\mathbf{r}_j = (x_1, y_1, z_1)$ are the Cartesian coordinates of the jth atom. Hence, there are $3N$ Cartesian coordinates, which can be labeled as r_i, $(i = 1, 2, \ldots, 3N)$. The dynamics of the system is uniquely defined by the Hamiltonian (Section 8.1.3), which is the total energy (kinetic and potential) of the system.

$$H(\mathbf{r}) = T(\mathbf{r}) + V(\mathbf{r}) \tag{9.78}$$

Near the equilibrium conformation ($\mathbf{r}_0$), the potential energy function is given by a Taylor expansion (4.35)

$$V(\mathbf{r}) = V(\mathbf{r_0}) + \sum_{i=1}^{3N} \left.\frac{\partial V}{\partial r_i}\right|_{\mathbf{r}_0} (r_i - r_{i0})$$

$$+ \frac{1}{2}\sum_{i=1}^{3N}\sum_{j=1}^{3N} \left.\frac{\partial^2 V}{\partial r_i\, \partial r_j}\right|_{\mathbf{r}_0} (r_i - r_{i0})\ (r_j - r_{j0}) + \cdots. \tag{9.79}$$

The potential energy of the reference structure can be defined to be zero, that is, $V\ (\mathbf{r}_0) = 0$ and, since $\mathbf{r}_0$ is defined to be the minimum energy conformation, the second term on the right also disappears. Furthermore, if the displacement is assumed to be small, the terms beyond the second order can be neglected. Hence, the potential energy function is approximated as

$$V(\mathbf{r}) = \frac{1}{2}\sum_{i=1}^{3N}\sum_{j=1}^{3N} \left.\frac{\partial^2 V}{\partial r_i\, \partial r_j}\right|_{\mathbf{r}_0} (r_i - r_{i0})\ (r_j - r_{j0}) \tag{9.80}$$

The kinetic energy function in (9.78) can be defined as

$$T\,(\mathbf{r}) = \frac{1}{2}\sum_{i=1}^{3N} m_i\, \dot{r}_i^2 \tag{9.81}$$

where m_i is the mass of the atom with coordinate r_i. It is to be noted that each mass is repeated three times. Using mass-weighted coordinates

$$X_i = m_i^{1/2}\,(r_i - r_{i0})$$

the Hamiltonian in (9.78) can be expressed as

$$H\,(X) = \frac{1}{2}\sum_{i=1}^{3N} \dot{X}_i^2 + \frac{1}{2}\sum_{i=1}^{3N}\sum_{j=1}^{3N} \left.\frac{\partial^2 V}{\partial X_i\, \partial X_j}\right|_{X_0} X_i X_j \tag{9.82}$$

(9.82) shows that the oscillatory motions described by the Hamiltonian are coupled. However, as we have seen earlier for coupled differential (9.69), the motions here can be "decoupled" by the use of appropriate normal mode coordinates. So, as in (9.71), the mass-weighted Cartesian coordinates and the normal coordinates are linearly related by

$$\mathbf{X} = \begin{bmatrix} X_1 \\ X_2 \\ \cdot \\ \cdot \\ \cdot \\ X_{3N} \end{bmatrix} = \mathbf{P}\,\mathbf{q} = \mathbf{P} \begin{bmatrix} q_1 \\ q_2 \\ \cdot \\ \cdot \\ \cdot \\ q_{3N} \end{bmatrix} \tag{9.83}$$

where $\mathbf{P}$ is a $3N \times 3N$ orthonormal transformation matrix that satisfies

$$\mathbf{P}^T\,\mathbf{P} = 1$$

Introduction of the new coordinates further simplifies the kinetic energy and potential energy terms as follows:

$$T(\mathbf{q}) = \frac{1}{2}\sum_{i=1}^{3N}\dot{X}_i^2 = \frac{1}{2}\dot{\mathbf{X}}^T\dot{\mathbf{X}} = \frac{1}{2}\dot{\mathbf{q}}^T\mathbf{P}^T\mathbf{P}\dot{\mathbf{q}} = \frac{1}{2}\dot{\mathbf{q}}^T\dot{\mathbf{q}} \tag{9.84}$$

and

$$V(\mathbf{q}) = \frac{1}{2}\sum_{i=1}^{3N}\sum_{j=1}^{3N}\left.\frac{\partial^2 V}{\partial X_i \partial X_j}\right|_{X_0} X_i X_j = \frac{1}{2}\mathbf{q}^T\mathbf{P}^T\mathbf{H}(X)\mathbf{P}\,\mathbf{q} \tag{9.85}$$

where

$$\mathbf{H}(X) = \left.\frac{\partial^2 V}{\partial X_i \partial X_j}\right|_{X_0}$$

is the Hessian matrix obtained from the second derivatives of the potential. If **P** diagonalizes $\mathbf{H}(X)$, (9.85) becomes

$$V(\mathbf{q}) = \frac{1}{2}\mathbf{q}^T\boldsymbol{\Lambda}\,\mathbf{q} \tag{9.86}$$

The Hessian **H** is a real symmetric matrix and, hence, diagonalizable by orthogonal transformation. The transformation matrix **P** is determined by solving the standard eigenvalue problem

$$\mathbf{H}\,\mathbf{P} = \boldsymbol{\Lambda}\,\mathbf{A} \tag{9.87}$$

Then, it can be found that **P** comprises the eigenvectors of **H** while the diagonal matrix contains the corresponding eigenvalues. That is,

$$\mathbf{P} = \left(\mathbf{p}_1, \ldots, \mathbf{p}_j, \ldots, \mathbf{p}_{3N}\right)$$

and

$$\boldsymbol{\Lambda} = \mathrm{diag}\left(\lambda_1, \ldots, \lambda_j, \ldots, \lambda_{3N}\right)$$

So, we have the Hamiltonian

$$H(\mathbf{q}) = \frac{1}{2}\dot{\mathbf{q}}^T\dot{\mathbf{q}} + \frac{1}{2}\mathbf{q}^T\boldsymbol{\Lambda}\,\mathbf{q} \tag{9.88}$$

which can also be written as

$$H(q) = \frac{1}{2}\sum_{j=1}^{3N}\dot{q}_j^2 + \frac{1}{2}\sum_{j=1}^{3N}\lambda_j q_j^2 \tag{9.89}$$

Clearly, the cross-term in the Hamiltonian expressed in (9.82) has disappeared in the transformed Hamiltonian (9.89) which now uncouples a set of independent harmonic oscillators. Each pair of eigenvector and associated eigenvalue ($\mathbf{p}_j$, λ_j) that represents such an oscillator, is a ***normal mode***. The normal mode oscillation is described by

$$q_j = B_j \cos\left(2\pi\nu_j t + \phi_j\right) \tag{9.90}$$

where the oscillation frequency

$$\nu_j = \frac{\lambda_j^{1/2}}{2\pi} \tag{9.91}$$

is the same for all the atoms in the jth state of the system. B_j and ϕ_j are respectively the maximum-amplitude and the phase determined from the initial conditions. It is to be noted that six of the modes have zero frequency and therefore called the null modes. These correspond to the three translational and three rotational rigid-body motions of the macromolecule. Hence, the dynamics of the system can be described by a linear combination of $3N - 6$ normal mode oscillations.

The magnitude of oscillation of each mode in thermal equilibrium can be determined based of a statistical thermodynamic theory that states that the average energy of each mode at temperature T is given by $k_BT/2$, where k_B is the Boltzmann constant. From (9.89), we have the potential energy of a single mode

$$V_j = \frac{\lambda_j \langle q_j^2 \rangle}{2}$$

Hence, the average square fluctuation of the q_j normal mode coordinates is found to be

$$\langle q_j^2 \rangle = \frac{k_B T}{\lambda_j} = \frac{k_B T}{(2\pi\nu_j)^2} \tag{9.92}$$

(9.92) clearly shows that the highest contribution to the atomic displacements comes from the lowest frequency normal modes, which therefore represent the most collective motions. This means that a relatively larger number of atoms undergo significant displacements in these modes.

9.4.3 Elastic network models

The basic biomolecular structure for NMA is provided by X-ray crystallography, NMR, or cryo-electron microscopy. Typically, the potential energy terms for atomic interactions in NMA are defined by standard MD force field (discussed in Section 9.3). This imposes a requirement for initial energy minimization without which negative frequency modes may emerge as a result of unstable equilibrium conditions. Besides, force fields containing terms related to van der Waal interactions cannot be well approximated to a quadratic form and therefore relatively difficult to compute.

To circumvent the problems inherent in traditional NMA, elastic network model (ENM)-based NMA considers the initial structure, usually imported from the PDB, as energy minimized. Besides, the complex biomolecular structures are reduced to a network of nodes and springs (Fig. 9.3) – the potential energy is assumed to be quadratic in the displacements, a characteristic of Hookean springs. Care is taken in the selection of the nodes so that the gross features of the molecular structure is

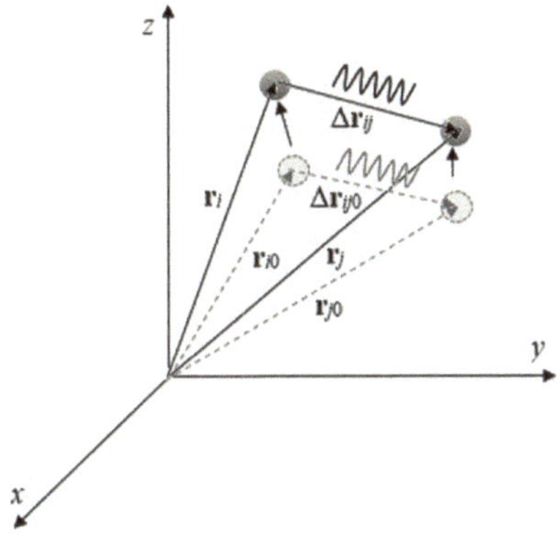

FIGURE 9.3

Interaction (spring) of the *i*th and *j*th nodes (C^{α}-atoms). Instantaneous positions are shown by dark arrows and equilibrium positions are shown by dashed arrows.

preserved by the network. A general approach is to represent each amino acid residue in a protein as a single node, coincident with the C^{α}-atom.

Two kinds of ENMs are widely used in NMA – the Gaussian network model (GNM) and the anisotropic network model (ANM). These are briefly introduced here.

Gaussian network model

In GNM, the equilibrium coordinates of a protein are identified by the position vectors $\mathbf{r}_{i0}$ of its C_{α} atoms in the PDB structure. Pairs of these atoms, considered as nodes, are connected by springs provided they are located within a cutoff distance (say, $\mathbf{r}_c$). The springs represent both bonded and nonbonded interactions between such pairs. The GNM is based on the assumption that the atoms (and hence the interatomic distances) undergo isotropic fluctuations in accordance with a Gaussian distribution. In this case, the potential energy function to be used in NMA is defined as

$$V_{\text{GNM}}(\mathbf{r}) = -\frac{1}{2}\sum_{ij}\gamma_{ij}\left(\Delta\mathbf{r}_{ij} - \Delta\mathbf{r}_{ij0}\right)\cdot\left(\Delta\mathbf{r}_{ij} - \Delta\mathbf{r}_{ij0}\right)\Gamma_{ij} \tag{9.93}$$

where γ_{ij} is the spring constant for interaction between the atom pair (i, j), and the *ij*th element of the connectivity matrix, known as the Kirchhoff matrix, of interresidue contacts is given by

$$\Gamma_{ij} = \begin{cases} -1 & i \neq j,\ \mathbf{r}_{ij} \leq \mathbf{r}_c \\ 0 & i \neq j,\ \mathbf{r}_{ij} > \mathbf{r}_c \\ -\sum_{i,i\neq j}\Gamma_{ij} & i = j \end{cases} \tag{9.94}$$

Applying statistical mechanics, we can obtain the relation

$$\overline{\mathbf{C}} = 3k_{\mathrm{B}}\,T\,\boldsymbol{\Gamma}^{-1} \tag{9.95}$$

where the covariance matrix

$$\overline{\mathbf{C}} = \begin{bmatrix} \langle(\Delta\mathbf{r}_1)^2\rangle & \langle\Delta\mathbf{r}_1\cdot\Delta\mathbf{r}_2\rangle & \cdots & \cdots & \langle\Delta\mathbf{r}_1\cdot\Delta\mathbf{r}_N\rangle \\ \langle\Delta\mathbf{r}_2\cdot\Delta\mathbf{r}_1\rangle & \langle(\Delta\mathbf{r}_2)^2\rangle & \cdots & \cdots & \langle\Delta\mathbf{r}_2\cdot\Delta\mathbf{r}_N\rangle \\ \vdots & \vdots & \vdots & \vdots & \vdots \\ \langle\Delta\mathbf{r}_N\cdot\Delta\mathbf{r}_1\rangle & \langle\Delta\mathbf{r}_N\cdot\Delta\mathbf{r}_2\rangle & \cdots & \cdots & \langle(\Delta\mathbf{r}_N)^2\rangle \end{bmatrix} \tag{9.96}$$

In GNM, the spring constant is assumed to be the same for all interacting pairs, $\gamma_{ij} = \gamma$. Hence, the overall interaction potential is given by

$$\begin{gathered} V_{\mathrm{GNM}}(\mathbf{r}) = \frac{1}{2}\sum_{ij}\gamma\left(\Delta\mathbf{r}_{ij} - \Delta\mathbf{r}_{ij0}\right)\cdot\left(\Delta\mathbf{r}_{ij} - \Delta\mathbf{r}_{ij0}\right) \\ \text{if } \left|\Delta\mathbf{r}_{ij0}\right| = \left|\mathbf{r}_{j0} - \mathbf{r}_{i0}\right| \le \mathbf{r}_c \end{gathered} \tag{9.97}$$

Anisotropic network model

In ANM, most of the assumptions are similar to those in GNM except that here the interatomic interactions are considered to be anisotropic. Hence, the potential energy is given by

$$V_{\mathrm{ANM}}(\mathbf{r}) = \frac{1}{2}\sum_{ij}\gamma_{ij}\left(\Delta\mathbf{r}_{ij} - \Delta\mathbf{r}_{ij0}\right)\cdot\left(\Delta\mathbf{r}_{ij} - \Delta\mathbf{r}_{ij0}\right) \tag{9.98}$$

but, in this case, γ_{ij} is not same for all atom pairs (i, j). Evidently, while GNM modes contain information only about the magnitude of the fluctuations, the ANM modes retain the directional information as well.

NMA has been widely used in the investigation of biomolecular structure and dynamics. Some of its applications include (1) domain motion of the GroEL chaperone subunit (Skaerven et al., 2011), (2) mechanisms of antibody–antigen recognition (Keskin, 2007), (3) retrieval of protein dynamics information from the PDB (Wako and Endo, 2017), (4) dynamics of the ribosomal tunnel (Kurkcuoglu et al., 2009), and (5) ligand binding (Ma, 2005). However, this is only a tiny fraction of the globally published literature in the area.

Exercise 9

1. Utilizing the definition of inner product and cross product in $\mathbb{R}^3$, calculate the products of two quaternions p and q, and show that the set of quaternions is closed under multiplication.

2. Let us consider a unit quaternion q and define an operator on a vector $\mathbf{v} \in \mathbb{R}^3$ as

$$\mathcal{R}_q(\mathbf{v}) = q\,\mathbf{v}\,q^*$$

Using the basic algebraic properties of a quaternion, show that the operator $\mathcal{R}_q(\mathbf{v})$ changes neither the length nor the direction of $\mathbf{v}$.
3. Let the potential energy of a diatomic molecule be

$$U(r) = \varepsilon\left\{\left(\frac{\sigma}{r}\right)^{12} - 2\left(\frac{\sigma}{r}\right)^{6}\right\}$$

where r is the interatomic separation. Find the value of r for which U is the minimum. What is the value of $U_{\min}$?
4. Write down the expressions for the nuclear kinetic energy and the electron kinetic energy operators of the molecule H_2^+ and from these expressions quantitatively justify Born–Oppenheimer approximation.
5. How is the Maxwell–Boltzmann distribution of molecular speed relevant in computation of MD trajectory?
6. Consider a coupled oscillator system with two objects, each of mass m, and three springs, each with spring constant k, as shown in Fig. 9.4. The displacements of the objects from equilibrium positions at any instant of time t are given in terms of coordinates $\mathbf{x} = (x_1, x_2)$. Write down the equations of motion for the two objects and find the solutions in terms of $\mathbf{x}$ and normal coordinates.
7. Write down the Hamiltonian for a system consisting of a set of n coupled simple harmonic oscillators. How is the Hamiltonian transformed so that it now represents a set of n uncoupled independent harmonic oscillators?

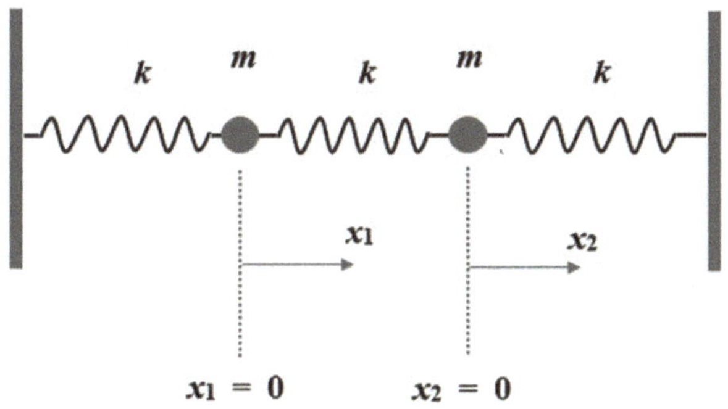

FIGURE 9.4

Exercise 9.6.

References

Keskin, O., 2007. Binding induced conformational changes of proteins correlate with their intrinsic fluctuations: a case study of antibodies. BMC Struct. Biol. 7, 31. Available from: https://doi.org/10.1186/1472-6807-7-31.

Kurkcuoglu, O., et al., 2009. Collective dynamics of the ribosomal tunnel revealed by elastic network modeling. Proteins 75 (4), 837–845.

Ma, J., 2005. Usefulness and limitations of normal mode analysis in modeling dynamics of biomolecular complexes. Structure 13, 373–380.

Skaerven, L., Martinez, A., Reuter, N., 2011. Principal component and normal mode analysis of proteins; a quantitative comparison using the GroEL subunit. Proteins 79, 232–243.

Wako, H., Endo, S., 2017. Normal mode analysis as a method to derive protein dynamics information from the Protein Data Bank. Biophys. Rev. 9, 877–893.

Further reading

Allen, M.P., 2004. Introduction to molecular dynamics simulation. In: Attig, N., et al., (Eds.), Computational Soft Matter: From Synthetic Polymers to Proteins, NIC Series, vol. 23. pp. 1–28. ISBN: 3-00-012641-4.

Atkins, P., Friedman, R., 2011. Molecular Quantum Mechanics. Oxford University Press.

Bahar, I., et al., 2010. Global dynamics of proteins: bridging between structure and function. Annu. Rev. Biophys. 39, 23–42.

Bahar, I., et al., 2010. Normal mode analysis of biomolecular structures: functional mechanisms of membrane proteins. Chem. Rev. 10 (3), 1463–1497.

Coutsias, E.A., Wester, M.J., 2019. RMSD and symmetry. Comput. Chem. 40 (15), 1496–1508.

Schlegel, H.B., 2011. Geometry optimization. WIREs 1 (5), 790–809.

Index

Note: Page numbers followed by "*b*" and "*t*" refer to boxes and tables, respectively.

D

E

F

G

H

N

Q

R

S

W

X

Z

FSC
www.fsc.org
MIX
Papier | Fördert gute Waldnutzung
FSC® C083411

Kolibri 360 GmbH
Ferdinand-Jühlke-Straße 7
99095 Erfurt, Deutschland
produktsicherheit@kolibri360.de